FUNDAMENTALS OF NETWORK BIOLOGY

Other World Scientific Titles by the Author

*Computational Ecology: Artificial Neural Networks and
Their Applications*
ISBN: 978-981-4282-62-8

Computational Ecology: Graphs, Networks and Agent-based Modeling
ISBN: 978-981-4343-61-9

Selforganizology: The Science of Self-Organization
ISBN: 978-981-4699-48-8

FUNDAMENTALS OF NETWORK BIOLOGY

WenJun Zhang

Sun Yat-sen University, China

World Scientific

NEW JERSEY · LONDON · SINGAPORE · BEIJING · SHANGHAI · HONG KONG · TAIPEI · CHENNAI · TOKYO

Published by

World Scientific Publishing Europe Ltd.

57 Shelton Street, Covent Garden, London WC2H 9HE

Head office: 5 Toh Tuck Link, Singapore 596224

USA office: 27 Warren Street, Suite 401-402, Hackensack, NJ 07601

Library of Congress Cataloging-in-Publication Data

Names: Zhang, WenJun, 1963–   author.

Title: Fundamentals of network biology / [by] WenJun Zhang (Sun Yat-sen University, China).

Description: New Jersey : World Scientific, [2018] | Includes bibliographical references.

Identifiers: LCCN 2017060719 | ISBN 9781786345080 (hc : alk. paper)

Subjects: LCSH: Biometry--Data processing. | Systems biology--Methodology. | Bioinformatics.

Classification: LCC QH323.5 .Z45 2018 | DDC 570.1/5195--dc23

LC record available at https://lccn.loc.gov/2017060719

British Library Cataloguing-in-Publication Data

A catalogue record for this book is available from the British Library.

First published 2018 (Hardcover)

Reprinted 2020 (in paperback edition)

ISBN 978-1-78634-894-4 (pbk)

For any available supplementary material, please visit
http://www.worldscientific.com/worldscibooks/10.1142/Q0149#t=suppl

Desk Editors: Dr. Sree Meenakshi Sajani/Jennifer Brough/Koe Shi Ying

Typeset by Stallion Press

Email: enquiries@stallionpress.com

Preface

Why have human cancers not been conquered so far?

Why does brain research emphasize the structure and function of neurons and brain networks?

Why did the successful cases of drug development meet a significant reduction in the past years?

Can the traditional Chinese medicine form a theoretical basis coincident with the modern science?

Why has ecological structure and some of the core ecological problems not yet been solved so far?

A lot of research has demonstrated that many problems in life sciences, medicine, and ecological science have been unable to be resolved relying on the local research of single factors, single molecules, and single reactions, and must be analyzed globally and systematically at the level of networks. In recent years, these methods and theories of network biology have been successfully used to solve problems such as ecological structure, coextinction, drug discovery, and so on, which have made significant progress and breakthroughs. Network science has begun to become the core theory and methodology for dealing with complex biological systems. Network science is mainly based on graph theory, topology, statistics, operational research, computational science, etc. It has been successfully applied in many areas, including statistical physics, particle physics, computer science, internet, biology, ecology, economics, sociology, etc. Network science focuses on the network structure, network dynamics and network functionality, etc.

The initial concept of network biology was first put forward by Barabasi and Oltvai in 2004. Since then network biology has developed rapidly. In recent years, I further expanded the scope and contents of network biology. So far, the scientific frame of network biology has been basically shaped. It is becoming a new frontier science that needs to be understood by university students and is a central research field for researchers.

Up till now, the research in network biology is mainly concentrated in such biological networks as protein–protein interactions, gene regulatory networks, metabolic networks, signaling networks, neural networks, food webs, etc. At the micro- and macro-level, all kinds of these biological networks offer a variety of sources for the development of theory and methods of network biology.

At present, studies of biological networks mainly focus on the following basic aspects: (1) Network structure: connection structure, degree distribution, network modules, patterns, motifs, robustness and stability of biological networks, etc. (2) Network dynamics: rule-based network evolution, biological adaptability of network evolution; dynamics of biological networks over time, among which network model for continuous variables can be used to analyze phase diagram, to conduct sensitivity analysis, and bifurcation analysis, etc. (3) Network control: positive and negative feedbacks, continuous/discrete time control, time delay control, etc. (4) Network design: design and formulation of biological networks. (5) Other topics.

In general, the scope and topics of network biology focus on, but are not limited to the following areas:

- Theories, algorithms, and programs of network analysis.
- Evolution, dynamics, optimization and control of biological networks.
- Network construction and link prediction.
- Network topology, topological analysis, relationship between topological structure and network functions, sensitivity analysis, network robustness, and stability.
- Network flow analysis.
- Design and formulation of biological networks.

- Ecological networks, food webs and natural equilibrium, coevolution, coextinction, biodiversity conservation.
- Metabolic networks, protein–protein interaction networks, biochemical reaction networks, gene networks, transcriptional regulatory networks, cell cycle networks, phylogenetic networks, network motifs, and modules.
- Physiological networks, social networks, epidemiological networks.
- Network regulation of metabolic processes, human diseases, and ecological systems.
- System complexity, self-organized systems, emergence of biological systems, agent-based modeling, neural network modeling and other network-based modeling, etc.
- Big data analytics of biological networks.

This book is the first comprehensive book on network biology. From this integrated and self-contained book, scientists, university teachers, and students will be provided with an in-depth and complete insight on knowledge, methodology, and recent advances of network biology. Matlab codes are presented in the book for easy use. It covers such contents as scientific fundamentals of network biology, construction and analysis of biological networks, methods for identifying crucial nodes in biological networks, link prediction, flow analysis, network dynamics and evolution, network simulation and control, ecological networks, social networks, molecular and cellular networks, network pharmacology and network toxicology, big data analytics, etc. From this unique and self-contained book, undergraduates, postgraduates, teachers, and researchers will thoroughly have an in-depth and complete insight on network biology. It is a valuable book for high-level undergraduates, postgraduates and scientists in the areas of biology, ecology, environmental sciences, medical science, computational science, applied mathematics, and social science, etc.

About the Author

Professor WenJun Zhang graduated from Northwestern Agricultural University in 1984, and earned his PhD degree in 1993. Professor Zhang was a postdoctoral fellow at Sun Yat-sen University during 1994–1996, and postdoctoral fellow and project scientist of International Rice Research Institute during 1997–2000. He was a visiting scientist to University of Arizona (1996) and Wageningen University (2002). He is the president of International Society of Network Biology and International Society of Computational Ecology. He is also the editor-in-chief of international journals *Network Biology, Computational Ecology and Software*, etc. Professor Zhang has published many papers and books. He is now working on ecology and biological science.

Acknowledgments

I am so grateful to the people who gave me guidance: Professors ShiZe Wang, HonSheng Shang, ZhenQi Li, DeXiang Gu, and KG Schoenly. I am also indebted to the people who contributed their valuable suggestions: Professors Yi Pang, RunJie Zhang, Yan Zhang, Jie Zhang, ChangLong Shu, and YanHong Qi. Special thanks is given to the anonymous reviewers for their comments and suggestions on this book. I thank Qi Zhang for her help in the manuscript preparation. This book is supported by The National Key Research and Development Program of China (2017YFD0201204), High-Quality Textbook Network Biology Project for Engineering of Teaching Quality and Teaching Reform of Undergraduate Universities of Guangdong Province (2015.6–2018.6), China, and Guangzhou Science and Technology Project (No. 201707020003).

Contents

Part 2　Crucial Nodes/Subnetworks/Modules, Network Types, and Structural Comparison　103

Part 1
Mathematical Fundamentals

Chapter 1

Fundamentals of Graph Theory

Graph theory is one of the most important fundamental sciences of network biology. Graph theory is a mathematical science with a long history (Chan *et al.*, 1982; Gross and Yellen, 2005; Zhang, 2012e). As early in 1736, Euler solved the well-known Königsberg Bridges Problem and published the first paper on graph theory. However, graph theory was widely known and began to develop rapidly until the late and mid-20th century.

1.1 Definitions and Concepts

A graph is a space made of some vertices and edges that connect vertices. Figure 1 shows a graph with 9 vertices and 14 edges. In other words, a graph X is an ordered triplet $(V(X), E(X), \varphi)$, or ordered pair $(V(X), E(X))$, in which $V(X)$ is the nonempty vertex set, with vertices as elements, $E(X)$ is the set of edges, with edges as elements, and φ is the incidence function that associates each edge (e) of X and un-ordered vertices (u, v), where $\varphi(e) = (u, v)$.

Two vertices associated with the same edge are called adjacent vertices. Two edges associated with the same vertex are called adjacent. Two edges associated with the same vertex are called adjacency edges. For example, the vertices v_1 and v_2 in Fig. 1 are adjacent; the edges e_6 and e_{10} are adjacent, and they are adjacency edges.

The edge with the same terminal vertex is called loop, i.e., the edge is a loop, given $\varphi(e) = (u, u)$. The edge with different initial

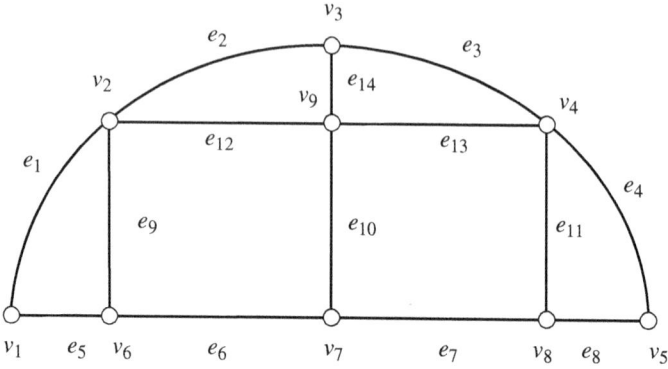

Fig. 1 A graph with nine vertices and 14 edges (Zhang, 2012e).

and terminal vertices is called link. There is no loop in Fig. 1, and all edges in Fig. 1 are links.

The vertex without linking any edge is called isolated vertex.

The graph in Fig. 1 can be mathematically represented as follows:

$$X = (V(X), E(X), \varphi)$$
$$V(X) = \{v_1, v_2, v_3, \ldots, v_9\}$$
$$E(X) = \{e_1, e_2, e_3, \ldots, e_{14}\}$$
$$\varphi(e_1) = v_1v_2, \quad \varphi(e_2) = v_2v_3, \quad \varphi(e_3) = v_3v_4, \quad \varphi(e_4) = v_4v_5,$$
$$\varphi(e_5) = v_1v_6, \quad \varphi(e_6) = v_6v_7, \quad \varphi(e_7) = v_7v_8, \quad \varphi(e_9) = v_2v_6,$$
$$\varphi(e_{10}) = v_7v_9, \quad \varphi(e_{11}) = v_4v_8, \quad \varphi(e_{12}) = v_2v_9,$$
$$\varphi(e_{13}) = v_4v_9, \quad \varphi(e_{14}) = v_3v_9$$

or

$$X = (V, E)$$
$$V = \{v_1, v_2, v_3, \ldots, v_9\}$$
$$E = \{e_1, e_2, e_3, \ldots, e_{14}\}$$
$$= \{(v_1, v_2), (v_2, v_3), (v_3, v_4), \ldots, (v_4, v_9), (v_3, v_9)\}$$

The graph without any vertex and edge is a null graph, denoted by ϕ.

1.1.1 *Finite graph and infinite graph*

A graph is a finite graph if it has a finite number of vertices (i.e., the vertex set V is a finite set) and a finite number of edges (i.e., the edge set E is a finite set). Figure 1 is a finite graph.

1.1.2 *Simple graph and planar graph*

The edges e_1 and e_2 are called parallel edges, given $\varphi(e_1) = \varphi(e_2)$. If a graph has neither loops nor parallel edges, it is called simple graph. Figure 1 is a simple graph. Figure 2(a) is not a simple graph. There is a loop (e_2) and two parallel edges (e_5, e_6) in Fig. 2(a).

In a graph X, delete all loops, so that every pair of adjacent edges have one link only, the resulting simple graph is the basic simple graph of graph X.

In a graph in which any two edges do not intersect each other but they intersect at the endpoints is called planar graph, or else it is a nonplanar graph. A planar graph can be drawn on a plane in a simple way. Figure 1 is a planar graph, and Fig. 2(b) is a nonplanar graph.

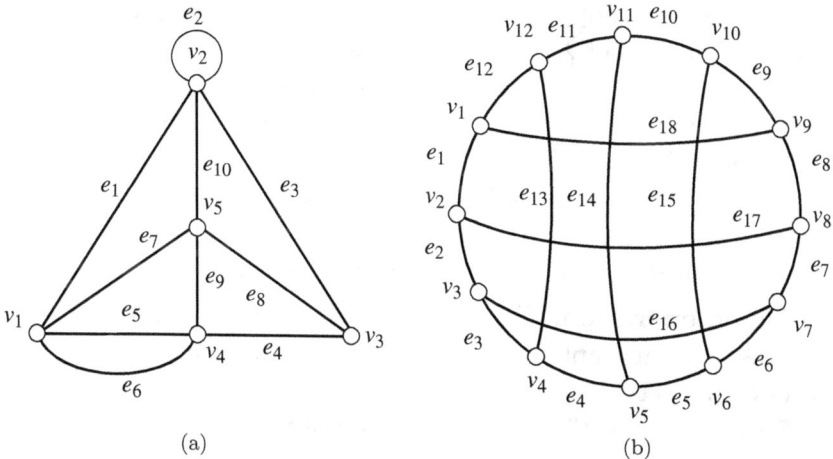

Fig. 2 Nonsimple graph (a) and nonplanar graph and (b) (Zhang, 2012e).

1.1.3 *Subgraph, proper subgraph, spanning subgraph, complementary graph*

Given $V(Y) \subseteq V(X)$ and $E(Y) \subseteq E(X)$, and φ_Y is the restriction of φ_X on $E(Y), Y$ is called the subgraph of X, denoted by $Y \subseteq X$. In this case, the vertex set and edge set of Y are subsets of the vertex set and edge set of X, respectively.

For example, in Fig. 1, the graph

$$Y = (V, E)$$
$$V = \{v_1, v_2, v_6\}$$
$$E = \{e_1, e_9, e_5\}$$

is a subgraph of Fig. 1.

Given $Y \subseteq X$ and $Y \neq X$, i.e., if the graph Y does not contain all edges of X, Y is called the proper subgraph of X. The Y above is the proper subgraph of Fig. 1.

Given $V(Y) = V(X)$, i.e., if the subgraph Y contains all vertices of X, Y is the spanning subgraph of X. For example, remove the edges e_7 and e_8, the rest of the graph is a spanning subgraph of Fig. 1.

Suppose $V' \subset V$ and V' is the vertex set, and the group of edges with two endpoints in V', the resulted subgraph is the V'-induced subgraph of X, which is expressed by $X[V']$. Similarly, suppose $E' \subset E$, and let E' is the edge set, and the group of vertices with edges in E', the resulted subgraph is the E'-induced subgraph of X, denoted by $X[E']$.

The set of all edges contained in X but not in Y is the complementary graph of Y.

Subgraphs can be operated using the following rules:

(1) **Union operation.** The union $X_1 \cup X_2$ of two subgraphs X_1 and X_2 of X is a subgraph made of all edges in which any edge belongs to X_1 or (and) X_2.

For example, in Fig. 1, the union of the two subgraphs

$$X_1 = (V, E)$$
$$V = \{v_1, v_2, v_6\}$$
$$E = \{e_1, e_9, e_5\}$$

and

$$X_2 = (V, E)$$
$$V = \{v_2, v_6, v_7, v_9\}$$
$$E = \{e_9, e_6, e_{10}, e_{12}\}$$

is

$$X_1 \cup X_2 = (V, E)$$
$$V = \{v_1, v_2, v_6, v_7, v_9\}$$
$$E = \{e_1, e_5, e_6, e_9, e_{10}, e_{12}\}$$

(2) **Intersection operation.** The intersection $X_1 \cap X_2$ of two subgraphs X_1 and X_2 of X is a subgraph made of all edges shared by X_1 and X_2. For example, the intersection of above subgraphs X_1 and X_2 is

$$X_1 \cap X_2 = (V, E)$$
$$V = \{v_2, v_6\}$$
$$E = \{e_9\}$$

The intersection of two subgraphs is a null graph if they do not share edges even if they share some vertices.

(3) **Complementary operation.** The complement $X_1 - X_2$ of two subgraphs X_1 and X_2 of X is a subgraph made of all edges belonging to X_1 but not X_2. For example, the complement of above subgraphs X_1 and X_2 is

$$X_1 - X_2 = (V, E)$$
$$V = \{v_1, v_2, v_6\}$$
$$E = \{e_1, e_5\}$$

(4) **Ring sum operation.** The ring sum $X_1 \oplus X_2$ of two subgraphs X_1 and X_2 of X is a subgraph made of all edges specific to either X_1

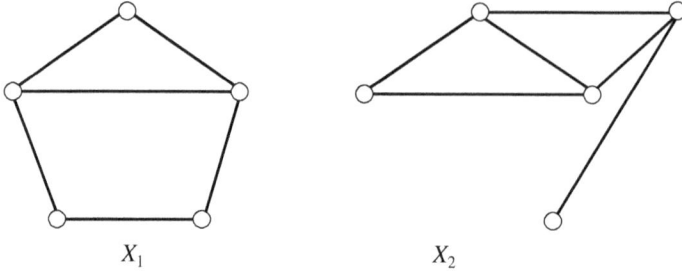

Fig. 3   Two subgraphs (Zhang, 2012e).

or X_2, and

$$X_1 \oplus X_2 = (X_1 \cup X_2) - (X_1 \cap X_2)$$

For example, the ring sum of above subgraphs X_1 and X_2 is

$$X_1 \oplus X_2 = (V, E)$$
$$V = \{v_1, v_2, v_6, v_7, v_9\}$$
$$E = \{e_1, e_5, e_6, e_{10}, e_{12}\}$$

The union, intersection and ring sum of two subgraphs X_1 and X_2 in Fig. 3 are indicated in Fig. 4.

1.1.4   *Complete graph and m-order complete graph*

A graph is a complete graph if there exists an edge between any pair of distinct vertices. Figure 1 is not a complete graph but it contains complete subgraphs, such as Y. Every induced subgraph of a complete graph is a complete subgraph.

A graph is called mth-order complete graph or mth-order regular graph if there are always m edges between any pair of distinct vertices. Figure 1 is not the mth-order complete graph, but it contains 1st-order complete subgraphs, such as Y.

$$X_1 \cup X_2 \qquad\qquad X_1 \cap X_2$$

$$X_1 \oplus X_2$$

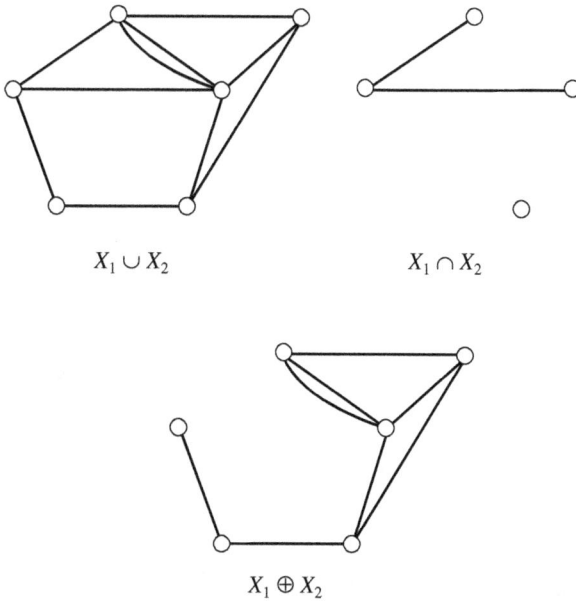

Fig. 4 The union, intersection and ring sum of two graphs X_1 and X_2 in Fig. 3 (Zhang, 2012e).

1.1.5 *Edge sequence, edge train, and path*

If k edges of graph X are naturally arranged and generate a finite sequence

$$e_1(v_1, v_2), e_2(v_2, v_3), e_3(v_3, v_4), \ldots, e_k(v_k, v_{k+1})$$

or

$$v_1 e_1 v_2 e_2 v_3 \ldots v_{k-1} e_k v_k, \quad k \geq 2$$

the sequence is called chain, or edge sequence. k is chain length. For example, an edge sequence in Fig. 1 is (Fig. 5)

$$e_5(v_1, v_6), e_9(v_6, v_2), e_{12}(v_2, v_9), e_{14}(v_9, v_3), e_3(v_3, v_4),$$
$$e_4(v_4, v_5), e_8(v_5, v_8), e_7(v_8, v_7)$$

The chain with distinct initial vertex and terminal vertex is called open chain, or else it is a closed chain. For example, the above chain is an open chain.

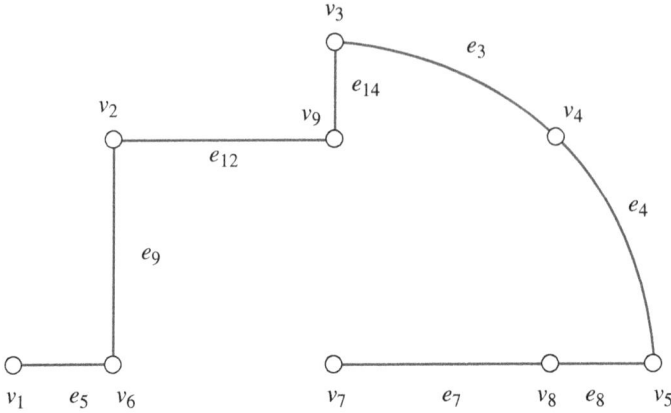

Fig. 5 A chain in Fig. 1 (Zhang, 2012e).

The chain without repeated edges is called simple chain, or edge train. A chain can be an open simple chain or closed simple chain. The chain above is an open simple chain.

The open simple chain without repeated vertices is called elementary chain, or path. The chain above is an elementary chain.

The elementary chain starting from initial vertex u to terminal vertex v exists if there exists at least a chain starting from initial vertex u to terminal vertex v.

Suppose the two endpoints of an elementary chain are the same vertex, the chain is called circuit. The circuit with length k is called k-circuit. According to the parity of k, a k-circuit can be odd circuit or even circuit. Graph X contains circuits if the number of edges is not less than the number of vertices.

1.1.6 *Connected graph, unconnected graph, and connected components*

Two vertices u and v in a graph X are said to be connected if there exists a path between u and v.

For a given graph X, let the vertex set to be

$$V = \bigcup_{i=1}^{m} V_i$$

Two vertices u and v in V_i are connected if and only if $u, v \in V_i$. The induced subgraphs $X[V_i]$, $i = 1, 2, \ldots, m$, are called the connected components of X. If $m = 1, X$ is a connected graph. That is, in the connected graph X, each pair of vertices is connected. Otherwise, X is an unconnected graph.

X is a connected graph if and only if for each of the classification that divides the vertex set V into two nonempty subsets V_1 and V_2, there always exists an edge, and one of its vertex is in V_1 and another vertex is in V_2. Figure 1 is a connected graph.

Suppose X is a connected graph, the maximal connected subgraph of X is just itself. If X is an unconnected graph, then each of its connected components is the maximal connected subgraph of X.

Assume the connected graph X has v vertices, its rank is defined as $v - 1$. If it is known the graph X has v vertices and m connected components, then its rank is $v - m$.

Suppose the connected graph X has v vertices and e edges, its nullity is defined as $e - v + 1$. And if X has m connected components, the nullity of X is defined as $e - v + m$.

In a connected graph, there exists shared vertex between any two longest elementary chains.

Given the connected graph X has $\omega(X)$ connected components, and the subgraph after deleting an edge e from X is $X - e$, we have

$$\omega(X) \leq \omega(X - e) \leq \omega(X) + 1$$

1.1.7 Separable graph, inseparable graph, bipartite graph, and disjoint graph

If there is a vertex in the connected graph X, and after removing the vertex the remaining graph becomes an unconnected graph, the vertex is called separation vertex.

If there is an edge in the connected graph X, and after removing the edge the remaining graph becomes an unconnected graph, the edge is called bridge.

The connected graph containing separation vertex is called separable graph, otherwise it is an inseparable graph. The maximal connected subgraph without separation vertex is called block. Figure 1 is an inseparable graph and also a block.

If the vertices of X can be divided into two subsets A and B, so that for each edge one vertex belongs to subset A and the other vertex belongs to subset B, then X is called bipartite graph, or bigraph. A graph is the bipartite graph if and only if it does not contain odd loops.

If the two graphs are separable from each other and they have not any shared vertex, they are called disjoint graphs.

1.1.8 *Degree of vertex*

The degree of a vertex refers to the number of edges associated with the vertex. A loop is equivalent to two edges. A vertex is an isolated vertex if its degree is zero. The degree of vertex v_1 in Fig. 1 is 2, and the degree of vertex v_9 is 4. There is no isolated vertex in Fig. 1.

Suppose the graph $X = (V, E), V = \{v_1, v_2, \ldots, v_n\}$, and $E = \{e_1, e_2, \ldots, e_m\}$, the sum of degree of vertex is $2m$.

1.1.9 *Directed graph and undirected graph*

Suppose all edges of graph X are directed, X is called directed graph, or else it is an undirected graph. For a directed edge from vertex v_i to v_j, v_i is said to be adjacent to v_j. The edges in Fig. 1 are undirected, it is thus an undirected graph.

For the directed graph, the above definitions should be adjusted correspondingly. For the directed graph, e.g., a vertex has outdegree $(d^+(v)$, the number of edges leaving the vertex) and indegree $(d^-(v)$, the number of edges pointing to the vertex), the degree of the vertex $d(v) = d^+(v) + d^-(v)$.

1.1.10 *Cutset and association set*

Remove a set of edges from the connected graph X, so that X is separated into two disjoint subgraphs, then the minimal edge set of this kind is called the cutset of X.

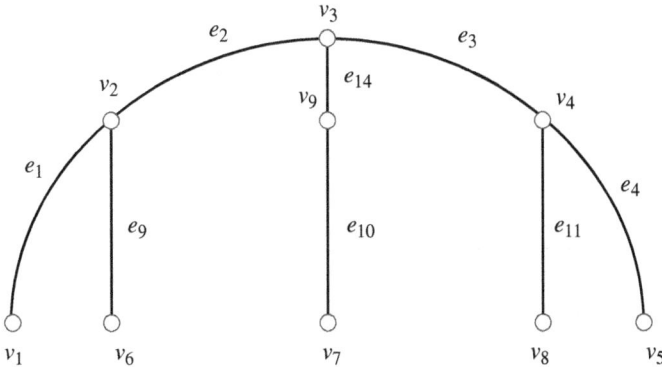

Fig. 6 A tree in Fig. 1 (Zhang, 2012e).

The set of edges associated with the same vertex is called association set.

1.1.11 *Tree and tree branch*

For a connected subgraph T of the connected graph X with v vertices, if T has $v - 1$ edges, contains all vertices of X, and does not contain any circuit, T is called the tree of X (Fig. 6). The edges in the tree are called tree branches. A connected graph X may contain multiple distinct trees.

1.1.12 *Isomorphism*

If there are two monogamy mappings, $\psi\colon V(X) \rightarrow V(Y)$, and $\varphi\colon E(X) \rightarrow E(Y)$, so that $\varphi_X(e) = uv$, if and only if $\varphi_Y(\varphi(e)) = \psi(u)\psi(v)$, then the mapping pair (ψ, φ) is called an isomorphism between graphs X and Y. In other words, two graphs X and Y are isomorphic if they have the same number of vertices and edges, and their vertices and edges are one-to-one mapped respectively, and the vertex–edge incidence relationships are maintained constant (Fig. 7). Automorphism refers to the isomorphism of a graph to itself.

If two graphs are isomorphic, the degree of their corresponding vertices is the same.

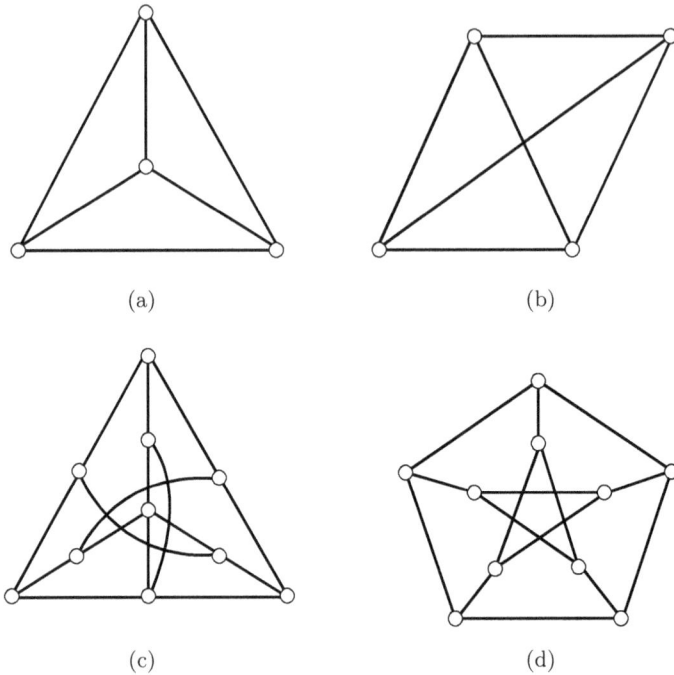

Fig. 7   Two pairs of isomorphic graphs (a) *vs.* (b) and (c) *vs.* (d) (Zhang, 2012e).

Given $V(X) = V(Y), E(X) = E(Y)$, and $\varphi_X = \varphi_Y$, the graphs X and Y are identical. Two identical graphs can be represented by a graph.

Divide the cut vertices of each graph in X and Y into two sets respectively, so that X and Y become two unconnected graphs. If the two unconnected graphs are isomorphic, the two graphs X and Y are called 1-isomorphic. If a graph is a separable graph, it may contain a 1-isomorphic graph.

1.2   Topological Definition of Graph

Graph theory is a branch of topology (Zhang, 2012e). Topology was firstly introduced in 1847 by Listing. It mainly treats the invariants and invariable properties of space in the topological transformation (homeomorphism). The theories and methods of topology (Spanier, 1966; Chan, 1987; Lin, 1998) are an important basis of graph theory.

1.2.1 *Homotopy, homotopy type*

Homeomorphism, or topologically invariant properties, is substantially homeomorphism classification (Zhang, 2012e).

Two mappings $f_0, f_1\colon X \to Y$ are called homotopic if there exists a continuous mapping $F\colon X \times I \to Y$ such that

$$F(x,0) = f_0(x), \quad F(x,1) = f_1(x), \quad \forall x \in X$$

F is called the homotopy from f_0 to f_1, denoted by $f_0 \simeq f_1$.

Generally, for $f \simeq g\colon X \to Y$, we can construct a function $F(x,t) = (1-t)f(x) + tg(x)$, $x \in X, t \in I$. In addition, $f \simeq g\colon X \to Y$, is equivalent if for any $x \in X$, $f(x)$ can be in Y continuously deformed to $g(x)$.

In the set of all mappings from X to Y, the homotopy relation, denoted by $\simeq$, is an equivalence relation from X to Y.

All mappings from X to Y can be classified into some equivalence categories based on homotopy relation, i.e., homotopy categories. The homotopy category of f is denoted by $[f]$.

Given $f_0 \simeq f_1\colon X \to Y$, and f_1 is a constant mapping, i.e., $f_1(X) = a \in Y$, f_0 is called null homotopy.

Theorem 1. *Let Y be a subspace of the Euclidean space R^n and $f, g\colon X \to Y$ are mappings. $f \simeq g$, if for each $x \in X$, $f(x)$ and $g(x)$ can be connected by a line segment in Y.*

Let A be a subspace of X, the ordered pair (X, A) is called space pair. If the mapping $f\colon X \to Y$ maps the subspace A of X onto the subspace B of Y, f is called the mapping between space pairs, denoted by $f\colon (X, A) \to (Y, B)$. Two mappings between space pairs $f, g\colon (X, A) \to (Y, B)$ are called to be homotopic if there exists a mapping between space pairs $F\colon (X \times I, A \times I) \to (Y, B)$, such that $F(x,0) = f(x), F(x,1) = g(x), x \in X$.

Let $f, g\colon (X, A) \to (Y, B)$ be the mappings between space pairs, such that $f|_A = g|_A$. f and g are called to be homotopic relative to A if there exists a mapping between space pairs, $F\colon (X \times I, A \times I) \to (Y, B)$, such that $F(x,0) = f(x), F(x,1) = g(x), x \in X$, and $F(a,t) = f(a) = g(a), a \in A, t \in I$, denoted by $f \simeq g$ rel A.

The topological spaces X and Y are called homotopy equivalent if there exist mappings $f\colon X \to Y$ and $g\colon Y \to Z$, such that $gf \simeq 1_X, fg \simeq 1_Y$, denoted by $X \simeq Y$.

Theorem 2. *The homotopy equivalence between topological spaces is an equivalence relation.*

The subspace A of X is called the retract of X if there exists a mapping $r\colon X \to A$, such that $r(a) = a, \forall a \in A$. Let $i\colon A \to X$ be an interior mapping, i.e., $i(a) = a, \forall a \in A$, then $ri = 1_A$. The mapping r is called retraction mapping. If there is a homotopy $ir \simeq 1_X$, the homotopy is called deformation retraction, and A is the deformation retract of X.

If A is the deformation retract of X, then $A \simeq X$, in which the interior mapping, $i\colon A \to X$ is a homotopy equivalence, and $r\colon X \to A$ is the homotopy inverse of i.

1.2.2 *Connectedness*

1.2.2.1 *Connectedness*

Let X be the union of the disjoint nonempty sets A and B of X, X is called unconnected space, otherwise connected space.

The continuous image of a connected space is a connected set. Connectedness of topological space is a topologically invariant property.

The subset A of a topological space X is called the connected subset of X if A, as a subspace, is connected.

Theorem 3. *Suppose the topological space X has a covering Θ, which consists of connected subsets, such that for any two members A and B in Θ, there exists finite number of members of $\Theta, A = \Theta_1, \Theta_2, \ldots, \Theta_n = B$, such that $\Theta_i \cap \Theta_{i+1} \neq \phi, i = 1, 2, \ldots, n - 1$, then X is connected.*

Let A be a connected subset of X and not be a proper subset of other connected sets, A is called the connected component of X. The connected component of X is a closed set. Different connected components are disjoint. X is the union of all of its connected components.

Topological space X is locally connected, if for any $x \in X$ and the neighborhood U_x of x, there exists a connected neighborhood V_x of x, such that $V_x \subset U_x$. Topological space X is locally connected if and only if the connected component of any open set of X is an open set. Local connectedness is a topologically invariant property.

1.2.2.2 *Path connectedness*

The mapping $f \colon I \to X$ that satisfies $f(0) = a, f(1) = b$ is called the path connecting a and b in X. X is said to be path connected if for any points a and b in the topological space X, there exists a path connecting them. a is called the origin of the path and b the end.

Path connectedness of topological space corresponds to the connectedness of graph.

Path connectedness is a topological invariant property. The continuous image of path-connected space is path connected.

The subset A of a topological space X is called path-connected subset, if A is, as a subspace, path connected. If A is a path-connected subset of X, but not a proper subset of other path-connected sets, A is called path-connected component. For the graph, the path-connected subset is the connected subgraph.

Every topological space is the union of its disjoint path-connected components.

A path-connected space must be a connected space. However, a connected space is not necessarily a path-connected space. Taking a graph as a topological space, then Theorem 3 holds also for graph.

Corollary (Zhang, 2012e). *Let the graph X be a topological space, there exists a covering Θ, consisting of connected subgraphs, such that for any two members A and B in Θ, there exist finite number of members of $\Theta, A = \Theta_1, \Theta_2, \ldots, \Theta_n = B$, such that $\Theta_i \cap \Theta_{i+1} \neq \phi, i = 1, 2, \ldots, n - 1$, then the graph X is a connected graph.*

Topological space X is locally path connected if for any $x \in X$, and the neighborhood U_x of x, there exists a path-connected neighborhood V_x of x, such that $V_x \subset U_x$. A locally path-connected space is also the locally connected space.

Topological space X is locally connected if and only if the connected component of any open set of X is an open set. Local connectedness is a topologically invariant property.

$X_1 \times X_2 \times \cdots \times X_n$ is connected (path connected, locally connected, locally path connected) if and only if each X_i is connected (path connected, locally connected, locally path connected).

1.2.3 *Simplicial complex, polyhedron, graph*

1.2.3.1 *Simplex*

The $n+1$ points, $a^0, a^1, \ldots$, and a^n in Euclidean space R^m are said to be geometry-independent, if the vectors $a^1 - a^0, a^2 - a^0, \ldots, a^n - a^0$ are linearly independent.

Let $a^0, a^1, \ldots, a^n$, be geometry-independent point set in Euclidean space R^m. The set

$$\sigma_n = \left\{ \sum_{i=0}^{n} \lambda_i a^i | \lambda_i \geq 0, \; \sum_{i=0}^{n} \lambda_i = 1 \right\}$$

is assigned the subspace topology of R^m. It is called n-dimensional simplex, denoted by $(a^0, a^1, \ldots, a^n)$, or σ_n. $a^0, a^1, \ldots$, and a^n are called vertices of n-dimensional simplex. The point x of σ_n can be uniquely expressed as

$$x = \sum_{i=0}^{n} \lambda_i a^i$$

If $\lambda_i > 0, i = 1, 2, \ldots, n$, then x is called the interior of σ_n, otherwise boundary point. The simplex extended from the subset of the vertex set $\{a^0, a^1, \ldots, a^n\}$ is called the facet of σ_n.

Zero-dimensional simplex is a point, two-dimensional simplex is a line segment, three-dimensional simplex is a triangle, and four-dimensional simplex is a tetrahedron, $\ldots$

The simplex $\sigma_n = (a^0, a^1, \ldots, a^n)$ is the minimum convex set that contains the vertices $a^0, a^1, \ldots$, and a^n, and it is a compact, closed and connected space in R^m.

Two two-dimensional simplexes are linearly homeomorphic.

1.2.3.2 *Simplicial complex*

The simplicial complex K is a set containing a finite number of simplexes in R^m and satisfies the conditions

(1) If $\sigma_n \in K$, then any facet of $\sigma_n \in K$.
(2) If $\sigma_n, \tau_m \in K$, then $\sigma_n \cap \tau_m$ is either empty or a shared facet.

The maximum dimension of simplexes in K is called the dimension of the complex K (i.e., dim K). Each zero-dimensional simplex of K is called the vertex of K. For example, the complex $K = \{(a^0 a^1 a^2), (a^0 a^1), (a^1 a^2), (a^0 a^2), (a^0 a^3), (a^1 a^4), (a^0), (a^1), (a^2), (a^3), (a^4)\}, \dim K = 2, a^0, a^1, a^2, a^3,$ and a^4 are the vertices of K.

The subset L of K is called the subcomplex of K if L satisfies the condition (1) (and naturally (2)). (K, L) is a simplex pair. If L and M are subcomplexes of K, $L \cap M$ and $L \cup M$ are also the subcomplexes of K.

1.2.3.3 *Polyhedron*

The union of all simplexes of the complex K

$$|K| = \bigcup_{\sigma \in K} \sigma$$

is assigned the subspace topology of R^m, which is called polyhedron of the simplex K. The polyhedron $|L|$ of subcomplex L of K is called subpolyhedron of the polyhedron $|K|$.

The polyhedron $|K|$ of complex K is a compact and closed subspace of R^m.

(K, f) is called the simplex partition of topological space X if K is a simplicial complex and $f: |K| \rightarrow X$ is a homeomorphism. Topological space X is called polytope, or curved polyhedron.

1.2.3.4 *Abstract complex, graph*

Abstract complex $\mathcal{K}$ is a family consisting of a set of a finite number of elements (abstract vertices), $a^0, a^1, \ldots, a^n$, together with the subsets $(a^{i0}, a^{i1}, \ldots, a^{ir})$ (i.e., abstract simplexes), such that every subset that satisfies every abstract simplex is still abstract simplex.

One-dimensional abstract complex $\mathcal{K}$ is called graph.

Chapter 2

Graph Algorithms

This chapter discusses the matrix representation and algorithms of graphs. There are numerous graph algorithms. Only basic and well-known graph algorithms are discussed.

2.1 Matrix Representation of Graphs

Generally, a graph can be represented by a matrix, and for some specific graphs by several matrices. Graph matrix contains all topological information and properties of the graph (Chan *et al.*, 1982; Li, 1982; Gross and Yellen, 2005; Zhang, 2012e).

2.1.1 *Undirected graph*

2.1.1.1 *Incidence matrix*

Incidence matrix describes the vertex–edge relationship of graph (Zhang, 2012e).

For an undirected graph X with v vertices and e edges, its incidence matrix $A = (a_{ij})_{v \times e}$, $a_{ij} = 1$, if $(v_i, v_k) = e_j$, and $a_{ij} = 0$, if $(v_i, v_k) \neq e_j$; $i = 1, 2, \ldots v$; $j = 1, 2, \ldots, e$.

For example, the incidence matrix A of a graph is as follows:

$$A = (a_{ij})_{5 \times 10} = \begin{bmatrix} 1 & 0 & 0 & 0 & 1 & 1 & 1 & 0 & 0 & 0 \\ 1 & 1 & 1 & 0 & 0 & 0 & 0 & 0 & 0 & 1 \\ 0 & 0 & 1 & 1 & 0 & 0 & 0 & 1 & 0 & 0 \\ 0 & 0 & 0 & 1 & 1 & 1 & 0 & 0 & 1 & 0 \\ 0 & 0 & 0 & 0 & 0 & 0 & 1 & 1 & 1 & 1 \end{bmatrix}$$

In the incidence matrix, the number of 1s in a row is the degree of the vertex that denotes the row. There are two 1s in a column, which means the two vertices associated with the edge corresponding to the column.

For a connected graph, each row in the incidence matrix contains at least one 1. If there is only one 1 in some row, the edge of the column containing the 1 is a suspended edge that is not in any circuit.

If an incidence matrix A has the following form:

$$A = \begin{bmatrix} A_{11} & 0 \\ 0 & A_{22} \end{bmatrix}$$

the graph is an unconnected graph X, and it has at least two maximal connected subgraph X_1 and X_2. For example, the incidence matrix of the unconnected graph in Fig. 1 is

$$\begin{bmatrix}
1 & 0 & 0 & 1 & 1 & 0 & 0 & 0 & \vdots & 0 & 0 & 0 \\
1 & 1 & 0 & 0 & 0 & 0 & 0 & 1 & \vdots & 0 & 0 & 0 \\
0 & 1 & 1 & 0 & 0 & 1 & 0 & 0 & \vdots & 0 & 0 & 0 \\
0 & 0 & 1 & 1 & 0 & 0 & 1 & 0 & \vdots & 0 & 0 & 0 \\
0 & 0 & 0 & 0 & 1 & 1 & 1 & 1 & \vdots & 0 & 0 & 0 \\
0 & 0 & 0 & 0 & 0 & 0 & 0 & 0 & \vdots & 1 & 1 & 1 \\
0 & 0 & 0 & 0 & 0 & 0 & 0 & 0 & \vdots & 1 & 1 & 1
\end{bmatrix}$$

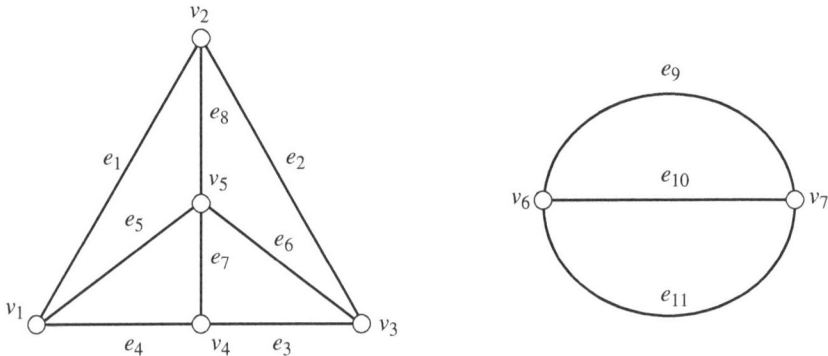

Fig. 1   An unconnected graph $X(X_1, X_2)$ (Zhang, 2012e).

The rank of incidence matrix of a connected graph with v vertices is $v - 1$. For a graph with v vertices and p maximal connected subgraphs, the rank of its incidence matrix is $v - p$.

Remove the row corresponding to the vertex v_i from the incidence matrix of a connected graph X with v vertices and e edges, a $(v - 1) \times e$ matrix can be obtained, which is called fundamental incidence matrix that corresponds to the vertex v_i. The rank of fundamental incidence matrix is $v - 1$.

If there is a circuit in a fundamental incidence matrix, the column vectors corresponding to edges in the circuit are linearly dependent.

The elementary transformation can be operated to incidence matrix with the additive rules $(0 + 0 = 0, 0 + 1 = 1, 1 + 0 = 1, 1 + 1 = 0)$ and multiplication rules $(0 \bullet 0 = 0, 0 \bullet 1 = 0, 1 \bullet 0 = 0, 1 \bullet 1 = 1)$. By deleting a row from the incidence matrix, we obtain an incidence submatrix.

A graph has a unique incidence matrix. In the sense of homomorphism, an incidence matrix corresponds to a unique graph.

2.1.1.2 *Circuit matrix*

Circuit matrix describes the edge–circuit relationship of a graph.

An undirected graph X with v vertices, e edges, and c circuits, is a circuit matrix $C = (c_{ij})_{c \times e}$, $c_{ij} = 1$, if the edge e_j is in the circuit c_i, and $c_{ij} = 0$, if the edge e_j is not in the circuit c_i; $i = 1, 2, \ldots, c$; $j = 1, 2, \ldots, e$.

For example, the circuit matrix of the graph in Fig. 2 is

$$C = (c_{ij})_{7 \times 6} = \begin{bmatrix} 1 & 0 & 0 & 1 & 0 & 1 \\ 0 & 1 & 0 & 0 & 1 & 1 \\ 0 & 0 & 1 & 1 & 1 & 0 \\ 1 & 1 & 1 & 0 & 0 & 0 \\ 1 & 1 & 0 & 1 & 1 & 0 \\ 1 & 0 & 1 & 0 & 1 & 1 \\ 0 & 1 & 1 & 1 & 0 & 1 \end{bmatrix}$$

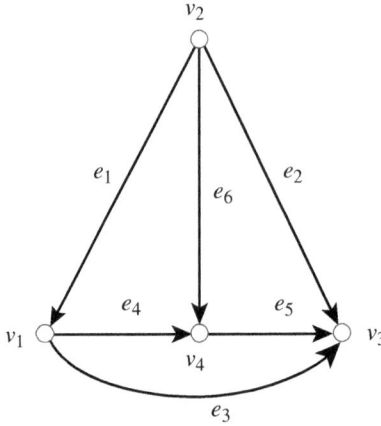

Fig. 2   A directed graph (Zhang, 2012e).

For an undirected connected graph X with v vertices and e edges, the rank of its circuit matrix is $e - v + 1$. There are $e - v + 1$ circuits in X.

Suppose A and C are incidence matrix and circuit matrix of an undirected and connected graph X, respectively, and the edges represented by columns of two matrices are the same, we have $AC^{\mathrm{T}} = 0$, $CA^{\mathrm{T}} = 0$.

For an undirected and connected graph X with v vertices and e edges, the fundamental circuit matrix of its tree T is $C = (c_{ij})_{(e-v+1) \times e}$, in which the rows are $e - v + 1$ fundamental circuits of the tree and the columns are e circuit branches. $c_{ij} = 1$, if the edge e_j is in the fundamental circuit c_i, and $c_{ij} = 0$, if the edge e_j is not in the fundamental circuit c_i; $i = 1, 2, \ldots, e - v + 1$; $j = 1, 2, \ldots, e$. The rank of fundamental circuit matrix is $e - v + 1$.

2.1.1.3   *Cutset matrix*

Cutset is the minimal edge set that makes a connected graph unconnected. Cutset matrix describes vertex–edge relationship of a graph.

For an undirected and connected graph X with v vertices and e edges, its cutset matrix is $Q = (q_{ij})_{k \times e}$, where k is the total number of cutsets. $q_{ij} = 1$, if the edge e_j belongs to the cutset k_i, and $q_{ij} = 0$, if the edge e_j does not belong to the cutset k_i; $i = 1, 2, \ldots, k$; $j = 1, 2, \ldots, e$.

For example, there are seven cutsets in the graph of Fig. 2, and the cutset matrix is

$$Q = (q_{ij})_{7 \times 6} = \begin{bmatrix} 1 & 0 & 1 & 1 & 0 & 0 \\ 1 & 1 & 0 & 0 & 0 & 1 \\ 0 & 1 & 1 & 0 & 1 & 0 \\ 0 & 0 & 0 & 1 & 1 & 1 \\ 0 & 1 & 1 & 1 & 0 & 1 \\ 1 & 0 & 1 & 0 & 1 & 1 \\ 1 & 1 & 0 & 1 & 1 & 0 \end{bmatrix}$$

An undirected and connected graph X with v vertices and e edges, has a fundamental cutset matrix is $Q = (q_{ij})_{(v-1) \times e}$, which responds to a tree and each row responds to a fundamental cutset and each column responds to an edge. $q_{ij} = 1$, if the edge e_j belongs to the fundamental cutset k_i, and $q_{ij} = 0$, if the edge e_j does not belong to the fundamental cutset k_i; $i = 1, 2, \ldots, k$; $j = 1, 2, \ldots, e$. The rank of fundamental cutset matrix Q is $v - 1$.

If Q and C are cutset matrix and circuit matrix of an undirected and connected graph respectively, and the edges represented by columns of two matrices are the same, then $QC^T = 0$, $CQ^T = 0$.

For an undirected and connected graph X with v vertices and e edges, an algorithm to calculate the cutset matrix Q of X is (Chan *et al.*, 1982; Zhang, 2012e) as follows:

(1) Choose a tree T from a graph X and write out the fundamental cutset matrix Q_a of tree T.
(2) Make all possible ring sum operations in Q_a and construct a new matrix Q_b by adding original rows of Q_a together with the new rows generated from ring sum operations.
(3) Eliminate redundant rows (i.e., the rows of the circuits with disjoint edges) in Q_b.
(4) Construct the cutset matrix Q from the left rows of Q_b.

The algorithm for cutset matrix Q is the same as that of the algorithm for circuit matrix C.

2.1.2   Directed graph

A directed graph can be represented by its fundamental graph together with arrows on it (Zhang, 2012e).

2.1.2.1   Incidence matrix

For a directed graph X with v vertices and e edges, its incidence matrix $A = (a_{ij})_{v \times e}$. $a_{ij} = 1$, if $(v_i, v_k) = e_j$; $a_{ij} = -1$, if $(v_k, v_i) = e_j$, and $a_{ij} = 0$, if $(v_i, v_k) \neq e_j$, $(v_k, v_i) \neq e_j$; $i = 1, 2, \ldots, v$; $j = 1, 2, \ldots, e$.

For example, the incidence matrix A of the graph in Fig. 2, is

$$A = (a_{ij})_{4 \times 6} = \begin{bmatrix} -1 & 0 & 1 & 1 & 0 & 0 \\ 1 & 1 & 0 & 0 & 0 & 1 \\ 0 & -1 & -1 & 0 & -1 & 0 \\ 0 & 0 & 0 & -1 & 1 & -1 \end{bmatrix}$$

The rank of incidence matrix of a directed graph with v vertices is $v - 1$. For a directed graph with v vertices and p maximal connected subgraphs, the rank of its incidence matrix is $v - p$.

2.1.2.2   Adjacency matrix

Adjacency matrix describes vertex–vertex relationship of a graph. Adjacency matrix represents some graph properties, e.g., the loop of a vertex (whether the corresponding diagonal element is 1), whether the edges occur in pairs (i.e., whether the matrix is symmetric), and the outdegree and indegree of a vertex (row sum and column sum of the matrix), etc. Moreover, the properties of a graph can be approached by matrix operations.

For an undirected graph X with v vertices and e edges, its adjacency matrix is $D = (d_{ij})_{v \times v}$. $d_{ij} = 1$, if v_i and v_j are adjacent, and $d_{ij} = 0$, if v_i and v_j are not adjacent; $i, j = 1, 2, \ldots, v$.

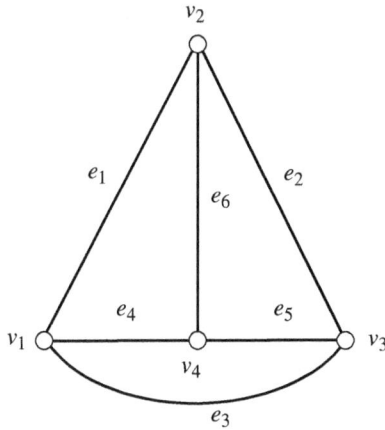

Fig. 3 An undirected connected graph X (Zhang, 2012e).

For example, the adjacency matrix of the graph in Fig. 3 is

$$D = (d_{ij})_{4\times 4} = \begin{bmatrix} 0 & 1 & 1 & 1 \\ 1 & 0 & 1 & 1 \\ 1 & 1 & 0 & 1 \\ 1 & 1 & 1 & 0 \end{bmatrix}$$

The number of 1's in a row or a column of adjacency matrix is the degree of the corresponding vertex.

If an adjacency matrix is a symmetric matrix with all diagonal elements as 0, then there is no loop in the graph.

In the elementary transformation of the adjacency matrix, permuting a row means that the corresponding column must be also permuted. Suppose there exists a permutation matrix P, such that

$$D_2 = P^{-1}D_1P$$

then the graphs corresponding to D_2 and D_1 are isomorphic.

If an adjacency matrix D has the following form:

$$D = \begin{bmatrix} D_{11} & 0 \\ 0 & D_{22} \end{bmatrix}$$

the graph is an unconnected graph, and it has at least two maximal connected subgraphs.

Adjacency matrix can be generalized to the multigraphs with parallel edges and weighted graphs. For example, the value of element d_{ij} can be the type of the edge from v_i to v_j, or be the weight w_{ij} of the edge, and if there is no edge from v_i to v_j, then let $d_{ij} = 0$.

2.1.2.3   *Circuit matrix*

For a directed and connected graph X with v vertices, e edges and c circuits, its circuit matrix is $C = (c_{ij})_{c \times e}$. $c_{ij} = 1$, if the edge e_j belongs to the circuit c_i, and in the same direction of the circuit; $c_{ij} = -1$, if the edge e_j belongs to the circuit c_i, and in the opposite direction of the circuit; $c_{ij} = 0$, if the edge e_j does not belong to the circuit c_i; $i = 1, 2, \ldots, c$; $j = 1, 2, \ldots, e$.

Given some rule for circuit direction, for example, the circuit matrix of the graph in Fig. 3 is

$$C = (c_{ij})_{7 \times 6} = \begin{bmatrix} 1 & 0 & 0 & 1 & 0 & -1 \\ 0 & -1 & 0 & 0 & 1 & 1 \\ 0 & 0 & -1 & 1 & 1 & 0 \\ 1 & -1 & 1 & 0 & 0 & 0 \\ 1 & -1 & 0 & 1 & 1 & 0 \\ 1 & 0 & 1 & 0 & -1 & -1 \\ 0 & 1 & -1 & 1 & 0 & -1 \end{bmatrix}$$

The rank of both circuit matrix and fundamental circuit matrix of a directed and connected graph X with v vertices and e edges is $e - v + 1$.

If A and C are incidence matrix and circuit matrix of a directed and connected graph X, respectively, and the edges represented by columns of two matrices are the same, we have $AC^{\mathrm{T}} = 0$, $CA^{\mathrm{T}} = 0$.

2.1.2.4   *Cutset matrix*

A directed and connected graph X with v vertices and e edges, its cutset matrix is $Q = (q_{ij})_{k \times e}$, where k is the total number of cutsets. $q_{ij} = 1$, if the edge e_j belongs to the cutset k_i, and in the same direction of the cutset; $q_{ij} = -1$, if the edge e_j belongs to the cutset k_i, and in the opposite direction of the cutset; $q_{ij} = 0$ if the edge e_j does not belong to the cutset k_i; $i = 1, 2, \ldots, k$; $j = 1, 2, \ldots, e$.

If Q and C are cutset matrix and circuit matrix of a directed and connected graph X, respectively, and the edges represented by columns of two matrices are the same, we have $QC^{\mathrm{T}} = 0$, $CQ^{\mathrm{T}} = 0$.

2.1.2.5 *Walk matrix and reachability matrix*

Suppose the adjacency matrix of a graph X with n vertices is D, the symbols $\vee$ and $\wedge$ represent the following defined matrix operations. If $D = (d_{ij})$, $C = (c_{ij})$, we have

$$D \vee C = (a_{ij}), \quad a_{ij} = d_{ij} \vee c_{ij}$$

$$D \wedge C = (b_{ij}), \quad b_{ij} = \bigvee_{k=1}^{n} d_{ik} \wedge c_{kj}$$

Let $D^{(m)} = D \wedge D \wedge \cdots \wedge D$. Consider the matrix

$$F = D \vee D^{(2)} \vee \cdots \vee D^{(n)}$$

$f_{ij} = 1$, if and only if there exists a path (walk) from v_i to v_j in X. F is called walk matrix of graph X.

Let $P = I \vee F$, where $I_{n \times n}$ is the unit matrix, P is called reachability matrix of graph X.

For example, calculate the walk matrix F and reachability matrix P of the graph in Fig. 4.

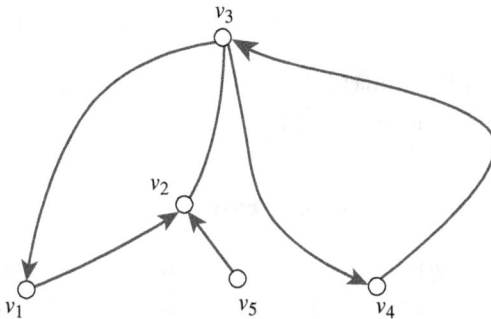

Fig. 4   A directed graph (Zhang, 2012e).

Suppose the adjacency matrix of the graph is D, we have

$$D = \begin{bmatrix} 0 & 1 & 0 & 0 & 0 \\ 0 & 0 & 1 & 0 & 0 \\ 1 & 0 & 0 & 1 & 0 \\ 0 & 0 & 1 & 0 & 0 \\ 0 & 1 & 0 & 0 & 0 \end{bmatrix}$$

$$D^{(2)} = \begin{bmatrix} 0 & 0 & 1 & 0 & 0 \\ 1 & 0 & 0 & 1 & 0 \\ 0 & 1 & 1 & 0 & 0 \\ 1 & 0 & 0 & 1 & 0 \\ 1 & 0 & 1 & 0 & 0 \end{bmatrix}$$

Thus, the walk matrix F and reachability matrix P are as follows:

$$F = D \vee D^{(2)} \vee D^{(3)} \vee D^{(4)} \vee D^{(5)} = \begin{bmatrix} 1 & 1 & 1 & 1 & 0 \\ 1 & 1 & 1 & 1 & 0 \\ 1 & 1 & 1 & 1 & 0 \\ 1 & 1 & 1 & 1 & 0 \\ 1 & 1 & 1 & 1 & 0 \end{bmatrix}$$

$$P = I \vee F = \begin{bmatrix} 1 & 1 & 1 & 1 & 0 \\ 1 & 1 & 1 & 1 & 0 \\ 1 & 1 & 1 & 1 & 0 \\ 1 & 1 & 1 & 1 & 0 \\ 1 & 1 & 1 & 1 & 1 \end{bmatrix}$$

The reachability matrix P shows that the graph in Fig. 4 is a undirected and connected graph.

2.2 Computer Storage of Graph

Binary matrix may be used to store the undirected graph. The adjacency matrix of an undirected graph is a symmetric matrix, so just the upper triangular matrix need to be stored (Chan *et al.*, 1982; Zhang, 2012e).

In addition to conventional matrix, a graph can also be stored as pairs of vertices. For example, the graph in Fig. 2 can be stored in pairs of vertices $(1, 2)$, $(2, 3)$, $(3, 1)$, $(1, 4)$, $(4, 3)$, $(2, 4)$.

The principle for storing a graph with two linear arrays is to define two arrays with the same dimension, A_1 and A_2; an element of A_1 stores a vertex of an edge and the corresponding element of A_2 stores another vertex of the edge. The third array should be used if edge weight needs to be stored. For example, the representation of two linear arrays of the graph in Fig. 2 is

$$A_1 = (1, 2, 3, 1, 4, 2)$$
$$A_2 = (2, 3, 1, 4, 3, 4)$$

Successor listing is often used in the graphs with many vertices and fewer edges. Define two arrays with the same dimension, A_1 and A_2, where A_1 stores vertices and A_2 stores between-vertex adjacency relation. For example, $A_1(i) = v_i$ $A_2(i+1) = v_{i+1}$, then the elements of A_2 starting from v_ith element are vertices adjacent to v_i. Thus the elements from $A_2(v_i)$ to $A_2(v_{i+1})$ are the vertices associated with the vertex i. The successor listing of the graph in Fig. 2 is

$$A_1 = (1, 4, 6, 9, 12)$$
$$A_2 = (2, 3, 4, \ 1, 3, 4, \ 1, 2, 4, \ 1, 2, 3)$$

Adjacency vertices listing is usually used in DFS algorithm. It is applicable to the situation with small ratio of edges *vs.* vertices. Two arrays are used in this representation. The one-dimensional array R marks the degree of every vertex, and two-dimensional array $P = (p_{ij})_{v \times d}$ marks the vertices adjacent to each vertex, where v is the number of vertices, p_{ij} is the labeled number of jth vertex adjacent to the vertex i in which the vertices adjacent to vertex i can be ordered arbitrarily. The one-dimensional array R and two-dimensional array P of the graph in Fig. 3 is

$$(3, 3, 3, 3)$$

$$\begin{bmatrix} 2 & 3 & 4 \\ 1 & 3 & 4 \\ 1 & 2 & 4 \\ 1 & 2 & 3 \end{bmatrix}$$

2.3 Graph Algorithms

The graph without any circuit is called acyclic graph. The connected acyclic graph is the tree (Zhang, 2012e). A tree is called the spanning tree of a graph X if the tree contains all vertices of X. A connected graph must contain a spanning tree. The complementary subgraph of a tree is called cotree of the tree. The branches of cotree are called chords. If T is the tree of a graph X the subgraph after removing the edges of T from X is the cotree of T. Adding a chord to a tree of the connected graph X will form a closed path that has a chord only and the others are all tree branches. The single-chord circuit generated in this way is called fundamental circuit. The number of fundamental circuits is equivalent to the number of chords.

Corresponding to an edge that belongs to a cutset of a tree in connected graph X, if there is a tree branch only and the others are all chords, the single-branch cutest is called fundamental cutset. The number of fundamental cutsets is equivalent to the number of tree branches.

2.3.1 *Tree algorithms*

Some theorems on tree are described as follows (Zhang, 2012e):

Theorem 1. *A graph is a tree if and only if there is one and only one path between any two edges.*

Theorem 2. *There exists at least one tree in a connected graph.*

Theorem 3. *The tree of a connected graph with v vertices must contain $v - 1$ edges.*

Theorem 4. *Given a connected graph X with v vertices, and $X_i \subset X$ has $v - 1$ edges and there is not any circuit in it, then X_i is a tree of X.*

Theorem 5. *The tree of a connected graph X with v vertices and e edges must contain and only contains $v-1$ tree branches and $e-v+1$ chords.*

Theorem 6. *If e is a link of a graph X, then the number of spanning trees of X is $\tau(X) = \tau(X - e) + \tau(X \bullet e)$, where $X \bullet e$ refers to the graph that eliminates e from X and makes two endpoints of e the same point.*

Theorem 7. *Let the incidence matrix of a connected graph X be B, the number of spanning trees of X is $\det(BB^{\mathrm{T}})$.*

2.3.1.1 DFS algorithm

Depth First Search (DFS) algorithm is used to obtain a tree from a graph. It is one of the most important algorithms in graph theory (Tarjan, 1972).

A graph X is stored using Adjacency Vertex Listing. The ID number of starting vertex to be searched is 1. If T is the set of edges in the tree, k is the sequence number, B is the set of edges not in the tree, v is the vertex being checked, w is the vertex to be checked, and $n(i)$ is the ID number of each vertex. The procedures of DFS algorithm are as follows (Zhang, 2012e, 2016h):

(1) Let $v = 1, k = 1, j = 1, n(1) = 1$.
(2) Search the incidence edge that is not yet checked as follows:

> First, take the first edge of v, being not yet checked, and set it to be (v, w). Reach the vertex w from this edge. The direction of the edge (v, w) is from v to w. Return to (3).

> If such an edge was not found after each of the incident edges of v has been checked, return to (4).

(3) If w is the vertex being not yet visited, i.e., $n(w)$ has not yet been determined, send the edge (v, w) into T, and let $v = w$, $k = k + 1$, $n(w) = k$.

> If w is the vertex that has been visited (i.e., $n(w) \neq 0$), send the edge (v, w) into B, return to the vertex v, and let $j = j + 1$, return to (2).

(4) Determine the edge (u, v) that orients to vertex v in T. Find out this edge and return to vertex u, let $v = u$ and return to (2). If such an exists does not exist, terminate the calculation.

The Matlab function for the DFS algorithm, DFS.m, is as follows (Zhang, 2016h):

```
%DFS algorithm to obtain a tree in a graph.
%d: adjacency matrix of the graph; Adjacency matrix is
  d=(dij)n*n,where
%n is the number of vertices in the graph. dij=1 if vi and vj
  are adjacent,
%and dij=0, if vi and vj are not adjacent; i,j=1,2,...,n.
%tree: string of a tree and all parameters and vectors.
%k: number of edges on the tree; l: number of edges not on
  the tree.
%t1[], b1[]: start vertices; t2[], b2[]: end vertices
%t1[],t2[]: set of edges on the tree; b1[],b2[]: set of edges
  not on the tree.
%num[]: DFS labels of vertices.
function [tree,k,l,t1,t2,b1,b2,num]=DFS(d)
n=size(d,1);
r=zeros(1,n);
r=sum(d);
e=max(r);
p=zeros(n,e);
for i=1:n
m=0;
for j=1:n
if (d(i,j)~=0) m=m+1;p(i,m)=j; end
end; end
num=zeros(1,n);
t1=zeros(1,n);
t2=zeros(1,n);
b1=zeros(1,n*e);
b2=zeros(1,n*e);
k=1; l=1; v=1; num(1)=1;
for i=2:n
num(i)=0;
end
lab3=0;
while (n>0)
s=r(v);
while (n>0)
lab2=0;
for i=1:s
if (p(v,i)==0) continue; end
w=p(v,i);
p(v,i)=0;
for j=1:r(w)
if (p(w,j)==v) p(w,j)=0; break; end
end
lab1=0;
if (num(w)==0)
```

```
t1(k)=v;
t2(k)=w;
k=k+1;
num(w)=k;
v=w;
lab1=1; break;
else
b1(l)=v;
b2(l)=w;
l=l+1;
lab2=1; break;
end; end
if (lab1==1) break;
elseif (lab2==1) continue; end
if (num(v)~=1)
m=num(v)-1;
v=t1(m);
break;
end
lab3=1; break;
end
if (lab1==1) lab1=0; continue; end
if (lab3==1) break; end
end;
k=k-1;
l=l-1;
tree='A tree in the graph:\n';
for i=1:k
tree=strcat(tree,'(',num2str(t1(i)),',',num2str(t2(i)),')');
if (i~=k) tree=strcat(tree,','); end
end
tree=strcat(tree,'\nDFS labels of vertices (num[]): \n');
for i=1:n
tree=strcat(tree,num2str(num(i)));
if (i~=n) tree=strcat(tree,','); end
end
tree=strcat(tree,'\nStart vertices of the edges in the tree
   (t1[]): \n');
for i=1:k
tree=strcat(tree,num2str(t1(i)));
if (i~=k) tree=strcat(tree,','); end
end
tree=strcat(tree,'\nEnd vertices of the edges in the tree
   (t2[]): \n');
for i=1:k
tree=strcat(tree,num2str(t2(i)));
if (i~=k) tree=strcat(tree,','); end
end
tree=strcat(tree,'\nStart vertices of the edges not in the
   tree (b1[]): \n');
```

```
for i=1:l
tree=strcat(tree,num2str(b1(i)));
if (i~=1) tree=strcat(tree,',');  end
end
tree=strcat(tree,'\nEnd vertices of the edges not in the tree
   (b2[]): \n');
for i=1:l
tree=strcat(tree,num2str(b2(i)));
if (i~=1) tree=strcat(tree,',');  end
end
```

2.3.1.2   *Minty's algorithm*

Minty's algorithm (Minty, 1965; Zhang, 2012e, 2016h) can be used to obtain all trees in a graph. Suppose an arbitrary edge of a graph X is e_i. Classify all trees into two categories based on e_i, in which one category contains e_i and another one does not contain e_i. Find out two subgraphs X_1 and X_2 from X, and add e_i in X_1, and eliminate e_i in X_2. Every tree in X_1 is added with e_i, which produces the first category of trees in X, and all trees in X_2 belong to the second category of trees in X. Choose another edge, repeat above procedures to get two subgraphs from X_1 and X_2 respectively. In such a way, two new subgraphs can be obtained each time. Delete this subgraph if the graph becomes a loop. After removing all edges, all edges of the subgraph constitute a tree. All trees are obtained after all subgraphs are handled.

Chan *et al.* (1982) proposed a revision based on Minty's algorithm. The Matlab function for the revised Minty algorithm, Minty.m, is as follows:

```
%Revised Minty algorithm to obtain all trees in a graph.
function trees=Minty(d)
%d: adjacency matrix of the graph; Adjacency matrix is d=(dij)n*n,where n
is the number of vertices in the graph. dij=1 if vi and vj are adjacent, and
dij=0, if vi and vj are not adjacent; i, j=1,2,..., n.
%trees: string of all trees.
n=size(d,1);
e=sum(sum(d~=0))/2;
d1=zeros(1,e);
d2=zeros(1,e);
num=0;
for i=1:n-1
for j=i+1:n
if (d(i,j)~=0)
```

```
num=num+1;
d1(num)=i;
d2(num)=j;
end
end; end
trees='';
edge=zeros(1,e);
vmem=zeros(n*e,n);
emem=zeros(n*e,e);
tree=zeros(1,e);
vert=zeros(1,n);
for i=1:e
edge(i)=1;
end
for i=1:n
vert(i)=0;
end
k=1;
f=1;
s=0;
while (n>0)
lab1=0; lab2=0;
for j=1:e
if (edge(j)~=1) continue; end
l=j;
edge(j)=0;
m=0;
for i=1:e
if (edge(i)~=0) m=m+1; end
end
if (m>=(n-1))
for i=1:e
emem(f,i)=edge(i);
end
for i=1:n
vmem(f,i)=vert(i);
end
f=f+1;
end
edge(l)=-1;
v1=d1(l);
v2=d2(l);
if (vert(v1)==0)
if (vert(v2)==0)
vert(v1)=k;
vert(v2)=k;
k=k+1;
lab1=1; break;
end
vert(v1)=vert(v2);
```

```
elseif (vert(v2)==0) vert(v2)=vert(v1);
else
l=vert(v1);
m=vert(v2);
if ((l-m)==0) break; end
if ((l-m)>0)
t=m;
m=l;
l=t;
end
for i=1:n
if ((vert(i)-m)==0) vert(i)=l; end
if ((vert(i)-m)>0) vert(i)=vert(i)-1; end
end
k=k-1;
end;
for i=1:n
if (vert(i)~=1) lab2=1; break; end
end
if (lab2==1) break; end
s=s+1;
l=1;
for i=1:e
if (edge(i)==-1)
tree(l)=i;
l=l+1;
end; end
trees=strcat(trees,'All edges of tree No.',num2str(s),':\n');
for i=1:l-1
trees=strcat(trees,'(',num2str(d1(tree(i))),',',num2str(d2(tree(i))),')');
if (i~=l-1) trees=strcat(trees,','); end
end
trees=strcat(trees,'\n');
fprintf(trees)
end
if ((lab1==1) | (lab2==1)) continue; end
if (f==1) break; end
f=f-1;
for i=1:e;
edge(i)=emem(f,i);
end
k=0;
for i=1:n
vert(i)=vmem(f,i);
if (vmem(f,i)>=k) k=vmem(f,i); end
end
k=k+1;
end
```

As an example, we used DFS algorithm and the adjacency matrices of tumor pathways (Huang and Zhang, 2012; Li and Zhang, 2013; Zhang, 2016e): the calculated tree for p53 pathway is (1,52), (52,4), (4,5), (5,2), (2,8), (2,10), (2,12), (2,14), (5,3), (5,6), (5,7), (7,9), (4,28), (52,11), (52,13), (52,15), (52,17), (52,19), (52,30), (52,48), (48,16), (16,18), (18,50), (50,20), (50,22), (50,24), (24,47), (47,26), (47,32), (32,40), (40,42), (42,38), (38,41), (40,43), (47,33), (47,34), (47,35), (35,37), (47,36), (47,39), (47,44), (47,45), (47,46), (50,51), (51,49), (49,21), (49,23), (49,25), (49,27), (16,29), (29,31).

For Ras tumor pathway, the calculated tree is $(1, 2), (2, 3), (3, 5)$, (5,4), (4,6), (4,8), (5,7), (5,9), (9,11), (11,13), (13,15), (15,17), (17,35), (35,33), (33,32), (32,31), (31,28), (28,26), (26,23), (23,21), (23,29), (23,30), (30,27), (32,34), (5,10), (10,12), (12,14), (12,19), (19,16), (16,18), (5,22), (22,20), (22,24), (5,25).

Using revised Minty algorithm and the adjacency matrix of p53 tumor pathway, the calculated trees (three trees are listed here) are as follows:

Tree No. 1
(1,52), (2,5), (2,8), (2,10), (2,12), (2,14), (3,5), (4,5), (4,28), (4,52), (5,6), (5,7), (7,9), (11,52), (13,52), (15,52), (16,18), (16,29), (16,48), (17,52), (18,50), (19,52), (20,50), (21,49), (22,50), (23,49), (24,47), (24,50), (25,49), (26,47), (27,49), (29,31), (30,52), (32,40), (32,47), (33,47), (34,47), (35,37), (35,47), (36,47), (38,41), (38,42), (39,47), (40,42), (40,43), (44,47), (45,47), (46,47), (48,49), (48,52), (49,51).

Tree No. 2
(1,52), (2,5), (2,8), (2,10), (2,12), (2,14), (3,5), (4,5), (4,28), (4,52), (5,6), (5,7), (7,9), (11,52), (13,52), (15,52), (16,18), (16,29), (16,48), (17,52), (18,50), (19,52), (20,50), (21,49), (22,50), (23,49), (24,47), (24,50), (25,49), (26,47), (27,49), (29,31), (30,52), (32,40), (32,47), (33,47), (34,47), (35,37), (35,47), (36,47), (38,41), (38,42), (39,47), (40,42), (40,43), (44,47), (45,47), (46,47), (48,49), (48,52), (50,51).

Tree No. 3
(1,52), (2,5), (2,8), (2,10), (2,12), (2,14), (3,5), (4,5), (4,28), (4,52), (5,6), (5,7), (7,9), (11,52), (13,52), (15,52), (16,18), (16,29), (16,48),

(17,52), (18,50), (19,52), (20,50), (21,49), (22,50), (23,49), (24,47), (24,50), (25,49), (26,47), (27,49), (29,31), (30,52), (32,40), (32,47), (33,47), (34,47), (35,37), (35,47), (36,47), (38,41), (38,42), (39,47), (40,42), (40,43), (44,47), (45,47), (46,47), (48,49), (48,52), (51,52).

2.3.1.3   *The shortest tree algorithm*

Suppose a graph X has v vertices and e edges, and between-vertex weights matrix of the graph is $a = (a_{ij})$, $ij = 1, 2, \ldots, v$, where a_{ij} is the weight between the vertices i and j. A method to find the shortest tree is (Lu and Lu, 1995; Zhang, 2012e, 2017d):

(1) Use a spanning tree arbitrarily.
(2) Add an edge of a cotree to form a circuit. In the circuit if there is an edge which is longer than the edge added, then replace the longer edge with the new added edge and thus achieve a new tree. Repeat this process until there are no more longer edges.

To find the shortest tree, we usually use Kruskal algorithm. In Kruskal algorithm, first we check the edges of X from smaller weight edge to larger weight edge, and add these edges to T based on the principle of not generating any loop until the number of edges of T is equal to the number of edges of $X - 1$. The procedures are as follows (Chan *et al.*, 1982; Zhang, 2012e, 2017d):

(1) Order the links in the edge set, from smaller weight edge to larger weight edge, as $e_1, e_2, \ldots, e_v$.
(2) Let $T = \{e_1\}, i = 1$, and $j = 2$.
(3) If $i = v - 1$, print T, and terminate calculation, otherwise return to (4).
(4) If a circuit is generated after e_i is added to T, let $j = j + 1$, return to (4), otherwise return to (5).
(5) Let $T = T \cup \{e_i\}, j = j + 1$, $i = i + 1$, return to (3).

The following are Matlab codes, Kruskal.m, for Kruskal algorithm:

```
%Kruskal algorithm to calculate the shortest tree in a graph.
a=input('Input the file name of between-vertex weights matrix of the
weighted graph (e.g., adj.xls, etc. The matrix is a=(aij)v*v, where v is the
```

number of vertices in the graph. aij is the weight between the vertices i
and j, i, j=1,2,..., v): ','s');

```
a=xlsread(a);
v=size(a,1);
cc=0;
t=zeros(v); t1=zeros(v); b=zeros(1,100000);
k=1;
for i=1:v-1
for j=i+1:v
if (a(i,j)>0)
b(k)=a(i,j);
kk=1;
for l=1:k-1
if (b(k)==b(l))
kk=0;
break;
end; end
k=k+kk;
end; end; end
k=k-1;
for i=1:k-1
for j=i+1:k
if (b(j)<b(i))
cc=b(j);
b(j)=b(i);
b(i)=cc;
end; end; end
m=0;
for l=1:k
if (m==v) break; end
for i=1:v-1
for j=i+1:v
if (a(i,j)==b(l))
t(i,j)=b(l);
t(j,i)=b(l);
for i1=1:v;
for j1=1:v;
t1(i1,j1)=t(i1,j1);
end; end
while(v>0)
in=1;
c=0;
for i1=1:v
kk=0;
for j1=1:v
if (t1(i1,j1)>0)
kk=kk+1;
c=j1;
end; end
if (kk==1)
```

```
t1(i1,c)=0;
t1(c,i1)=0;
in=0;
end
end
if (in~=0) break; end
end
in=0;
for i1=1:v-1
for j1=i1+1:v
if (t1(i1,j1)>0)
in=1;
break;
end; end; end
if (in~=0)
t(i,j)=0;
t(j,i)=0;
else m=m+1; end
end; end; end; end
fprintf(['The shortest tree:' '\n']);
tree=t
```

Using Kruskal algorithm and the adjacency matrix of tumor pathway p53 (Huang and Zhang, 2012; Li and Zhang, 2013; Zhang, 2017d), the calculated shortest tree in the p53 pathway is $(2, 5)$, $(3, 5)$, $(4, 5)$, $(5, 6)$, $(5, 7)$, $(2, 8)$, $(7, 9)$, $(2, 10)$, $(2, 12)$, $(2, 14)$, $(16, 18)$, $(4, 28)$, $(16, 29)$, $(29, 31)$, $(35, 37)$, $(32, 40)$, $(38,41)$, $(38,42)$, $(40,42)$, $(40,43)$, $(24,47)$, $(26,47)$, $(32, 47)$, $(33, 47)$, $(34, 47)$, $(35, 47)$, $(36, 47)$, $(39, 47)$, $(44, 47)$, $(45, 47)$, $(46, 47)$, $(16, 48)$, $(21, 49)$, $(23, 49)$, $(25, 49)$, $(27, 49)$, $(48, 49)$, $(18, 50)$, $(20, 50)$, $(22, 50)$, $(24, 50)$, $(49, 51)$, $(1, 52)$, $(4, 52)$, $(11, 52)$, $(13, 52)$, $(15, 52)$, $(17, 52)$, $(19, 52)$, $(30, 52)$, $(48, 52)$.

2.3.2   *Connectedness algorithms*

2.3.2.1   *Connectedness*

Graph connectedness can be tested by using vertex-fusion algorithm. In the algorithm, for the adjacency matrix of a graph, starting from a vertex, fusing all its adjacent vertices and then fusing the new added vertices adjacent to it, until no new adjacent vertex is added, a connected component is thus obtained. When fusing the vertices v_i and v_j, add row j of adjacency matrix to row i, and add column j to column i, and then delete row j and column j. By doing so, all connected components can be obtained (Chan *et al.*, 1982; Zhang, 2012e, 2016g).

The following are the Matlab functions connDet.m, for calculating graph connectedness The program will display the number of connected components and all vertices of each connected component (Zhang, 2012e, 2016g).

```
function conn=connDet(d)
%Calculate graph connectedness. g[]: if vertex i belongs to j-th connected
%component then g[i]=j.
%d[][]: adjacency matrix, d=(dij)v*v, where v is the number of vertices
in the
%graph. dij=1 if vi and vj are adjacent, and dij=0, if vi and vj are not
adjacent;
%i, j=1,2,..., v.
%conn: string containing number of connected components and all vertices
%of each connected component.
v=size(d,1);
g=zeros(1,v+1);
for i=1:v
g(i)=0;
end
s=1;
t=1;
while (v>0)
g(t)=s;
a1=0;
for j=1:v
a1=a1+d(j,t);
end
m=t+1;
while (v>0)
for h=m:v
if ((d(t,h)==0) | (g(h)~=0)) continue; end
g(h)=s;
for i=m:v
if (d(i,h)==0) continue; end
d(i,t)=1;
d(t,i)=1;
end; end
a2=0;
for j=1:v
a2=a2+d(j,t);
end
if ((a1-a2)<0)
a1=a2;
continue;
end
if ((a1-a2)>=0) break; end
end
lab=0;
for k=m:v
if (g(k)==0) break; end
```

```
if (k==v) lab=1; break; end
end
if (lab==1) break; end
t=k;
s=s+1;
end
conn='';
conn=strcat(conn,'Number of connected components in the graph:
',num2str(s),'\n');
conn=strcat(conn,'Vertex----Belonged connected component\n');
for i=1:v
conn=strcat(conn,num2str(i),'----', num2str(g(i)),'\n');
end
conn=strcat(conn,'\n');
```

Using the data of adjacency matrices of tumor pathways, p53, RAS, VEGF, PPAR (Zhang, 2016e), the results of network connectedness detection show that there is only one connected component in these tumor pathways.

Use the data of adjacency matrices of world's 54 human races and populations (nodes) (Zhang and Qi, 2014; Zhang, 2015e). The nodes (i.e., vertices) Nos. 20 and 21 belong to a component, and the remaining 52 nodes belong to another component.

2.3.2.2   *Vertex connectivity*

Connectivity is a property of graph connectedness (Chan *et al.*, 1982; Zhang, 2012e). Removing a minimal set of vertices such that the graph is unconnected or trivial graph, then the number of vertices in the set is called vertex connectivity.

If the connectivity of a graph is K, the minimal degree (i.e., the degree of the vertex that has minimal associated edges) is d, and the number of vertices and edges is n and e, respectively, we have $1 \leq K \leq d \leq 2e/n$. For the regular graph, all vertices have the same degree, $d = 2e/n$. The graph has a maximal connectivity, if $K = d$. The connectivity of the tree is minimal.

2.3.2.3   *Connectedness of directed graph*

The vertex v is called reachable in the graph X starting from vertex u, if there exists a directed path (u, v) in X. Two vertices are called strongly connected (bilaterally connected; Fig. 5)

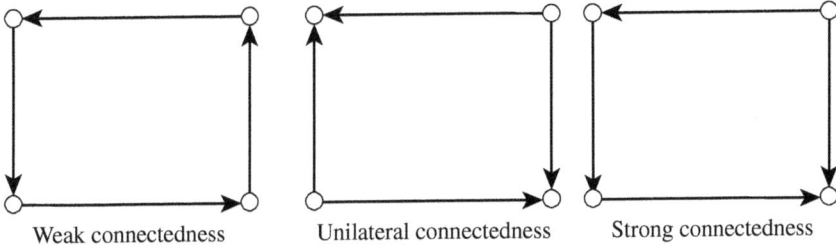

Weak connectedness   Unilateral connectedness   Strong connectedness

Fig. 5 Types of connectedness (Zhang, 2012e).

if they are reachable for each other (Zhang, 2012e). Strongly connectedness is an equivalence relation with respect to the vertex set of X. The directed subgraphs $X(V_1), X(V_2), \ldots, X(V_m)$, induced by a classification $(V_1, V_2, \ldots, V_m)$ of vertex set $V(X)$ which is based on the relation of strongly connectedness, are called strongly connected components of X. A directed graph X is strongly connected if it has and must have only one strongly connected component.

There are also unilaterally connectedness, weak connectedness, etc. (Fig. 5). Suppose X is a directed graph, u and v are any two vertices in X.

(1) If v is reachable from u, or u is reachable from v, then X is unilaterally connected.

(2) If for every pair of vertices (u, v), at least v is reachable from u, or u is reachable from v, then X is weak connected.

(3) If for every pair of vertices (u, v), there exists a vertex w such that u and v are reachable from w, then X is pseudo-strongly connected.

Theorem 8. *A directed graph X is strongly connected if and only if X is connected and every block of X is strongly connected.*

Theorem 9. *A directed graph X is strongly connected if and only if every edge of X is in a directed circuit.*

Theorem 10. *A directed graph X is unilaterally connected if and only if every edge of X is in a directed path.*

Equivalently, there are some conclusions as follows:

A directed graph X is strongly connected if and only if there exists a circuit in X that passes through every vertex at least one time.

A directed graph X is unilaterally connected if and only if there exists a path in X that passes through every vertex at least one time.

A strongly connected graph must be a unilaterally connected graph. A unilaterally connected graph must be a weak connected graph.

2.3.2.4   *Block, cut vertex, bridge*

Theorem 11. *Two edges belong to the same block if and only if there exists a circuit that contains the two edges.*

Theorem 12. *A connected graph X is a block if and only if for any three vertices u, v and w in X, there exists a path from u to w and the path does not contain v.*

Theorem 12 reveals that there is no bottleneck in a block. The vertex v is a bottleneck if any path from u to w must go through v. In this case v is a cut vertex.

The following is Matlab function, cutVertex.m, for calculating cut vertices. The bridges and blocks are easily found by discriminating cut vertices from all vertices in the graph (Zhang, 2016j).

```
%Matlab function to obtain cut vertices in a graph.
%d: adjacency matrix of the graph; Adjacency matrix is d=(dij)n*n,where n
%is the number of vertices in the network. dij=1 if vi and vj are
   adjacent, and
%dij=0, if vi and vj are not adjacent; i, j=1,2,..., n.
%cutset: string of cutvertex set.
   function cutset=cutVertex(d)
   n=size(d,1);
   r=sum(d);
   e=max(r);
   num=zeros(1,n);
   t1=zeros(1,n);
   t2=zeros(1,n);
   b1=zeros(1,n*e);
   b2=zeros(1,n*e);
   lw=zeros(1,n);
   cut=zeros(1,n);
   [tree,et,eb,t1,t2,b1,b2,num]=DFS(d);
   for i=1:n;
```

```
cut(i)=0;
lw(i)=num(i);
end
for i=1:eb
v1=b1(i);
v2=b2(i);
if (lw(v1)>=num(v2)) lw(v1)=num(v2); end
end
for i=1:et
v1=t1(et-i+1);
v2=t2(et-i+1);
if (lw(v2)<=lw(v1)) lw(v1)=lw(v2); end
end
s=0;
for i=1:et
v1=t1(i);
v2=t2(i);
if (v1==1) s=s+1; end
if ((lw(v2)>=num(v1)) & (v1~=1)) cut(v1)=v1; end
end
if (s>=2) cut(1)=1; end
cutset='Cutvertex set:\n{';
for i=1:n
if (cut(i)~=0)
cutset=strcat(cutset,num2str(cut(i)));
if (i~=n) cutset=strcat(cutset,','); end
end
end
cutset=strcat(cutset,'}\n');
```

The Matlab function, DFS.m, used in the program above, is described in the previous section.

Using the algorithm and the adjacency matrices of tumor networks (Huang and Zhang, 2012; Li and Zhang, 2013; Zhang, 2016e), the set of cut nodes for p53 tumor pathway is $\{5, 32, 38, 40, 42, 47, 49, 50, 52\}$, for Ras tumor pathway is $\{2, 3, 4, 5, 10, 12, 16, 19, 22, 23\}$, and for MARK is $\{6, 13, 15, 21, 23, 25, 31, 33, 36, 58\}$. These cut nodes are crucial nodes in the tumor pathways.

2.3.3 *The shortest path*

2.3.3.1 *Floyd algorithm*

The shortest path problem dedicates to find a path with minimal weight sum (i.e., shortest path) in a weighted graph (Zhang, 2012e, 2016b). The shortest path refers to the one between two given vertices, or the one between a given vertex to every other vertex. The weights in weighted graph are cost, distance, etc. Floyd algorithm can be used to address the shortest path problem.

In the Floyd algorithm (1962), each time insert a vertex, and compare the weight sum of the path between any two vertices with the weight sum of the path going through the two nodes and inserted vertex, if the former is larger, then replace the former path with the path having inserted the vertex (Zhang, 2012e). Suppose the graph X has v vertices, and the weight matrix is $d = (d_{ij})$, where $d_{ij} = 1$ for unweighted graph and $d_{ij} = w_{ij}$ for weighted graph (w_{ij} is the weight for the edge v_i to v_j), if v_i and v_j are adjacent, and $d_{ij} = 0$, if v_i and v_j are not adjacent; $i, j = 1, 2, \ldots, v$. The algorithm is (Floyd, 1962; Zhang, 2012e)

(1) Let $k = 1$.
(2) Let $i = 1$.
(3) Calculate

$$d_{ij} = \min(d_{ij}, d_{ik} + d_{kj}), \quad j = 1, 2, \ldots, v$$

(4) Let $i = i + 1$, if $i \leq v$, then return to (3).
(5) Let $k = k + 1$, if $k \leq v$, then return to (2) or else terminate the calculation.

The following are Matlab codes, Floyd.m, for Floyd algorithm:

```
d=input('Input the file name of adjacency matrix of the weighted graph
(e.g., adj.txt, adj.xls, etc. Adjacency matrix is d=(dij)v*v, where v is the
number of vertices in the graph. dij=1 for unweighted graph and dij=wij for
weighted graph (wij is the weight for the edge vi to vj), if vi and vj are
adjacent, and dij=0, if vi and vj are not adjacent; i, j=1,2,..., m):','s');
d=load(d);
%d: weighted adjacency matrix; distances: matrix of distances between
    different
%vertices; paths: string of paths and distances between any of two vertices.
inf=1e+50;
v=size(d,1);
a=zeros(v);
b=zeros(1,v*(v-1)/2);
e=zeros(1,v*(v-1)/2);
h=zeros(1,v*(v-1)/2);
distances=zeros(v);
```

```
for i=1:v
for j=1:v
if ((d(i,j)==0) & (i~=j)) d(i,j)=inf; end
end; end
for i=1:v
for j=1:v
if (d(i,j)~=inf) a(i,j)=j; end
end; end
for i=1:v
for j=1:v
for k=1:v
c=d(j,i)+d(i,k);
if (c<d(j,k)) d(j,k)=c; a(j,k)=i; end
end; end; end
paths='';
for p=1:v
for q=1:v
if (p==q) continue; end
u=a(p,q);
m=1;
b(1)=u;
while (v>0)
m=m+1;
b(m)=a(b(m-1),q);
if (q==b(m)) break; end
if (b(m)==b(m-1)) break; end
if (m>v) break; end
end
n=1;
e(1)=u;
while (v>0)
n=n+1;
e(n)=a(p,e(n-1));
if (p==e(n)) break; end
if (e(n)==e(n-1)) break; end
if (n>v) break; end
end
for i=1:m+n-1
if (i==1) h(i)=p; end
if ((i<=n) & (i>1)) h(i)=e(n-i+1); end
if ((i>n) & (i<(m+n-1))) h(i)=b(i-n+1); end
if (i==(m+n-1)) h(i)=q; end
end
paths=strcat(paths,'The shortest path from',num2str(p),'to',
  num2str(q),':\n');
for i=1:m+n-1
if ((h(i)~=0) & (d(p,q)~=inf))
if ((h(i)==h(i+1)) & (i<m+n-1)) continue; end
if (i<m+n-1) paths=strcat(paths,num2str(h(i)),'->'); end
if (i>=m+n-1) paths=strcat(paths,num2str(h(i)),'\n'); end
```

```
end; end
if (d(p,q)∼=inf)paths=strcat(paths,'Distance=',num2str(d(p,q)),'\n');
distances(p,q)=d(p,q); end
if (d(p,q)==inf) paths=strcat(paths,'No path','\n'); end
end; end
disp('Distances matrix')
distances
disp('Shortest paths')
fprintf(paths)
```

I used the PPAR tumor pathway to find shortest paths (Huang and Zhang 2012; Li and Zhang, 2013; Zhang, 2016b). The adjacency matrix (unweighted network) of PPAR tumor pathway is as follows:

```
0 0 0 0 0 1 0 0 0 0 0 0 0 0 0 0 0 0 0 0 0 0 0 0 0 0 0 0 0
0 0 0 0 0 1 0 0 0 0 0 0 0 0 0 0 0 0 0 0 0 0 0 0 0 0 0 0 0
0 0 0 0 0 0 0 0 0 1 0 0 1 0 0 0 0 0 0 0 0 0 0 0 0 0 0 0 0
0 0 0 0 0 1 0 0 0 0 0 0 0 0 0 0 0 0 0 0 0 0 0 0 0 0 0 0 0
0 0 0 0 0 0 0 0 0 0 0 0 0 0 0 0 0 0 0 0 0 0 0 0 0 1 1 1 0
1 1 0 0 0 0 0 1 0 0 0 0 0 0 0 0 0 0 0 0 0 0 0 0 0 0 0 0 0
0 0 0 1 0 0 0 0 1 0 0 0 0 0 0 0 0 0 0 0 0 0 0 0 0 0 0 0 0
0 0 0 0 0 1 0 0 0 1 0 0 0 0 0 0 0 0 0 0 0 0 0 0 0 0 0 0 0
0 0 0 0 0 0 1 0 0 0 1 0 0 0 0 0 0 0 0 0 0 0 0 0 0 0 0 0 0
0 0 1 0 0 0 0 1 0 0 0 0 0 0 0 0 0 0 0 0 0 0 0 0 0 0 0 1 0
0 0 0 0 0 0 0 0 1 0 0 0 1 0 0 0 0 0 0 0 0 0 0 0 0 0 0 0 0
0 0 0 0 0 0 0 0 0 0 0 0 0 0 1 0 0 1 0 0 0 0 0 0 0 0 0 0 0
0 0 1 0 0 0 0 0 0 0 1 0 0 0 0 0 0 1 0 0 0 0 0 0 0 0 0 1
0 0 0 0 0 0 0 0 0 0 0 0 1 0 0 0 1 0 0 0 0 0 0 0 0 0 0 0 0
0 0 0 0 0 0 0 0 0 0 0 0 0 0 0 0 0 0 0 0 0 0 1 0 0 0 0 0
0 0 0 0 0 0 0 0 0 0 0 0 0 1 0 0 0 1 0 0 0 0 0 0 0 0 0 0 0
0 0 0 0 0 0 0 0 0 0 0 1 0 0 0 0 0 1 0 0 0 0 0 0 0 0 0 0 0
0 0 0 0 0 0 0 0 0 0 0 0 1 0 0 1 0 0 0 1 0 0 0 1 0 0 0
0 0 0 0 0 0 0 0 0 0 0 0 0 0 0 0 1 0 0 0 1 0 0 0 0 0 0 0
0 0 0 0 0 0 0 0 0 0 0 0 0 0 0 0 0 0 1 0 0 0 0 0 0 1 1 1
0 0 0 0 0 0 0 0 0 0 0 0 0 0 0 0 0 0 0 1 0 0 0 0 1 0 0 0 0
0 0 0 0 0 0 0 0 0 0 0 0 0 0 1 0 0 0 0 0 0 0 0 1 0 0 0
0 0 0 0 0 0 0 0 0 0 0 0 0 0 0 0 0 0 0 0 0 1 0 0 1 0 0 0
0 0 0 0 0 0 0 0 0 0 0 0 0 0 0 0 0 0 1 0 0 0 1 1 0 1 0 0
0 0 0 0 1 0 0 0 0 0 0 0 0 0 0 0 0 0 0 1 0 0 0 1 0 0 0
0 0 0 0 1 0 0 0 0 1 0 0 0 0 0 0 0 0 0 1 0 0 0 0 0 0 0
0 0 0 0 0 0 0 0 0 0 0 0 1 0 0 0 0 0 0 1 0 0 0 0 0 0 0
```

Running the Matlab program, the distances matrix is as follows:

```
 0  2  4  9  5  1  8  2  7  3  6  9  5  8  9  7 10  6 10  5  9  8  8  7  6  4  6
 2  0  4  9  5  1  8  2  7  3  6  9  5  8  9  7 10  6 10  5  9  8  8  7  6  4  6
 4  4  0  5  3  3  4  2  3  1  2  5  1  4  5  3  6  2  6  3  5  4  4  3  4  2  2
 9  9  5  0  8  8  1  7  2  6  3  8  4  7  8  6  9  5  9  6  8  7  7  6  7  7  5
 5  5  3  8  0  4  7  3  6  2  5  6  4  5  4  4  6  3  5  2  4  3  3  2  1  1  3
 1  1  3  8  4  0  7  1  6  2  5  8  4  7  8  6  9  5  9  4  8  7  7  6  5  3  5
 8  8  4  1  7  7  0  6  1  5  2  7  3  6  7  5  8  4  8  5  7  6  6  5  6  6  4
 2  2  2  7  3  1  6  0  5  1  4  7  3  6  7  5  8  4  8  3  7  6  6  5  4  2  4
 7  7  3  2  6  6  1  5  0  4  1  6  2  5  6  4  7  3  7  4  6  5  5  4  5  5  3
 3  3  1  6  2  2  5  1  4  0  3  6  2  5  6  4  7  3  7  2  6  5  5  4  3  1  3
 6  6  2  3  5  5  2  4  1  3  0  5  1  4  5  3  6  2  6  3  5  4  4  3  4  4  2
 9  9  5  8  6  8  7  7  6  6  5  0  4  1  6  2  1  3  2  4  3  5  4  4  5  5  5
 5  5  1  4  4  4  3  3  2  2  1  4  0  3  4  2  5  1  5  2  4  3  3  2  3  3  1
 8  8  4  7  5  7  6  6  5  5  4  1  3  0  5  1  2  2  3  3  4  4  4  3  4  4  4
 9  9  5  8  4  8  7  7  6  6  5  6  4  5  0  4  6  3  5  4  4  1  3  2  3  5  5
 7  7  3  6  4  6  5  5  4  4  3  2  2  1  4  0  3  1  4  2  4  3  3  2  3  3  3
10 10  6  9  6  9  8  8  7  7  6  1  5  2  6  3  0  4  1  5  2  5  3  4  5  6  6
 6  6  2  5  3  5  4  4  3  3  2  3  1  2  3  1  4  0  4  1  3  2  2  1  2  2  2
10 10  6  9  5  9  8  8  7  7  6  2  5  3  5  4  1  4  0  5  1  4  2  3  4  6  6
 5  5  3  6  2  4  5  3  4  2  3  4  2  3  4  2  5  1  5  0  4  3  3  2  1  1  1
 9  9  5  8  4  8  7  7  6  6  5  3  4  4  4  4  2  3  1  4  0  3  1  2  3  5  5
 8  8  4  7  3  7  6  6  5  5  4  5  3  4  1  3  5  2  4  3  3  0  2  1  2  4  4
 8  8  4  7  3  7  6  6  5  5  4  4  3  4  3  3  3  2  2  3  1  2  0  1  2  4  4
 7  7  3  6  2  6  5  5  4  4  3  4  2  3  2  2  4  1  3  2  2  1  1  0  1  3  3
 6  6  4  7  1  5  6  4  5  3  4  5  3  4  3  3  5  2  4  1  3  2  2  1  0  2  2
 4  4  2  7  1  3  6  2  5  1  4  5  3  4  5  3  6  2  6  1  5  4  4  3  2  0  2
 6  6  2  5  3  5  4  4  3  3  2  5  1  4  5  3  6  2  6  1  5  4  4  3  2  2  0
```

Some of the shortest paths are achieved as follows:

```
Shortest paths
Shortest path from1 to2:
1->6->2
Distance=2
Shortest path from1 to3:
1->6->8->10->3
Distance=4
Shortest path from1 to4:
1->6->8->10->13->11->9->7->4
Distance=9
Shortest path from1 to5:
1->6->8->10->26->5
```

```
Distance=5
Shortest path from1 to6:
1->6
Distance=1
Shortest path from1 to7:
1->6->8->10->13->11->9->7
```

2.3.3.2   *Dijkstra algorithm*

To find between-vertex shortest path of an undirected graph, Dijkstra algorithm can be used (Dijkstra, 1959). First, define the weight matrix $D = (d_{ij})$ of an undirected graph X, where d_{ij} is the weight of the edge e_{ij}. $d_{ij} = 0$ if $i = j$; $d_{ij} > 0$ if there exists an edge e_{ij}, and $d_{ij} = \infty$ if there is no edge e_{ij}. Suppose the two vertices are A (starting vertex) and B (terminal vertex), then the Dijkstra algorithm is as follows (Zhang, 2012e):

(1) Label v_A as $v_A = 0$, and the other vertex v_i as $v_i = \infty, i \neq A$.
(2) Label the unlabeled vertex v_j as

$$v_j = (j_{\text{old}}, i_{\text{old}} + d_{ij})$$

(3) Find the minimum of labels and take the label as the fixed label of the vertex; return to (2), until B is labeled. The shortest path and its length are thus achieved.

The following is the Matlab function, Dijkstra.m, for Dijkstra algorithm:

```
function [ss,pat,distances,paths]=Dijkstra(d)
%d: weighted adjacency matrix; ss: total number of paths; pat: number of
   paths
%passing through each vertex; distances: matrix of distances between
different
%vertices; paths: string of paths and distances between any of two vertices.
v=size(d,1);
p=zeros(1,v); w=zeros(1,v); a=zeros(1,v); b=zeros(1,v);
pat=zeros(1,v);
distances=zeros(v);
for i=1:v
for j=1:v
if ((d(i,j)==0) & (i~=j)) d(i,j)=inf; end
end; end
paths='';
su=0;
```

```
for j=1:v-1
for k=j+1:v
for i=1:v
p(i)=0; w(i)=0;
a(i)=inf;
end
a(j)=0; w(j)=1; n=j; h=0;
while (v>0)
ma=inf;
for i=1:v
if (w(i)==1) continue; end
iv=d(n,i)+a(n);
if (iv<a(i)) a(i)=iv; b(i)=n; end
if (a(i)>ma) continue; end
ma=a(i); h=i;
end
w(h)=1;
if (h==k) break; end
n=h;
end
sd=a(k); p(1)=k; c=k;
for i=2:v
if (c==j) break; end
p(i)=b(c); c=b(c);
end
paths=strcat(paths,'The shortest path from',num2str(j),'to',
   num2str(k),':\n');
for i=v:-1:1
if ((p(i)~=0) & (sd~=inf))
if (i>1) paths=strcat(paths,num2str(p(i)),'->'); end
if (i<=1) paths=strcat(paths,num2str(p(i)),'\n'); end
end; end
for i=v:-1:1
for h=1:v
if (p(i)==h) pat(h)=pat(h)+1; break; end
end; end
if (sd~=inf) paths=strcat(paths,'Distance=',num2str(sd),'\n');
   distances(j,k)=sd;
distances(k,j)=sd; end
if (sd==inf) paths=strcat(paths,'No path','\n'); su=su+1; end
end; end
ss=v*(v-1)/2-su;
```

2.3.4   *Circuit algorithm*

2.3.4.1   *Paton's fundamental circuit finding algorithm*

The following is the algorithm of fundamental circuit set developed by Paton (1969) (Zhang, 2017b). Suppose the vertex set of a graph X

is $V = \{1, 2, \ldots, n\}$, its adjacency matrix is D, the set of the vertices already on the tree is T, and the set of the vertices to be tested is S. Let $1 \in T, S = V$, and the vertex 1 be the tree root, then

(1) If $T \cap S = \phi$ terminate calculation.
(2) If $T \cap S \neq \phi$, choose a vertex in $T \cap S$.
(3) Sequentially test every edge associated with the vertex v; if there is no edge to be tested, then remove v from S and return to (1).
(4) If there exists an edge (v, w) to be tested, test whether the vertex w is in T or not.
(5) If $w \in T$, find out the edge (v, w) and the fundamental circuit generated by the unique path (in the tree) that links v and w; remove the edge (v, w) from the graph and return to (3).
(6) If $w \notin T$, add the edge (v, w) to the tree and the vertex w to T; remove the edge (v, w) from the graph and return to (3).

The Matlab function for calculating fundamental circuit set, fundCircuit.m, is as follows:

```
function [num,n,circuits]=foundCircuit(d)
%d: weighted adjacency matrix; num: total number of fundamental circuits; n:
number of fundamental circuits containing each vertex; circuits: string of
all circuits
v=size(d,1);
l=zeros(1,v); vp=zeros(1,v); ts=zeros(1,v); circuit=zeros(1,v*(v-1)/2);
n=zeros(1,v);
num=0;
for i=1:v
l(i)=-1;
end
circuits='';
t=1;
while (v>0)
its=1;
ts(1)=t; l(t)=0;
while (v>0)
if (its==0) break; end
r=ts(its); lm=l(r)+1;
for w=1:v
if (d(r,w)<=0) continue; end
if ((d(r,w)>0) & ((l(w)+1)==0))
ts(its)=w;
its=its+1;
vp(w)=r; l(w)=lm;
d(r,w)=0; d(w,r)=0;
```

```
continue; end
num=num+1; a=vp(w);
m=1;
circuit(1)=r; j=r;
while_(v>0)
j=vp(j);
m=m+1;
circuit(m)=j;
if (j==a) break; end
end
m=m+1;
circuit(m)=w;
circuits=strcat(circuits,'Number of fundamental circuit:',
   num2str(num),'\n');
circuits=strcat(circuits,'Fundamental circuit: ');
for j=1:m
circuits=strcat(circuits,num2str(circuit(j)),'->');
end
circuits=strcat(circuits,num2str(circuit(1)),'\n');
for i=1:v
for j=1:m
if (circuit(j)==i) n(i)=n(i)+1; break; end
end; end
d(r,w)=0; d(w,r)=0;
end
its=its-1;
end
la=0;
for t=t:v
if (l(t)==-1) la=1; break; end
end
if (la==1) continue; end
break;
end
```

Using Paton algorithm and the adjacency matrix of tumor pathway p53 (Huang and Zhang, 2012; Li and Zhang, 2013; Zhang, 2017b), the calculated fundamental circuits of p53 pathway were

```
Number of fundamental circuit:1
Fundamental circuit:47->26->50->24->47
Number of fundamental circuit:2
Fundamental circuit:47->26->50->51->52->48->47
Number of fundamental circuit:3
Fundamental circuit:37->47->35->37
Number of fundamental circuit:4
Fundamental circuit:16->18->50->51->52->48->16
Number of fundamental circuit:5
Fundamental circuit:31->29->16->18->50->51->52->48->31
```

```
Number of fundamental circuit:6
Fundamental circuit:49->51->52->48->49
Number of fundamental circuit:7
Fundamental circuit:28->52->4->28
Number of fundamental circuit:8
Fundamental circuit:2->14->52->8->2
Number of fundamental circuit:9
Fundamental circuit:2->14->52->10->2
Number of fundamental circuit:10
Fundamental circuit:2->14->52->12->2
Number of fundamental circuit:11
Fundamental circuit:5->2->14->52->4->5
Number of fundamental circuit:12
Fundamental circuit:5->2->14->52->7->5
Number of fundamental circuit:13
Fundamental circuit:9->52->7->9
```

2.3.4.2   *Chan's circuit matrix algorithm*

For an undirected and connected graph X with v vertices and e edges, an algorithm to calculate the circuit matrix C of X is (Chan *et al.*, 1982)

(1) Arbitrarily choose a tree T from a graph X and write out the fundamental circuit matrix C_a of the tree T.
(2) Perform all possible ring sum operations in C_a and construct a new matrix C_b by adding original rows of C_a together with the new rows generated from ring sum operations.
(3) Eliminate redundant rows, i.e., the rows of the circuits with disjoint edges in C_b.
(4) Construct the circuit matrix C from the remaining rows of C_b. The algorithm can also be used to test the graph planarity (Zhang, 2012e).

2.3.5   *Matching problem*

Matching problem is an important part of graph theory (Zhang, 2012e). It has important applications in the optimal assignment problem.

Let M be a subset of the edge set E of a graph X. If any two edges of M are disjoint in X, M is called a matching of X. The two vertices of an edge of M are called matched in M. If an edge of the matching M is associated with vertex v, then M saturates vertex v, otherwise v is called M-unsaturated. The matching M is called the optimum matching if every vertex of X is M-saturated. M is called a maximum matching of X, if there is no other matching M' in X, such that $|M'| > |M|$. Every optimum matching is a maximum matching.

In a graph X, the M-augmenting path refers to a staggered path that both initial vertex and terminal vertex are M-unsaturated.

Theorem 13. *The matching M of graph X is a maximum matching if and only if there is no M-augmenting path in X.*

2.3.5.1 *Maximum matching*

Discuss an algorithm to find maximum matching of a bipartite graph, i.e., Hungarian Algorithm (Zhang, 2012e).

An example for maximum matching problem of a bipartite graph is that, suppose there are m tree species with which to green n hills, we want to find a plan that optimizes biodiversity conservation. Here, any of m tree species can be planted on one or more hills. Not all tree species will certainly be planted on any hill. The problem is whether we can assign each tree species to a hill for planting (Zhang, 2012e).

We denote m tree species and n hills as $G = \{g1, g2, \ldots, gm\}$ and $H = \{h1, h2, \ldots, hn\}$, respectively. gi and hj are adjacent if and only if the tree species gi can be planted on the hill hj. The problem is thus to find a maximum matching of graph X.

The principle of Hungarian Algorithm is, starting from any matching M of a graph X, search for M-augmenting path for all M-unsaturated vertices in X. If there is no M-augmenting path, then M is the maximum matching. If there exists a M-augmenting path C, then exchange the M edges and non-M edges in C to obtain a matching $M1$ which has one more edge than M. Repeat the above process on $M1$. Suppose $X = (G, H, E)$ is a bipartite graph, where $G = \{g1, g2, \ldots, gm\}, H = \{h1, h2, \ldots, hn\}$, arbitrarily find an initial matching M of X, and then

(1) Let $S = \phi, T = \phi$, return to (2).
(2) If M saturates all vertices of $G - S$, then M is the maximum matching of bipartite graph X. Otherwise, arbitrarily find a M-unsaturated vertex $u \in G - S$, let $S = S \cup \{u\}$, return to (3).
(3) Let $N(S) = \{v | u \in S, uv \in E\}$. If $N(S) = T$, return to (2). Otherwise, find $h \in N(S) - T$. If h is M-saturated, return to (4), otherwise return to (5).
(4) Suppose $gh \in M$, then let $S = S \cup \{g\}, T = T \cup \{h\}$, return to (3).
(5) $u - h$ path is a M-augmenting path, denoted by C, and let $M = M \oplus C$, return to (1), where, $M \oplus C = M \cup C - M \cap C$.

The Matlab codes, maxMatch.m, of Hungarian Algorithm for maximum matching problems are as follows. Data is $a = (a_{ij})_{v*n}$, where v is the number of elements in G and n is the number of elements in H.

```
a=input('Input the excel file name of data matrix for maximum matching
problem (e.g., a.xls, etc. Data is a=(aij)v*n,where v is the number of
elements in G, and n is the number of elements in H):','s');
a=xlsread(a);
v=size(a,1); n=size(a,2);
c=zeros(n,5); m=zeros(v,n);
g=zeros(1,v); h=zeros(1,n); hh=zeros(1,n);
cp=0;
for i=1:v
for j=1:n
if (a(i,j)~=0)
m(i,j)=1;
break;
end; end
if (m(i,j)~=0) break; end
end
while (v>0)
g=zeros(1,v); h=zeros(1,n);
for i=1:v
cd=1;
for j=1:n
if (m(i,j)~=0) cd=0; break; end
end
if (cd~=0) g(i)=-n-1; end
end
cd=0;
while (v>0)
gi=0;
```

```
for i=1:v
if (g(i)<0)
gi=i;
break;
end; end
if (gi==0)
cd=1;
break;
end
g(gi)=-g(gi);
k=1;
for j=1:n
if ((a(gi,j)~=0) & (h(j)==0))
h(j)=gi;
hh(k)=j;
k=k+1;
end; end
if (k>1)
k=k-1;
for j=1:k
cp=1;
for i=1:v
if (m(i,hh(j))~=0)
g(i)=-hh(j);
cp=0;
break;
end; end
if (cp~=0) break; end
end
if (cp~=0)
k=1;
j=hh(j);
while (v>0)
c(k,2)=j;
c(k,1)=h(j);
j=abs(g(h(j)));
if (j==(n+1)) break; end
k=k+1;
end
for i=1:k
if (m(c(i,1),c(i,2))~=0) m(c(i,1),c(i,2))=0;
else m(c(i,1),c(i,2))=1; end
end
break;
end; end
end
if (cd~=0) break; end
end
fprintf(['Maximum matching:' '\n']);
m
```

Suppose there are six objects in G and five areas in H. The data matrix for maximum matching is

$$
\begin{array}{ccccc}
1 & 0 & 1 & 0 & 0 \\
0 & 1 & 0 & 1 & 0 \\
0 & 1 & 1 & 0 & 1 \\
1 & 0 & 1 & 0 & 0 \\
0 & 1 & 0 & 1 & 1 \\
1 & 0 & 1 & 0 & 1
\end{array}
$$

Using the algorithm, we obtained the maximum matching matrix as the follows:

$$
\begin{array}{ccccc}
1 & 0 & 0 & 0 & 0 \\
0 & 1 & 0 & 0 & 0 \\
0 & 0 & 1 & 0 & 1 \\
0 & 0 & 1 & 0 & 0 \\
0 & 0 & 0 & 1 & 0 \\
0 & 0 & 0 & 0 & 0
\end{array}
$$

And the maximum matching is: (object 1, area 1), (object 2, area 2), (object 3, area 3, area 5), (object 4, area 3), (object 5, area 4).

2.3.5.2   *Optimum matching*

The procedures for optimum matching are as follows (Zhang, 2012e). Suppose the graph $X = (U, V, C, D)$ is a complete bipartite and weighted graph, L is a label for initial available vertex. Let

$$
L(p) = \max\{D(pq)|q \in V\}, \quad p \in U
$$
$$
L(q) = 0, \quad q \in V
$$

M is a matching of XL.

(1) If every vertex of U is M-saturated, then M is the optimum matching, otherwise find a M-unsaturated vertex $w \in U$, let $S = \{w\}, T = \phi$, return to (2).
(2) Let $NL(S) = \{v|w \in S, wv \in XL\}$. If $NL(S) = T$, then XL has no optimum matching, otherwise return to (3); or else return to (4).

(3) Adjust the label of available vertex. Calculate

$$IL = \min\{L(p) + L(q) - D(pq)|p \in S, \quad q \in V - T\}$$

to obtain the new available vertex label

$$G(v) = L(v) - IL, \quad v \in S$$
$$G(v) = L(v) + IL, \quad v \in T$$
$$G(v) = L(v), \quad \text{others}$$

Let $L = G, XL = XG$, once again give a matching M of XL, return to (1).

(4) Choose $q \in NL(S) - T$, if q is M-saturated, return to (5), otherwise return to (6).

(5) If $pq \in M$, then let $S = S \cup \{p\}, T = T \cup \{q\}$, return to (2).

(6) The $w - q$ path in XL is the M-augmenting path, denoted by P, and let $M = M \oplus P$, return to (1), where $M \oplus P = M \cup P - M \cap P$.

Chapter 3

Fundamentals of Network Theory

Graph is the mathematical terminology of network (Bohman, 2009). In a narrow sense, however, network is a type of graph. A network X refers to a directed graph (the fundamental directed graph of X) with two specific vertex subsets A and B, where A and B are disjoint and nonempty. The vertices in A are called source vertices and vertices in B are called sink vertices. The vertices not belonging to both A and B are called intermediate vertices. The set of intermediate vertices is denoted by I. For example, in the network of Fig. 1, $A = \{v_1, v_2, v_3\}$, $B = \{v_7, v_8\}$, $I = \{v_4, v_5, v_6\}$.

In this book or other studies, the terminology of graph and network is always used without a clear distinction between them (Zhang, 2012e). Therefore, networks share almost the same theoretical system

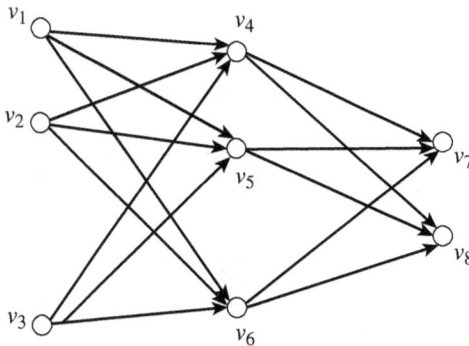

Fig. 1 A classific network (Zhang, 2012e).

and methodology with graphs. In the network, vertices and edges are usually called nodes and links (arcs), respectively. In some cases, network edges are assigned weights, and the network is thus a weighted graph.

3.1 Network Topology

There are generally six types of network topology, including cellular topology, bus topology, tree topology, star topology, ring topology, and mesh topology. Among these, the former five are mainly layouts for various information transmission (Zhang, 2016), and the last one, mesh topology, is the most popular network topology for biological networks.

3.1.1 *Cellular topology*

Cellular topology is commonly used in wireless LANs. It is characterized by point-to-point and multi-point transmission through wireless transmission medium (microwave, satellite, infrared, etc.) and is widely used in urban networks, campus networks, corporate networks, etc. (Kovalevsky, 2004).

3.1.2 *Bus topology*

With a single trunk cable as the transmission medium, all the sites are directly connected to the transmission bus through the corresponding interface. In the bus topology, all nodes share a data channel, and any node can send information along the bus in two directions. The bus topology is characterized by simple and flexible structures, and the failure of a site does not affect other sites (Barranco *et al.*, 2011).

3.1.3 *Tree topology*

In essence, tree topology is an expansion of the star topology. In a complex network, the hierarchical arrangement of star topology induces the tree topology. In principle, tree topology is also the extension of bus topology. Tree topology produces hierarchical branches in the network of bus topology. In a network of tree topology, the information originating from any node can be received by the remaining nodes. In such networks, all connections are fixed and the structure is symmetric, with a certain fault tolerance. Some of the larger LANs, campus networks, etc., are tree-like networks (Li et al., 2012).

3.1.4 Star topology

Star topology is the most common structure of LAN. The star topology has a central node which is connected with the surrounding nodes. Each node and the central node are connected in the point-to-point way, and the surrounding nodes are arranged radially around the central node. The communication between any two nodes must be routed through the central node. The central node performs the control strategy of centralized communication. Thus the central node is quite complex, the burden is heavy, and it is easy to form communication bottleneck. The entire network will be paralyzed if the central node fails (Cena and Valenzano, 2013).

3.1.5 Ring topology

Ring topology networks are also known as token ring networks. Each node in the ring topology is self-connected to its starting and end point to form a closed ring. Each node is connected to a closed ring road through a loop interface. Any node on the loop can send a "token" to request communication with other nodes. If the request is approved, it can send information to the loop. The ring line is public, so the information issued in the ring transmits in the direction of the loop. The data in the ring topology can be either bidirectional or unidirectional.

Ring topology is suitable for information collection system, information processing system, and automation system with stricter time restriction. However, communication token is easy to lose in ring topology.

3.1.6 *Mesh topology*

In the mesh topology, each node is connected to other nodes. The mesh topology has a higher reliability, but its structure is complex and difficult to manage and maintain (Marina *et al.*, 2010).

3.2 Random and Complex Networks

3.2.1 *Random network*

Suppose the frequency of the vertices with degree k is p_k, vertex degree is therefore a random variable. The probability distribution of vertex degree is called degree distribution.

In a random network, vertex degree is binomially distributed, and its limit distribution is Poisson distribution. In the random network, the majority of vertices have the same degree with the averaged. It is the simplest network.

In the formation of Erdos–Renyi random graph, each time a pair of vertices of n vertices are randomly (uniformly distributed) connected and added to the graph (Erdos and Renyi, 1959), the random graph demonstrates small-world effect and is an approximation of the real world (Zhang, 2012e).

Another method for generating a random graph, proposed by Watts and Strogatz (1998), is (1) produce a complete graph in which each vertex has k adjacent edges; (2) for each vertex, the edge is reoriented with probability p; (3) generate a complete random graph if $p = 1$; the generated graph is likely similar to a small-world network, if $0 < p < 1$.

3.2.2 *Complex networks*

So far, numerous research works have been done for simple networks. Many results and methods have been obtained from these studies. However, the network systems in the past years have become more and more complex. Occasionally, there are millions of vertices and edges in a complex network. Graph theory, combinatorial optimization, statistics, stochastic processes, etc., are quickly becoming the scientific basis and effective tools for studying complex networks.

Dependent on network sources, the complex networks can be divided into four categories, namely social networks, information networks, technological networks, and biological networks (Zhang, 2012e).

(1) Social networks reflect the interaction patterns of human population. Social networks include friend networks, business networks, family networks, WWW, etc.
(2) Information networks refer to literature indexing networks, computer networks, etc.
(3) Technological networks include electrical systems, power networks, energy grids, transportation networks, banking networks, telephone networks, Internet, etc.
(4) Biological networks include metabolic networks, protein interaction networks, gene regulatory networks, food webs, neural networks, the blood circulatory system, etc.

3.2.2.1 *Measurement of complex networks*

Watts and Strogatz (1998) found that a complex network could be measured with two topological indices, clustering coefficient (the average number of adjacent edges of a vertex, C) and the length of characteristic path (the median of average shortest distances between two vertices, L). A complex system with shorter characteristic path and larger clustering coefficient is considered to be a small-world network. Small-world network is an important tool to approach complex systems (Zhang, 2012e).

Suppose the total number of vertices is n, the number of adjacent vertices of vertex i is n_i. The clustering coefficient of the network is

$$C = (1/n) \sum_{i=1}^{n} C_i$$

where C_i is the clustering coefficient of vertex i

$$C_i = 2/[n_i(n_i - 1)] \sum_{j=1}^{n_i} \sum d_j$$

where $d_i = 1$, if $e_{ij} \in E_i$, otherwise $d_i = 0; 0 \leq C \leq 1$. For a vertex v and an arbitrary vertex j, the distance between the two vertices is d_{vj}. For d_{vj} of each vertex v, calculate the average d_v. The length of characteristic path L is the median of all d_v's.

Clustering coefficient and length of characteristic path should be divided by the corresponding values of the complete network (that is, the network in which every vertex has the same number of adjacent edges) with the same size (i.e., the same number of vertices and edges) to stabilize the network.

In the complex network, degree distribution is typically a power-law distribution, and the network is called the scale-free network (Barabasi and Albert, 1999). Many complex networks, such as Internet, metabolic networks, communication networks, etc., are scale-free networks. Scale-free property is an important feature of complex networks. The possible mechanism for scale-free property include the following: (1) with the addition of new vertices, the network continues to expand; (2) new vertices tend to connect to already better connected vertices (Barabasi and Albert, 1999).

Some complex networks are exponentially degree-distributed, such as energy grids and railway networks, while others are complex networks with the exponent power-law degree distribution, such as cooperation networks, etc.

Actually, the degree distribution of networks is a continuum (Zhang, 2016f). Power-law distribution is just a typical model. Various distribution models and indices can be used to measure the network complexity (Huang and Zhang, 2012; Zhang, 2012d, 2012e; Zhang and Zhan, 2011).

According to the research, the average length of network paths is almost unchanged if some vertices are randomly removed from the complex network. However, it will grow rapidly as vertices are removed in a targeted manner.

A study of the topology complex networks will help facilitate deep understanding of the network dynamics, for example, the influence of species loss on ecosystems. Although the dynamics of complex networks are extremely important, the research is still seldom due to a lack of relevant knowledge.

In general, at present, there are not enough mathematical tools, methods, and performance indicators to treat complex networks. Further studies on network dynamics and behaviors are required (Newman *et al.*, 2006; Ferrarini, 2011a, 2011b, 2013a, 2013b; Nedorezov, 2012; Zhang, 2012c, 2015b, 2016f).

3.2.2.2 *Generation and calculation of complex networks*

Barabasi and Albert (1999) presented a method to generate scale-free random graphs. Suppose there are n vertices. For every pair of vertices i and j, an edge is generated with probability p. When a vertex is connected and added, all edges are then generated in the way similar to the generation of random graph. The new vertex traverses all existing vertices but is connected and added with a fixed probability. However the probability values are different. The drawback of this algorithm is that the vertex connected and added earlier tends to be the attractor more likely, which is different from the real world. To this end, they gave a fitness value for each vertex The rate of a vertex for accepting a new connection is positively proportional to its fitness value. At the start of the calculation, each vertex is assigned a uniform distributed fitness value. When the vertex is assigned a new edge, it increases the fitness value. With the increase of the connected edges, the fitness value grows and a positive feedback loop is formed (Zhang, 2012e).

The algorithm proposed by Cancho and Sole (2001) can generate a variety of complex networks with diverse degree distributions. In the algorithm, the optimize the energy function $E(\lambda)$ is

$$E(\lambda) = \lambda d + (1 - \lambda)\rho$$
$$\rho = m/(n - 1)$$

where $0 \le \lambda \le 0.25$, and m is the average degree of vertices. Reorganize the random network and change λ. As λ increases, the resultant network is subsequently a random network, exponential network, and scale-free network. If $\lambda > 0.25$, the resultant network is a star network (Zhang, 2012e).

3.3   Network Analysis

Between-node relationship is a binary relationship. The methodology of network analysis provides a formalism to represent, to measure, and to simulate relational structures (Butts, 2009). Weighting edges and representing multi-edges relationship with super-edge are allowed to improve binary relationship (Wasserman and Faust, 1994). In addition, the dynamic changes of the relationship may be treated as a time series (Newman *et al.*, 2006; Zhang, 2012c, 2015b, 2016f).

The topological structure of network changes with the addition of nodes and links (Zhang, 2012c, 2013b, 2015b, 2016f, 2016r). That is, to explain the topological structure of a network, we first need to describe how the network came into being.

Occasionally, a network will not evolve in a continuous way during its formation. A small change in control variables would result in sudden and substantial structure changes (phase transitions) of network, e.g., the sudden changes in connection structure, which is called "explosive diffusion" (Achlioptas *et al.*, 2009; Bohman, 2009). A network may experience several phase transitions during its formation in which the size of the giant component will change. Here, the component refers to a set of nodes in which any of pairs of nodes are connected with links. Suppose there are m randomly connected links. The network may have many small components if $m < n/2$. If m is slightly larger than $n/2$, there will be a unique component which contains many nodes, while other components are still very small. With the addition of new edges the giant component will gradually increase, and finally cover all nodes (Bollobás, 2001). Occurrence and evolution of giant component is an important part of network dynamics.

The size of node set may significantly affect the resultant network. In order to avoid misleading conclusions, the definition of the node set should include all the different entities, and they are relevant to the relationship being studied (Butts, 2009). Moreover, the classification of entities is also important (Butts, 2009). For example, an ecologist wants to examine inter animal interactions. To construct a network, he (she) will take samples in some designated area, record all the interactions, and may also record the environment–animal relationships (Zhang, 2007, 2010, 2011a, 2011b, 2011c, 2012b, 2012e, 2013a, 2014b, 2015c, 2015d; Zhang and Li, 2015a, 2015b; Zhang *et al.*, 2014b, 2014d).

3.4 Basic Algorithms

In this section, the maximum flow and minimum cost flow algorithms are described. A transportation network is a connected directed network with n nodes (vertices). The node v_0 is the source node, and it has no terminal link (edge). The node v_n is the sink node, and it has no initial link. The weight c_{ij} of directed link e_{ij} refers to the maximal transportation capacity of the link; f_{ij} is the flux of the link, $0 \leq f_{ij} \leq c_{ij}$, and the influx sum and outflux sum of node v_j are the same (Chan *et al.*, 1982; Zhang, 2012e, 2017c)

$$\sum_{i=0}^{n-1} f_{ij} = \sum_{k=1}^{n} f_{jk}, \quad j = 1, 2, \ldots, n-1$$

The outflux sum of source node is equal to the influx sum of sink node.

3.4.1 *Maximum flow*

The maximum flow problem is

$$\max \sum_{j=1}^{n} f_{0j}$$

Max-flow and min-cut theorem (Ford–Fulkerson Theorem). In a transportation network, the value of maximum flow is equal to the capacity of a minimum cutting of the network.

Based on Ford–Fulkerson Theorem, Ford and Fulkerson (1957). proposed an algorithm to find maximum flow. The principle of the algorithm is, starting from a given flow, recursively construct a series of flows with increasing values and terminate at the maximum flow. Once every new flow f is generated, if there exists an f-incremental path g, then find the path g, construct a revised flow

f' which is based on g, and take it as the next flow of the series. If there is no f-incremental path g, terminate calculation, and f is the maximum flow.

The main procedures of Ford–Fulkerson algorithm is as follows (Chan *et al.*, 1982; Zhang, 2012e, 2017):

(1) Labeling

    (a) Label the source node vs with $(+, +\infty)$, $ds = +\infty$.

    (b) Choose a labeled node x. For all unlabeled adjacent nodes y of x, handle them using the following rules

        If $yx \in E$ and $fyx > 0$, let $dy = \min\{fyx, dx\}$, and label y with (x, dy).

        If $xy \in E$ and $fxy < Cxy$, let $dy = \min\{Cxy - fxy, dx\}$, and label y with $(x+, dy)$.

    (c) Repeat step (b) until the sink node vt has been labeled or no more nodes can be labeled. If vt is labeled, then there exists an augmenting chain, and so continue to procedure (2) to adjust the process; if vt is not labeled, labeling process is not able to be conducted, and f is thus the maximum flow.

(2) Adjusting

    (a) Determine adjusting magnitude $d = dvt$, and let $u = vt$.

    (b) If the node u is labeled by $(v+, du)$, then replace fvu with $fvu + d$, and if the node u is labeled by (v, du), then replace fvu with $fvu + d$.

    (c) If $v = vs$, then remove all labels and return the procedure (1) for labeling again; otherwise let $u = v$, return (b).

        If the computation terminates, let the set of labeled nodes be S, and cutset (S, Sc) is the minimum cut, and the maximum flow is $Mf = C(S, Sc)$.

The following theorems are based on max-flow and min-cut theorem (Ford and Fulkerson, 1956, 1957; Chan *et al.*, 1982; Zhang, 2012e).

Theorem 1. *Suppose X is a network with v_0 as the source node and v_n as the sink node, and each link has a unit capacity*

(1) *The value of maximum flow in X is equal to the maximal number of directed (v_0, v_n) paths with unrepeated links in X.*

(2) *The capacity of minimum cutset in X is equal to the minimal number of such links; all directed (v_0, v_n) paths in X will be damaged if these links are eliminated.*

Theorem 2. *A graph X is k-edge connected if and only if any two different vertices in X are linked by a path with at least k unrepeated edges.*

Theorem 3. *Suppose v_0 and v_n are two vertices of a graph X, then the maximal number of the (v_0, v_n) paths with unrepeated edges is equal to the minimal number of such edges; if these edges are eliminated, then all (v_0, v_n) paths in X will be damaged.*

The following are Matlab codes, maxiFlow.m, for Ford–Fulkerson algorithm. The Matlab algorithm needs the user to load an excel file that stores the Two Array Listing data of the form (d_{1i}, d_{2i}, c_i), where d_{1i}, d_{2i}, and c_i are start nodes and end node of the link i, and the flow capacity of the link i, respectively, $i = 1, 2, \ldots, e$.

```
%Ford--Fulkerson algorithm for maximum flow in the network (graph).
%v: the number of nodes; Data are stored in Two Array Listing.
%d1(1-e), d2(1-e): start and end nodes of links; d(1-e): flow capacity of
    links.
%Nodes are sequentially numbered in the network.
d=input('Input the excel file name of Two Array Listing data (e.g., adj.xls,
    etc.
Data is d=(d1i, d2i, ci), where d1i, d2i, and ci are start node and end node
of the link i, and the flow capacity of the link i, respectively,
i=1,2,...,e): ','s');
d=xlsread(d);
e=size(d,1);
d1=d(:,1); d2=d(:,2); dd=d(:,3);
v=max(max([d1 d2]));
c=zeros(v); f=zeros(v);
no=zeros(1,v); d=zeros(1,v);
for i=1:v
for j=1:v
for k=1:e
if ((d1(k)==i) & (d2(k)==j))
c(i,j)=dd(k);
break;
end; end; end; end
for i=1:v
```

```
for j=1:v f(i,j)=0;
end; end
for i=1:v
no(i)=0;
d(i)=0;
end
pr=1;
while(v>0)
no(1)=v+1;
d(1)=1e+30;
while(v>0)
pr=1;
for i=1:v
if (no(i)~=0)
for j=1:v
if ((no(j)==0) & (f(i,j)<c(i,j)))
no(j)=i;
d(j)=c(i,j)-f(i,j);
pr=0;
if (d(j)>d(i)) d(j)=d(i); end
end
if ((no(j)==0) & (f(j,i)>0))
no(j)=-i;
d(j)=f(j,i);
pr=0;
if (d(j)>d(i)) d(j)=d(i); end
end
end
end; end
if ((no(v)~=0) |(pr~=0)) break; end
end
if (pr~=0) break; end
dv=d(v);
s=v;
while (v>0)
if (no(s)>0) f(no(s),s)=f(no(s),s)+dv; end
if (no(s)<0) f(no(s),s)=f(no(s),s)-dv; end
if (no(s)==1)
for i=1:v
no(i)=0;
d(i)=0;
end
break;
end
s=no(s);
end; end
mf=0;
for j=1:v
mf=mf+f(1,j);
end
```

```
fprintf(['Maximum flow matrix:' '\n']);
f
fprintf(['Maximum flow=:' num2str(mf) '\n']);
fprintf(['Labels for minimum cut:' '\n']);
no
```

As an example, suppose there are eight nodes and 11 links in a network. The data are as follows:

Node	Node	Flow capacity
1	2	5
1	3	1
1	4	3
2	4	2
2	6	3
3	5	1
3	8	3
4	7	2
5	8	7
6	7	3
7	8	4

Using the algorithm above, the maximum flow matrix is achieved as follows:

$$
\begin{matrix}
0 & 2 & 1 & 2 & 0 & 0 & 0 & 0 \\
0 & 0 & 0 & 0 & 0 & 2 & 0 & 0 \\
0 & 0 & 0 & 0 & 0 & 0 & 0 & 1 \\
0 & 0 & 0 & 0 & 0 & 0 & 2 & 0 \\
0 & 0 & 0 & 0 & 0 & 0 & 0 & 0 \\
0 & 0 & 0 & 0 & 0 & 0 & 2 & 0 \\
0 & 0 & 0 & 0 & 0 & 0 & 0 & 4 \\
0 & 0 & 0 & 0 & 0 & 0 & 0 & 0
\end{matrix}
$$

The maximum flow is 5, and the labels for minimum cut are 9, 1, 0, 1, 0, 2, 6, 0.

3.4.2 *Minimum cost flow*

Minimum cost flow problem refers to that in the maximum flow problem above, set transportation cost a_{ij} on each link, try to achieve the flux $\sum f_{0j}$ with v_0 as the source node and v_n as the sink node, and minimize the total cost (Zhang, 2017c)

$$\min \sum_{i,j} a_{ij} f_{ij}$$

Given a network $X = (V, E, C)$ with n nodes, v_i, $i = 1, 2, \ldots, n$. In the maximum flow problem, assign each link a cost a_{ij}. With v_0 as the source node and v_n as the sink node, calculate the flow $\sum f_{0j}$, and minimize the total cost (Chan *et al.*, 1982; Zhang, 2012e, 2017c)

$$\min \sum_{i,j} a_{ij} f_{ij}$$

Take the initial available flow f as the zero flow. The algorithm to solve minimum cost flow problem is (Chan *et al.*, 1982; Zhang, 2012e, 2017c)

(1) Generate a weighted directed graph $Xf = (V, Ef, F)$, for any $e_{ij} \in E, Ef$ and F are defined as

$$e_{ij} \in Ef, F(e_{ij}) = b_{ij}, \qquad \text{if } f_{ij} = 0$$
$$e_{ji} \in Ef, F(e_{ji}) = b_{ij}, \qquad \text{if } f_{ij} = C_{ij}$$
$$e_{ij} \in Ef, F(e_{ij}) = b_{ij}, e_{ji} \in Ef, F(e_{ji}) = -b_{ij}, \quad \text{if } 0 < f_{ij} < C_{ij}$$

(2) Find the shortest path h from source node vs to sink node vt in the weighted directed graph $Xf = (V, Ef, F)$ (Zhang, 2016a–2016r). If there exists a shortest path h, return (3), otherwise f is the maximum flow with minimum cost, terminate the algorithm.

(3) Enhance flow. The same procedures as finding maximum flow, let

$$d_{ij} = c_{ij} - f_{ij}, \qquad \text{if } e_{ij} \in h^+$$
$$d_{ij} = f_{ij}, \qquad \text{if } e_{ij} \in h^-$$
$$d = \min\{d_{ij} | e_{ij} \in h\}$$

and define the flow $f = \{f_{ij}\}$ as

$$f_{ij} = f_{ij} + d, \quad \text{if } e_{ij} \in h^+$$
$$f_{ij} = f_{ij} - d, \quad \text{if } e_{ij} \in h^-$$
$$f_{ij} = f_{ij}, \qquad \text{otherwise}$$

If *Mf* is not less than the desired flow, then reduce *d*, such that *Mf* is equal to the desired flow, by doing so *f* is the minimum cost flow, and terminate the algorithm, otherwise return (1).

The following are Matlab codes, minCost.m, for the minimum cost flow algorithm. The Matlab algorithm needs the user to load an excel file that stores the Two Array Listing data of the form $(d_{1i}, d_{2i}, c_i, b_i)$, where d_{1i}, d_{2i}, c_i, and b_i are start node and end node of the link *i*, flow capacity of the link *i*, and cost per unit flow of link *i*, respectively, $i = 1, 2, \ldots, e$.

```
%Data are stored in Two Array Listing.
da=input('Input the excel file name of Two Array Listing data (e.g.,
    adj.xls, etc.
Data is d=(d1i, d2i, cci, bbi),where d1i, d2i, cci, bbi are start node and
end node of the link i, flow capacity of the link i, and cost of unit flow
of link i, respectively, i=1,2,¡−,e): ','s');
da=xlsread(da);
e=size(da,1);
d1=da(:,1); d2=da(:,2); cc=da(:,3); bb=da(:,4);
v=max(max([d1 d2]));
c=zeros(v);
a=c; b=c;
p=zeros(1,v); s=p;
for i=1:v
for j=1:v
for k=1:e
if ((d1(k)==i) & (d2(k)==j))
c(i,j)=cc(k);
b(i,j)=bb(k);
break;
end; end; end; end
mf=0;
mf0=inf;
f=zeros(v);
while (v>0)
for i=1:v
for j=1:v
if (j~=i) a(i,j)=inf; end
end; end
for i=1:v
for j=1:v
if ((c(i,j)>0) & (f(i,j)==0)) a(i,j)=b(i,j);
elseif ((c(i,j)>0) & (f(i,j)==c(i,j))) a(j,i)=-b(i,j);
elseif (c(i,j)>0)
a(i,j)=b(i,j);
a(j,i)=-b(i,j);
end
```

```
end; end
for i=2:v
p(i)=inf;
s(i)=i;
end
for k=1:v
d=1;
for i=2:v
for j=1:v
if (p(i)>(p(j)+a(j,i)))
p(i)=p(j)+a(j,i);
s(i)=j;
d=0;
end; end; end
if (d~=0) break; end
end
if (p(v)>=inf) break; end
dv=inf;
m=v;
while (v>0)
dvt=-inf;
if (a(s(m),m)>0) dvt=c(s(m),m)-f(s(m),m);
elseif (a(s(m),m)<0) dvt=f(m,s(m)); end
if (dv>dvt) dv=dvt; end
if (s(m)==1) break; end
m=s(m);
end
d=0;
if ((mf+dv)>=mf0)
dv=mf0-mf;
d=1;
end
m=v;
while (v>0)
if (a(s(m),m)>0) f(s(m),m)=f(s(m),m)+dv;
elseif (a(s(m),m)<0) f(m,s(m))=f(m,s(m))-dv; end
if (s(m)==1) break; end
m=s(m);
end
if (d~=0) break; end
mf=sum(f(1,:));
end
mmf=sum(sum(b.*f));
fprintf(['Maximum flow with minimum cost:' '\n']);
f
fprintf(['Maximum flow with minimum cost=' num2str(mf) '\n']);
fprintf(['Minimum cost=' num2str(mmf) '\n']);
```

Suppose there are five nodes and seven links in a network. The data are as follows:

From-node	To-node	Flow capacity	Cost per unit flow of the link
1	2	7	3
1	4	5	6
2	3	7	3
2	5	3	4
3	4	6	5
3	5	4	2
4	5	8	4

Using the algorithm above, we achieve the maximum flow matrix with minimum cost as follows:

$$
\begin{matrix}
0 & 7 & 0 & 5 & 0 \\
0 & 0 & 4 & 0 & 3 \\
0 & 0 & 0 & 0 & 4 \\
0 & 0 & 0 & 0 & 5 \\
0 & 0 & 0 & 0 & 0
\end{matrix}
$$

The maximum flow with minimum cost is 12 and the minimum cost is 103.

Chapter 4

Other Fundamentals

Network biology is based on a wide range of mathematical areas. It is impossible to fully discuss all of them in a book. Some of the mathematical fundamentals are described below. Details can be found in Zhang (2010, 2016r).

4.1 Bayes' Rule

Bayes' rule is a fundamental theory in network science. Bayes' rule states that the probability that a postulate A will be true is positively proportional to the multiplication of the postulate's prior probability and the conditional probability of information, I, being observed given H is true.

Known a discrete sample space S, and suppose $A_i \in S$, $i = 1, 2, \ldots$, where $\cup A_i = S$, and $A_i \cap A_j = \phi$, $i \neq j$, Bayes' rule is expressed as

$$p(A_i/I) = p(I/A_i)p(A_i) \bigg/ \sum p(I/A_j)p(A_j)$$

For the continuous sample space, Bayes' rule is

$$p(A/I) = p(I/A)p(A) \bigg/ \int p(I/A)p(A)\,dA$$

4.1.1 *Model selection*

Bayes' rule can be used in the selection of models (Mackey, 1992). Suppose there are several models, and the data or information I is known, the posterior probability of model A_i, is given by Bayes' rule

$$p(A_i/I) = p(I/A_i)p(A_i)/p(I)$$

where $p(A_i)$ is the prior probability of the model A_i; $p(I/A_i)$ is the evidence of the model A_i, which is expressed as

$$p(I/A_i) = \int p(I/w, A_i)p(w/A_i)\,dw$$

where $w = (w_1, w_2, \ldots, w_n)$ is a weight vector.

4.1.2 *Bayesian learning*

Without the information or samples I, let $p(w)$ be a prior distribution of network weights, where $w = (w_1, w_2, \ldots, w_n)$. When the information or samples are known, the posterior distribution is

$$p(w/I) = p(I/w)p(w) \left/ \int p(I/w)p(w)\,dw \right.$$

Without any prior information, the prior distribution $p(w)$ and the conditional distribution $p(I/w)$ can be represented by exponential distributions (Yan and Zhang, 2000)

$$p(w) = \exp(-cf(w))/h(c)$$
$$p(I/w) = \exp(-ag(I))/q(a)$$

And the posterior distribution of network weights is obtained

$$p(w/I) = \iint p(w/c, a, I)p(c, a/I)\,dcda$$

4.2 Linear Regression

4.2.1 *Single-variable linear regression*

A single-variable linear regression, with the dependent variable y and the independent variable x, is

$$y = a + bx$$

where

$$b = \left(\sum (xy) - \left(\sum x \sum y \right) / n \right) \Big/ \left(\sum x^2 - \left(\sum x \right)^2 / n \right)$$

$$a = \sum y / n - b \sum x / n$$

and n: number of samples.

4.2.2 *Multiple-variable linear regression*

A multiple-variable linear regression is (Qi, 2004)

$$y_i = b_{0_i} + b_{1_i} x_1 + b_{2_i} x_2 + \cdots + b_{mi} x_m, \quad i = 1, 2, \ldots, p$$

Given n samples and a data matrix $(x_{ij})_{n \times m}$, for the independent variable, and a data matrix $(y_{ij})_{n \times p}$, for the dependent variable. Let $L = (l_{ij})$, $i, j = 0, 1, \ldots, m$; $D = (d_{ij})$, $i = 0, 1, \ldots, m$; $j = 1, 2, \ldots, p$. Let $l_{00} = n$, and calculate

$$l_{0j} = \sum_{k=1}^{n} x_{kj}, \quad l_{i0} = \sum_{k=1}^{n} x_{ki}, \quad l_{ij} = \sum_{k=1}^{n} x_{ki} x_{kj},$$

$$i, j = 1, 2, \ldots, m$$

$$d_{0j} = \sum_{k=1}^{n} \sum y_{kj}, \quad d_{ij} = \sum_{k=1}^{n} x_{ki} y_{kj},$$

$$i = 1, 2, \ldots, m; \quad j = 1, 2, \ldots, p$$

The matrix for correlation coefficients is $b = L^{-1} * D$, where $b = (b_{ij})$, $i = 0, 1, \ldots, m$; $j = 1, 2, \ldots, p$.

4.3 Randomization, Bootstrap, and Monte Carlo Methods

Randomization, bootstrap, and Monte Carlo methods are usually used in network biology. Details of randomization, bootstrap, and Monte Carlo techniques are found in Gentle (2002), Manly (1997), Zhang (2010), etc.

4.3.1 *Random numbers*

4.3.1.1 *General random numbers*

Random numbers are always used in randomization, bootstrap, and Monte Carlo methods.

To generate a series of uniformly distributed random numbers (0–1), the Matlab codes are (Mathworks, 2015)

```
x=rand;
m=rand(5,6);
```

Randomly permutate the integer numbers, $1 - n$

```
n=10;
x=randperm(n);
```

However, the random number produced by Matlab generator is a pseudo-random number. The generator resets once Matlab is initiated. The generated sequence of random numbers is thus the same. The following codes record the present status (R) of the generator, and set the status to R

```
R=rand(''state'');
rand(''state'',R);
```

Different status of the generator can be set at any time

```
rand(''state'',sum(100*clock));
```

4.3.1.2 *Probability distributions and random numbers*

The random numbers of normal distribution (norm), χ^2 distribution (chi^2), t distribution (t), F distribution (f), β distribution (beta), uniform distribution (unif), exponential distribution (exp), etc., can be easily generated in Matlab environment. The commands of probability density, probability distribution, inverse probability distribution, mean and variance, and random number are pdf, cdf, inv, stat, and rnd, respectively. Various combinations of the corresponding commands will generate different probability distributions and random numbers (Zhang, 2010, 2016r)

```
x=5;
%Yield a density value of normal distribution with mean 5
%and standard deviation 30
p=normpdf(x,5,30);
%Yield a random number of normal distribution with mean 3
%and standard deviation 25
r=normrnd(3,25);
%Yield a value of probability distribution of t-distribution with
%degree of freedom 8
s=tcdf(x,8);
%Yield a 0.95 percentile of F-distribution with degrees of freedom
%3 and 9
g=finv(0.95,3,9);
```

The random numbers of a probability distribution can be generated on the basis of uniformly distributed random numbers (Gentle, 2002; Zhang, 2010). The methods include:

(1) **Inverse transformation.** Given the probability distribution $F(x)$, and its inverse function F^{-1}, generate an uniformly distributed random number on [0,1], $X \sim U(0,1)$, and take $X = F^{-1}(U)$.

(2) **Convolution.** Given the random variable $Y = X_1 + X_2 + \cdots + X_n$, where X_i, $i = 1, 2, \ldots, n$, are independent and they share the same probability distribution, first generate X_i, $i = 1, 2, \ldots, n$, and take $Y = X_1 + X_2 + \cdots + X_n$.

(3) **Acceptance–rejection.** Suppose the density function to be found is $f(x)$; first, get a density function $g(x)$, and a constant c, such that $f(x) \leq cg(x)$; second, generate a random number x of $g(x)$; take $r = cg(x)/f(x)$; finally, generate a uniformly distributed random number u, thus x is the desired random number if $ru < 1$, or else repeat the above procedure.

The Matlab codes for generating the random numbers of various probability distributions are (Zhang, 2010, 2016r)

```
%Generate a matrix of uniformly distributed random numbers
rand(3,4)
%Generate a matrix of normally distributed random numbers,
%N(3,25²)
random('norm',3,25,5,7);
%Generate a matrix of F-distributed random numbers with
%degrees of freedom 10 and 20
random('f',10,20,6,8)
%Generate a matrix of binomially distributed random numbers,
%B(5,0.3)
random('bino',5,0.3)
```

4.3.1.3 *Multivariable random numbers*

Multivariable random numbers are usually generated on the basis of single random variables (Gentle, 2002). Suppose $X = (X_1, X_2, \ldots, X_n)$ is a random vector, where X_i, $i = 1, 2, \ldots, n$, are independent and they share the same probability distribution with variance 1 and mean 0. Build a random vector, $Y = AX$, with covariance matrix AA^{T}, where $|A| \neq 0$. Find A, such that $AA^{\mathrm{T}} = \sum$, where $\sum$ is the desired covariance matrix.

4.3.2 *Randomization-based data partition*

Data partition includes cross-validation, jackknife, etc. Data partition is used to mine more information from a limited data set (Gentle, 2002; Zhang, 2010, 2016r).

4.3.2.1 *Cross-validation*

In the cross-validation, we randomly or systematically divide the data set (x_i, y_i), $i = 1, 2, \ldots, n$, into two parts, the training set (T) and the validation set (V). The training set is used to determine parameters in the fitted model $y = f(x)$, and the validation set is used to validate and test the model. Moreover, the training set and validation set can be exchanged to estimate fitting error in x

$$E(R(Y_0, f(x_0))) = \left(\sum_{i \in V} R(Y_i, f_1(x_i)) + \sum_{i \in T} R(Y_i, f_2(x_i)) \right) \Big/ n$$

where $f_1(x)$, $f_2(x)$: the fitted functions using training set T and V respectively, and $R(Y_0, f(x_0))$: prediction error.

A data set can be divided into several intersected subsets (Breiman, 2001), e.g., K subsets with similar sizes. Choose a subset as the validation set and the remaining subsets as the training set in order to calculate prediction error. The averaged error for K subsets is the prediction error to be estimated.

4.3.2.2 Jackknife method

A data set is systematically divided in jackknife method in order to obtain the estimates, e.g., variance or mean, from the data set (Gentle, 2002). Suppose the statistic T of a random sample X_1, $X_2, \ldots, X_n$, is the estimate of population parameter θ. Divide the data set into r subsets, each with m elements (m is always 1 or 2).

Remove subset i from the data set and calculate the estimate T_{-i} from the remaining subsets. The estimate of population parameter θ is

$$T' = \sum_{i=1}^{r} T_{-i}/r$$

Moreover, the jackknife T is

$$J(T) = T^{*} = \sum_{i=1}^{r} T_i^*/r = rT - (r-1)T'$$

where $T_i^* = rT - (r-1)\,T_{-i}$. If T_i^* are independent, the variance of T can be estimated from the jackknife variance $V(J(T))$

$$V(J(T)) = \sum_{i=1}^{r} (T_i^* - J(T))^2/(r(r-1))$$

or

$$V(J(T)) = \sum_{i-1}^{r} (T_i^* - T)^2/(r(r-1))$$

Suppose $m = 1$, and the deviation of T may be extended as (Gentle, 2002)

$$D(T) = \sum_{i=1}^{\infty} a_i/n^i$$

$D(T)$ is unbiased, if $a_i = 0$, $i = 1, 2, \ldots$; $D(T)$ has 2nd-order precision, if $a_1 \neq 0$. Jackknife estimate of deviation of T is thus

$$D(J(T)) = E(J(T)) - \theta = n \sum_{i=1}^{\infty} a_i/n^i - (n-1) \sum_{i=1}^{\infty} a_i/(n-1)^i$$

4.3.3 Bootstrap method

In bootstrap method, an observed sample is treated as a population and the population is resampled. Infer the statistic distribution according to the conditional distribution of the sample taken from the population (Gentle, 2002; Zhang, 2010, 2016r). Given the observed sample x_i, $i = 1, 2, \ldots, n$, and the population parameter θ, statistic T is used to estimate θ. First, the sampling distribution of T should be determined in order to achieve the confidence interval of θ. Statistic T is the functional of empirical cumulative distribution function P_n, i.e.,

$$T = T(P_n) = \theta(P_n) = \int g(x) dP_n(x)$$

In bootstrap method, the observed sample x_i, $i = 1, 2, \ldots, n$, is resampled to generate a resampling sample x_i^*, $i = 1, 2, \ldots, n$, and the corresponding statistic is T^*. The variance of T is

$$V(T) = V(T^*) = \sum (T^{*j} - T_{\text{bar}}^*)^2/(m-1)$$

where T^{*j}: the jth T^* of T, and T_{bar}^*: the average of T^{*^i} variance of m samples taken from P_n, with size n.

The confidence interval of θ can be derived [1] tionship between θ and T, e.g., $f(T, \theta)$, and the

(Gentle, 2002; Zhang, 2010, 2016r)

$$P(f_{\alpha/2} \le f(T, \theta) \le f_{1-\alpha/2}) = 1 - \alpha$$

in which $f_{\alpha/2}$ and $f_{1-\alpha/2}$ can be approximated with bootstrap method if the probability of population is not available. $f_{\alpha/2}$ and $f_{1-\alpha/2}$ are determined using the percentiles of Monte Carlo samples of $T^* - T_o$, where T_o is the T value of observed sample.

4.3.4 *Monte Carlo method*

Monte Carlo method is used to test the characteristics of statistical methods, approximate the distribution of a statistic with asymptotic approximation, and compute the expectation of some function of random variables (Manly, 1997; Zhang, 2010, 2016r).

First, Monte Carlo method is used in function approximation. For example, we want to estimate $F(x)$ from $f(x)$. Given a set of random inputs x_i, $i = 1, 2, \ldots, n$, the corresponding outputs are $y_i = f(x_i)$, $i = 1, 2, \ldots, n$. The mean and variance of $f(x)$ can thus be obtained.

Monte Carlo method is also used to estimate the variance, test statistic significance, etc. A procedure for statistical test is as follows (Gentle, 2002; Zhang, 2010, 2016r): generate a random sample from observed sample and compute the characteristic of the random sample; repeat the procedure n times (n randomizations); finally, test the null hypothesis, i.e., observed samples are from the same distribution. The statistic p, is

$$p = r/n$$

or

$$p = (r + 1)/(n + 1)$$

where r is the number of the characteristic of the random sample greater than that of observed sample.

Random sampling of a data set and subset generation from the data set are also applications of Monte Carlo method. In addition, missing data can be generated by Monte Carlo method (Gentle, 2002). A data matrix X is composed of the observed data block and the missing data block. We can generate m missing data and analyze the corresponding m complete data blocks.

4.4 Stochastic Process

A stationary stochastic process is the stochastic process that meets the condition (Zhang, 2010, 2016r)

$$p(x(t_i) = c_i | i = 1, 2, \ldots, n) = p(x(t_i + \tau) = c_i | i = 1, 2, \ldots, n),$$

$$\forall t_i, \ c_i, \ \tau \in R$$

A stochastic process is reversible if

$$p(x(t_i) = c_i | i = 1, 2, \ldots, n) = p(x(\tau - t_i) = c_i | i = 1, 2, \ldots, n),$$

$$\forall t_i, \ c_i, \ \tau \in R$$

4.5 Optimization Methods

4.5.1 *Steepest descent method*

The steepest descent method is an unconstrained optimization technique (Zhang, 2010, 2016r). Known an unconstrained optimization problem

$$\min f(x)$$

where $x = (x_1, x_2, \ldots, x_n)^T$, and $f(x)$ is the nonlinear function of x, calculate $t^i \in R$, such that

$$f(x^i + t^i p^i) = \min_t f(x^i + t p^i)$$

The searching direction p^i is determined by

$$p^i = -\nabla f(x^i) = -(\partial f / \partial x_1, \partial f / \partial x_2, \ldots, \partial f / \partial x_n) | x = x^i$$

The next point, $x^{i+1} = x^i + t^i p^i$, is thus achieved.

4.5.2 *Conjugate gradient method*

Suppose the objective function $f(x)$ is approximately a quadratic function in the neighborhood of the extreme point x^*

$$f(x) \approx a + b^T x + x^T A x / 2$$

Calculate

$$p_0 = -b - A x_0$$

$$g_i = b + A x_i$$

$$\beta_{i-1} = \|g_i\|^2 / \|g_{i-1}\|^2$$

$$p_i = -g_i + \beta_{i-1} p_{i-1}$$

such that

$$f(x^i + t^i p^i) = \min_t f(x^i + t p^i)$$

The next point, $x^{i+1} = x^i + t^i p^i$, is thus achieved. The iteration terminates, if $\|g_i\| \le \varepsilon$.

4.5.3 *Newton's method*

Suppose the objective function $f(x)$ can be expanded as a quadratic Taylor polynomial near the neighborhood of the point x^i

$$f(x) \approx f(x^i) + \nabla f(x^i)^T \Delta x + \Delta x^T A^i \Delta x / 2$$

where $\Delta x = x - x^i$. Calculate $t^i \in R$, such that

$$f(x^i + t^i p^i) = \min_t f(x^i + t p^i)$$

The searching direction p_i is determined by

$$p^i = -(A^i)^{-1} \nabla f(x^i)$$

The next point, $x^{i+1} = x^i + t^i p^i$, is finally achieved.

4.5.4 *DFP method*

Given an unconstrained optimization problem

$$\min f(x)$$

where $x = (x_1, x_2, \ldots, x_n)^{\mathrm{T}}$, and $f(x)$ is the nonlinear function of x. Suppose the initial values $x_0 = (x_{10}, x_{20}, \ldots, x_{n0})$, and the iterative error $e > 0$. $H_0 = I$, and I is the unit matrix. Search in the direction of $-H_0 f'(x_0)$ in order to determine λ_0

$$\min f(x_0 - \lambda H_0 f'(x_0)) = f(x_0 - \lambda_0 H_0 f'(x_0))$$

then the solution $x_1 = x_0 - \lambda_0 H_0 f'(x_0)$. Repeat the process, we obtain x_k. Let

$$g'_{k-1} = f'(x_k) - f'(x_{k-1}), \quad dx_{k-1} = x_k - x_{k-1}$$

Calculate the scaling matrix

$$H_k = H_{k-1} + dx_{k-1}dx_{k-1}/(g'_{k-1}dx_{k-1})$$
$$- H_{k-1}g'_{k-1}g'_{k-1}H_{k-1}/(g'_{k-1}H_{k-1}g'_{k-1})$$

and search in the direction of $-H_k f'(x_k)$ to determine λ_k

$$\min f(x_k - \lambda H_k f'(x_k)) = f(x_k - \lambda_k H_k f'(x_k))$$

The solution is $x_{k+1} = x_k - \lambda_k H_k f'(x_k)$. If $||dx_k|| < e$, x_{k+1} is the optimization solution; otherwise let $k = k+1$, repeat the process.

4.5.5 *Dynamic programming*

Dynamic programming is based on Bellman's principle (Bellman, 1957), which states that the optimal decision sequence of a multiphase decision process has such a property, i.e., regardless of what the initial phase, take the phase and state generated in the first decision as the initial condition, then the subsequent decisions must constitute the optimal sequences with respect to the corresponding problem (Norton, 1972; Li, 1982).

The procedures of the discrete and deterministic dynamic programming are as follows (Zhang, 2010, 2012e, 2016r).

First, divide the process into n phases, $k = 1, 2, \ldots, n$, and determine the state x_k of phase k, where x_k is an initial state of phase k.

Second, determine the decision variables in each phase. Assume $u_k(x_k)$ is the decision variable in phase k when its state is x_k. $u_k(x_k) \in D_k(x_k)$, where $D_k(x_k)$ is the permissible decision set of phase k. The decision function series from phase k to the end point, i.e., the substrategies are

$$P_{kn} = \{u_k(x_k),\ u_{k+1}(x_{k+1}), \ldots, u_n(x_n)\}$$

Third, determine the rules for state transition between phases. Given the state variable x_k of phase k, and using the decision variable u_k, then the state x_{k+1} of phase $k+1$ can be determined, i.e., $x_{k+1} = T_k(x_k, u_k)$.

Fourth, define the goal function. The goal function V_{kn} is used to assess the goodness of the procedure

$$V_{kn} = V_{kn}(x_k, u_k, x_{k+1}, \ldots, x_{n+1}), \quad k = 1, 2, \ldots, n$$

where the optimal value of V_{kn} is the optimal goal function $f_k(x_k)$. Formula of the goal function is

$$V_{kn} = v_k(x_k,\ u_k) + V_{k+1n}(x_{k+1}, \ldots, x_{n+1})$$

where $v_k(x_k, u_k)$ is the goal value of phase k. Goal function is the function of initial condition and strategies, so its calculation formula can be written as

$$V_{kn}(x_k,\ P_{kn}) = v_k(x_k,\ u_k) + V_{k+1n}(x_{k+1}, P_{k+1n})$$

where $P_{kn} = \{u_k(x_k), P_{k+1n}(x_{k+1})\}$

Finally, conduct the reverse sequence optimization

$$\text{opt}(P_{kn})V_{kn}(x_k,\ P_{kn}) = \text{opt}(u_k)\{v_k(x_k,\ u_k) + \text{opt}(P_{k+1n})V_{k+1n},$$
$$k = n,\ n-1, \ldots, 1$$
$$f_1(x_1) = \text{opt}(P_{1n})V_{1n}(x_1,\ P_{1n})$$

or

$$f_k(x_k) = \text{opt}(u_k \in D_k(x_k))\{v_k(x_k, \ u_k) + f_{k+1}(x_{k+1})\},$$
$$k = n, n - 1, \ldots, 1$$
$$f_{n+1}(x_{n+1}) = 0$$

where opt denotes minimization (min) or maximization (max). Starting from $k = n$, calculate ahead until $f_1(x_1)$ is obtained. And the optimal strategy and optimal value of goal function are obtained.

4.6 Functional Analysis

4.6.1 *Functional representation*

The functional relationship, $s = f(r)$, where $s = y(t)$, and $r = x(t)$, is called an operator. The functional relationship, $y(t) = f(t, x(\tau)|\tau \le t)$, is generally called a functional (Yan and Zhang, 2000; Zhang, 2010, 2016r). Usually, a linear functional is represented by

$$f(t, x(\tau)|\tau \le t) = \int g(t, \tau)x(\tau)d\tau$$

A common nonlinear functional is the nth order regular homogeneous functional

$$f(t, x(\tau)|\tau \le t) = \int\!\!\int \ldots \int g(t, \tau_1, \tau_2, \ldots, \tau_n)$$
$$\times x(\tau_1)x(\tau_2)\ldots x(\tau_n)d\tau_1 d\tau_2 \ldots d\tau_n$$

A functional, $f(t, x(\tau)|\tau \le t)$, can be expanded as a Volterra series

$$f(t, x(\tau)|\tau \le t) = \sum \int \ldots \int g(t, \tau_1, \tau_2, \ldots, \tau_n)x(\tau_1)$$
$$\times x(\tau_2)\ldots x(\tau_n)d\tau_1 d\tau_2 \ldots d\tau_n$$
$$= g_0(t) + \int g(t, \tau)x(\tau)d\tau$$

$$+ \iint g(t, \tau_1, \tau_2) x(\tau_1) x(\tau_2) d\tau_1 d\tau_2 + \cdots$$

$$+ \int \cdots \int g(t, \tau_1, \tau_2, \ldots, \tau_n) x(\tau_1) x(\tau_2) \ldots x(\tau_n)$$

$$\times d\tau_1 d\tau_2 \ldots d\tau_n + \cdots$$

which is equivalent to the three-tier feed-forward artificial neural network (Yan and Zhang, 2000).

4.6.2 *Functional analysis*

Some methods for functional analysis can be used in the analysis of mathematical models. Some principles and methods of functional analysis are described in the following (Chen, 1987; Rudin, 1991; Liu, 2000; Men and Feng, 2005; Zhang, 2010, 2016r).

Deflation Principle 1. Suppose (X, d) is a complete metric space, a mapping $T : X \to X$, such that

$$d(Tx, Ty) \le \theta d(x, y), \quad \forall x, y \in X$$

where $\theta \in (0, 1)$, then there is only one fixed point $x' \in X$, such that $Tx' = x'$.

Deflation Principle 2. Suppose (X, d) is a complete metric space, and $T : X \to X$, is a mapping. If there is a natural number n_0, such that

$$d(T^{n_0} x, T^{n_0} y) \le \theta d(x, y), \quad \forall x, y \in X$$

where $\theta \in [0, 1)$, then there is only one fixed point $x' \in X$, such that $Tx' = x'$.

According to Deflation Principle 1, take $x_0 \in X$, and perform iteration, $x_{n+1} = Tx_n$. If $\{x_n\}$ is sequential convergent, then the limit of $\{x_n\}$ is the fixed point x'. Moreover, the error to approximate x' with x_n is

$$d(x_n, x') \le \theta^n d(x_0, Tx_0) / (1 - \theta)$$

The nearer x_0 approaches Tx_0, the smaller the error is.

As an example, consider a differential equation

$$dx/dt = f(x,t), x|_{t=t_0} = x_0$$

where $f(x,t)$ is continuous on R^2, and satisfies with the condition

$$|f(x_1,t) - f(x_2,t)| \leq K|x_2 - x_1|$$

in respect to x. Let's show that the above problem has a unique solution in the neighborhood of t_0.

Take $\delta > 0$, such that $K\delta < 1$. Define an operator on $C[t_0 - \delta, t_0 + \delta]$

$$Tx(t) = \int_{t_0}^{t} f(x(\tau), \tau)d\tau + x_0$$

then T is the mapping of R^2 to itself, and

$$d(Tx_1, Tx_2) = \max_{|t-t_0|\leq\delta} \left| \int_{t_0}^{t} (f(x_1(\tau), \tau) - f(x_2(\tau), \tau))d\tau \right|$$

$$\leq \max_{|t-t_0|\leq\delta} \left| \int_{t_0}^{t} K|x_2(\tau) - x_1(\tau)|d\tau \right|$$

$$\leq K\delta \max_{|t-t_0|\leq\delta} |x_2(\tau) - x_1(\tau)|$$

$$= K\delta d(x_1, x_2)$$

The space $C[t_0 - \delta, t_0 + \delta]$ is complete, and $0 \leq K\delta < 1$. The existence and uniqueness of solution is thus shown by the Deflation Principle.

Suppose X and Z are normed spaces, and $S(T)$ is the linear subspace of X. If the mapping $T : S(T) \to Z$, satisfies the following conditions:

$$T(x + y) = Tx + Ty, \ T(\alpha x) = \alpha Tx, \quad \forall x, y \in S(T), \ \alpha \in K$$

then T is said to be a linear operator from inside X to inside Z. $S(T)$ is the domain of definition of T, then T is said to be a linear operator on X into Z, if $S(T) = X$. f is called linear functional if it is a linear operator on normed space X into number field K. For the linear operator T on normed space X into normed space Z, $\exists M \in K$, such

that $||Tx|| \leq M||x||, \forall x \in X$, T is called bounded linear operator.
T is a linear operator on normed space X into normed space Z, then
T is continuous if and only if T is bounded.

Hahn–Banach Theorem on Real Space. Suppose M is the linear
subspace of the linear real space X, $g : X \to R$, and

$$g(x+y) \leq g(x) + g(y), \quad g(\alpha x) = \alpha g(x), \quad \forall x, y \in X, \alpha \geq 0$$

In addition, f is a linear functional on M and

$$f(x) \leq g(x), \quad x \in M$$

then there is a linear functional $p(x)$ on X, such that

$$p(x) = f(x), \quad x \in M$$

$$-g(-x) \leq p(x) \leq g(x), \quad x \in X$$

Riesz Theorem 1. *Suppose f is the bounded linear functional on
$C[a, b]$, then there is a function of bounded variation $v(t)$ on $[a,b]$,
such that*

$$f(x) = \int_a^b x(t)\, dv(t), \quad x \in C[a, b]$$

$$||f|| = V(v)$$

*where $V(v)$ is the total variation of $v(t)$ on $[a,b]$. Further, based
on any function of bounded variation $v(t)$ on $[a,b]$, we can define a
bounded linear functional on $C[a,b]$ through above expression.*

Riesz Theorem 2. *Suppose H is a Hilbert space, and f is an arbitrary bounded linear functional on H, then there is only one $y_f \in H$,
such that*

$$f(x) = (x, y_f), \quad \forall x \in H$$

$$||f|| = ||y_f||$$

Hilbert–Schmidt Theorem. *Suppose T is a self-conjugate compact operator on Hilbert space H, then there is an orthonormal system*

$\{e_n\}$, *which is composed of eigenvectors corresponding to eigenvalues* $\{\lambda_n\}$, $\lambda_n \neq 0$, *such that*

$$x = \sum \alpha_n e_n + x_0 \ Tx_0 = 0$$

$$Tx = \sum \lambda_n \alpha_n e_n$$

and if $\{e_n\}$ *is infinite, then*

$$\lim_{n \to \infty} \lambda_n = 0$$

4.7   Algebraic Topology

Homology method is usually used to solve local minimum problem (Lin, 1998; Zhang, 2010, 2016r).

Using homotopy method to find the zero point of the nonlinear function $f(x)$, we need to get a related and simpler function $g(x)$.

First, try to obtain the zero point of $g(x)$, and gradually transit to get the zero point of $f(x)$. In homotopy method, we construct a homotopy function

$$H(t, x) = (1 - t)g(x) + t\,f(x)$$

where t is the parameter variable. In the training, t will gradually change from 0 to 1, and

$$H(0, x) = g(x), \quad \text{if } t = 0; \text{ zero point is easy to obtain}$$

$$H(1, x) = f(x), \quad \text{if } t = 1; \text{ zero point is what we need}$$

Trace the trajectory $x_0(t)$ of zero point of $H(t, x)$ when t changes from 0 to 1, and thus perform solution transit from $x_0(0)$ to $x_0(1)$ to obtain zero point.

4.8   Entropy of Systems

The probability model of a dynamic system can be developed according to the principle of maximum entropy and the principle of minimum relative information (Zhang, 2010, 2012e, 2016r).

4.8.1 *Maximum entropy*

If the prior distribution of a micro-state is not given, the principle of maximum entropy is expressed as the following optimum problem:

$$\max \ H = -\sum_{i=1}^{n} p_i \log p_i$$

$$\sum_{i=1}^{n} p_i f(x_i) = f^-(x_i)$$

$$\sum_{i=1}^{n} p_i = 1$$

where $x(t) = (x_1(t), x_2(t), \ldots, x_n(t))^{\mathrm{T}}$, and $f(x) = (f(x_1), f(x_2), \ldots, f(x_n))^{\mathrm{T}} \geq 0$, is the state function, which is an analogue of energy function; $f^-(x)$ is the mean of $f(x)$; and p_i is the occurrence probability of micro-state x_i. Using Lagrange multiplier method, the maximum entropy can be achieved as follows:

$$H_{\max} = \mu + \mu f(x)$$

where $\mu \geq 0$, is the Lagrange parameter ($1/\mu$ is the analogue of temperature in thermodynamics) and

$$L(\mu) = \sum_{i=1}^{n} \exp(-\mu f(x_i))$$

$$\mu = \log L(\mu)$$

$$p_i = \exp(-\mu f(x_i))/L(\mu), \quad i = 1, 2, \ldots, n$$

4.8.2 *Minimum relative information*

If the prior distribution of a micro-state, $(p_1^0, p_2^0, \ldots, p_n^0)$, is given, the principle of minimum relative information is represented by

(Zhang, 2010, 2012e, 2016r)

$$\min \ I = \sum_{i=1}^{n} p_i \log(p_i/p_i^0)$$

$$\sum_{i=1}^{n} p_i \, f(x_i) = f^-(x_i)$$

$$\sum_{i=1}^{n} p_i = 1$$

Minimum relative information minimizes the difference between $\{p_i\}$ and $\{p_i^0\}$. It is the generalization of principle of maximum entropy. The former is applicable to both continuous and discrete systems. For a continuous system, the principle of minimum relative information is represented by

$$\min \ I = \int p(x) \log(p(x)/p^0(x)) \, dx$$

$$\int p(x) f(x) \, dx = f^-(x)$$

$$\int p(x) \, dx = 1$$

The solution of this problem is

$$L(\mu) = \int p^0(x) \exp(-\mu f(x)) \, dx$$

$$p(x) = p^0(x) \exp(-\mu f(x))/L(\mu)$$

$$I_{\min} = -\mu f^-(x) - \log L(\mu)$$

4.8.3 *Minimum mean energy*

Minimum mean energy describes the convergence degree of a system to its limit state, given a certain degree of disorder. The principle of minimum mean energy is represented by (Zhang, 2010, 2012e, 2016r)

$$\min f^-(x) = \sum_{i=1}^{n} p_i \, f^-(x_i)$$

$$-\sum_{i=1}^{n} p_i \log p_i = E$$

$$\sum_{i=1}^{n} p_i = 1$$

And the solution is

$$L(\mu) = \sum_{i=1}^{n} \exp(-\mu f(x_i))$$

$$p_i = \exp(-\mu f(x_i))/L(\mu), \quad i = 1, 2, \ldots, n$$

4.8.4 *Probability distribution with maximum entropy*

The entropy of exponential distribution

$$p(x) = \lambda \exp(-\lambda x), x > 0, \quad \lambda > 0$$

$$p(x) = 0, \quad x \le 0$$

is $H = \log \lambda - 1$. Among all of the density functions with the same mean, the exponential distribution has the maximum entropy.

The entropy of the normal distribution is

$$H = -\int_{-\infty}^{\infty} p_0(x) \log p(x) dx$$

where $p_0(x)$ is an arbitrary density function having the same mean and variance with the normal distribution $p(x)$. Among all of the density functions with the same mean and variance, the normal distribution has the maximum entropy (Zhang, 2010, 2012e, 2016r).

Part 2

Crucial Nodes/Subnetworks/ Modules, Network Types, and Structural Comparison

Chapter 5

Identification of Crucial Nodes and Subnetworks/Modules

Crucial nodes are the nodes governing the structure and dynamics of a network (Junker, 2006). Missing or even small changes of crucial nodes may substantially change the network. Identification of crucial nodes is a fundamental procedure in network analysis (Pimm *et al.*, 1991; Montoya *et al.*, 2006; Butts, 2009; Ding, 2012; Zhang, 2012d, 2012e). In this chapter, I introduce some mathematical methods for identifying crucial nodes in networks.

Some networks (including some biological networks) consist of subnetworks/modules. In certain cases, we need to identify subnetworks/modules in a network. In this chapter, I introduce some methods for identifying subnetworks/modules.

5.1 Features of Crucial Nodes

Generally, crucial nodes should possess the following features (Zhang, 2012d, 2012e):

(1) Their existence is indispensable in preserving topological structure and functionality of a network.
(2) They determine network structure independently.
(3) Crucial nodes are closely related to other nodes in the network.

5.2　Indices and Methods of Crucial Nodes

5.2.1　*Node perturbation index*

Node perturbation index is defined as (Zhang, 2012d)

$$N_i = dN/dn/N$$

or

$$N_i = dN/dn$$

where N: a measure of network structure; n: state value or proportion of node i in the network. Theoretically, N_i distributes normally, i.e., $N_i \approx 0$ for most nodes, but crucial nodes have N_i much larger or less than 0.

The measure of network structure N can be total links, total number of nodes, network flow (Latham, 2006), degree distribution (Zhang, 2011c; Zhang and Zhan, 2011), aggregation index, coefficient of variation, entropy (Zhang and Zhan, 2011; Zhang, 2012e), and others (Paine, 1992; Power *et al.*, 1996; Dunne *et al.*, 2002; Montoya and Sole, 2003; Allesina *et al.*, 2005; Barabasi, 2009).

Alternatively, N_i can be expressed as

$$N_i = (N_0 - N_t)/N_0/n_0$$

or

$$N_i = (N_0 - N_t)/N_0$$

where N_t, N_0: a measure of network structure after and before node i is completely removed from the network, respectively; n_0: state value or proportion of the node in the network before the node is removed from the network. $N_i \approx 1$ if the functionality of the node is positively proportional to its state value or proportion in the network; $N_i \approx -1$, if the functionality of the node is negatively proportional to its state value or proportion in the network; $N_i \gg 1$, if the node is a crucial node.

N_i is a general index that can be further materialized in various ways.

5.2.2 *Criticality index*

Criticality index is defined as (Zhang, 2012d)

$$C_i = \sum_{c=1}^{n}(1 + C_{bc})/d_c + \sum_{e=1}^{m}(1 + C_{fe})/f_e$$

where C_i: value of criticality index of node i; n: the number of source nodes directing to target node i; d_c: the number of target nodes of the cth source node, and C_{bc}: the backward-oriented criticality index of the cth source node. Similarly, m: the number of target nodes of source node i; f_e: the number of source nodes of the eth-target node, and C_{fe}: the forward-oriented criticality index of the eth-target node.

The nodes with larger C_i tend to be crucial nodes. This index takes both forward- and backward-oriented between-node relationships into consideration, and only the nodes within the same network can be compared for their relative importance.

This index was defined based on the keystone index, etc. (Jordán *et al.*, 1999, 2006; Jordán, 2001; Zhang, 2012d, 2012e).

5.2.3 *Degree change index*

Degree change index is defined as (Zhang, 2012d)

$$D_i = \sum_{j=1}^{v}[|(O_{tj} - O_{0j})/O_{0j}| + |(I_{tj} - I_{0j})/I_j|]$$

or

$$D_i = \sum_{j=1}^{v}(|O_{tj} - O_{0j}| + |I_{tj} - I_{0j}|)$$

where D_i: value of degree change index of node i; v: total number of nodes in the network; O_{tj}, O_{0j}: out-degree of node j after and before node i is changed, respectively; I_{tj}, I_{0j}: in-degree of node j after and before node i is changed, respectively.

The nodes with larger D_i tend to be crucial nodes.

5.2.4 *Flow change index*

Flow change index is defined as (Zhang, 2012d)

$$F_i = \sum_{j=1}^{v}[|(FO_{tj}-FO_{0j})/FO_{0j}| + |(FI_{tj} - FI_{0j})/FI_j|]$$

or

$$F_i = \sum_{j=1}^{v}(|FO_{tj} - FO_{0j}| + |FI_{tj} - FI_{0j}|)$$

where FC_i: value of flow change index of node i; v: total number of nodes in the network; FO_{tj}, FO_{0j}: outflow of node j after and before node i is changed, respectively; FI_{tj}, FI_{0j}: influx of node j after and before node i is changed, respectively. The nodes with larger F_i tend to be crucial nodes.

Further, flow change index can also be defined as

$$F_k = \sum_{i}\sum_{j<i}|f_{ijt} - f_{ij0}|$$

where f_{ijt}, f_{ij0}: flow between node i and j after and before node k is changed. The nodes with larger F_k tend to be crucial nodes.

5.2.5 *Adjacency matrix index*

Following the definition of Zhang (2012e), suppose the adjacency matrix of a network with v nodes is $D = (d_{ij})_{v\times v}$. If $d_{ij} = d_{ji} = 0$, there is no connection from v_i to v_j; if $d_{ij} = -d_{ji}$, and $|d_{ij}| = 1$, there is only a directed connection from v_i to v_j; if $d_{ij} = d_{ji} = 1$, there is only an undirected connection from v_i to v_j; if $d_{ij} = d_{ji} = 2$, there are two parallel connections from v_i to v_j; if $d_{ii} = 3$, v_i has a loop; if $d_{ii} = 4$, then v_i is an isolated node; if $d_{ii} = 5$, v_i is an isolated node and it has a loop. $i, j = 1, 2, \ldots, v$. Adjacency matrix index is defined as (Zhang, 2012d)

$$A_k = \sum_{i}\sum_{j}|d_{ijt} - d_{ij0}|$$

where d_{ijt}, d_{ij0}: value of the element d_{ij} after and before node k is changed. The nodes with larger A_k tend to be crucial nodes.

5.2.6 *Branch flourishing index*

Branch flourishing index is (Zhang, 2012d)

$$B_i = \sum_{j \neq i} (n_{ij} \times ml_{ij})$$

where B_i: branch flourishing index of the node i; n_{ij}: the total number of paths between nodes i and j. ml_{ij}: the mean path length of all paths between nodes i and j, $j \neq i$; v: the total number of nodes in the network. The nodes with larger B_i tend to be crucial nodes.

5.2.7 *Connections and between-node connection strength*

Measures on connection strength and number of connections can be used to identify crucial nodes (Paine, 1980; Zhang, 2007, 2011c, 2012b; Ding, 2012). For the statistic networks (Zhang, 2012b), a node with more connections (d_i) and larger mean correlation (mc_i) tends to be a crucial node. A simple index is

$$CS_i = d_i \times mc_i$$

Another index, node importance index, for identifying crucial nodes in statistic networks is (Zhang, 2012d)

$$S_i = \sum_{j \neq i} d_{ij}$$

where S_i: node importance index of the node i; d_{ij}: the path strength between nodes i and j in the network, $j \neq i$; v: the total number of nodes in the network. d_{ij} can be defined in different ways, e.g.,

$$d_{ij} = \max_{n_{ij}} \prod_t |r_{kl}|$$

where r_{kl}: the correlation between nodes k and l in the path t between nodes i and j, $t = 1, 2, \ldots, n_{ij}$; n_{ij}: total number of paths between nodes i and j. The nodes with larger S_i tend to be crucial nodes.

5.2.8   *Centrality indices*

Centrality indices are widely used in network analysis (Scardoni and Laudanna, 2012; Zhang, 2012c, 2012d, 2012e; Shams and Khansari, 2014; Khansari *et al.*, 2016).

5.2.8.1   *Degree centrality*

Degree centrality, always called degree, is a local centrality based on neighborhood (Zhang, 2011c, 2012d, 2012e, 2016p). It represents the influence of a node on its neighborhoods. A node degree centrality is the sum of the weights of the links attached to that node. It represents the whole involvement of a node in the network (Opsahl *et al.*, 2010; Khansari *et al.*, 2016)

$$y_i = \sum_{j=1}^{v} w_{ij}$$

where w_{ij} is the weight of link from node i to j. For the nonweighted network, $w_{ij} = 1$ if there is a link between node i and j.

5.2.8.2   *Distance-based centralities*

(1) *Closeness centrality*

Closeness centrality is a distance-based measure. It measures how close a node is to the rest of nodes. It is based on the proximity principle and quantifies how short the minimal paths from a given node to all other nodes are (Wassermann and Faust, 1994). For a distance-based measure, different values (cost, connection strength, etc.) can be assigned to different links in weighted networks. It is a global centrality which represents the independence of a node in the network. Closeness centrality is defined as reciprocal of the sum of the node's geodesic distances to all other nodes (the distances of the shortest paths) in the network

$$y_i = 1 \left/ \sum_{j=1}^{v} d_{ij} \right.$$

where, d_{ij} is the weighted geodesic path between node i and j. Dijkstra algorithm (Dijkstra, 1959; Zhang, 2012d) is used to calculate the shortest path and corresponding distance between two nodes.

Another form of closeness centrality is (Zhang, 2012d)

$$y_i = (v - 1) \left/ \sum_{j=1}^{v} d_{ij} \right.$$

where $i \neq j$, and d_{ij} is the length of the shortest path between nodes i and j in the network. The smallest value of y_i will be for that trophic group that upon being removed will affect the majority of other groups.

(2) *Betweenness centrality*

Betweenness centrality is a distance-based measure. It represents the node's ability to control the data flow in the network (Freeman, 1978). It measures how central a given node is in terms of being adjacent to many shortest paths in the network (Navia *et al.*, 2010). This measure is the proportion of number of geodesic paths that pass through the given node to total number of geodesic paths between any pair of nodes in the network (Khansari *et al.*, 2016)

$$y_i = n_i/s$$

where n_i is the number of weighted geodesic paths which pass through node i, and s is the total number of weighted geodesic paths between any pair of nodes in the network.

Betweenness centrality is based on quantifying how often node i is on the shortest path between each pair of nodes j and k. The standardized centrality index for node i is

$$y_i = 2 \sum_{j \leq k} g_{jk}(i)/g_{jk}/[(v - 1)(v - 2)]$$

where $i \neq j$ and k, g_{jk} is the number of equally shortest paths between nodes j and k, $g_{jk}(i)$ is the number of these shortest paths to which node i is adjacent, and v is the total number of nodes. The denominator is twice the number of pairs of nodes without node i. If y_i is large for trophic group i, the loss of this node will have many rapidly spreading effects in the network.

(3) *Radiality centrality*

Radiality denotes the extent of access into the network provided by the node's neighbors (Valente and Foreman, 1998). A high radiality means that less time is needed for the infectious node to reach others in the network (Khansari *et al.*, 2016)

$$y_i = \sum_{j \in V} (\Delta + 1 - d(i, j))/g_{jk}$$

where Δ is the diameter of graph, V is the set of nodes in graph, $d(i, j)$ is the weighted geodesic path between nodes i and j.

(4) *Katz centrality*

Katz centrality is the generalization of degree centrality (Katz, 1953; Khansari *et al.*, 2016). It is the number of nodes that are accessible through a specific path

$$y_i = \alpha \sum_{j} w(i, j) y_j + \beta$$

where $w(i, j)$ is the element of weighted adjacency matrix. α is an attenuation factor, $0 \leq \alpha \leq 1$; $\beta > 0$.

(5) *Subgraph centrality*

The number of occurrences of a node in different subgraphs of a network is the subgraph centrality of the node. The measure is based on spectral properties and characterizes the contribution of a node in different possible subgraphs. Subgraph centrality of a node is the sum of closed paths with different lengths that launch from and terminate in that node (Estrada and Rodriguez, 2005; Khansari *et al.*, 2016)

$$y_i = \sum_{k=0}^{\infty} \mu_k(i)/k$$

where $\mu_k(i)$ is the number of closed walks of cost k starting and ending at node i.

(6) *Circuit centrality*

Circuit centrality is proposed to characterize the contribution of a node in different possible subgraphs (Zhang, 2016p). It is a distance-based measure and is based on the concept of subgraph centrality. This measure is the proportion of number of fundamental circuits (i.e., cycles, or closed paths, or loops, see Paton (1969) and Zhang (2012e)) that pass through the given node to total number of fundamental circuits in the network

$$y_i = n_i/s$$

where n_i is the number of fundamental circuits which pass through node i, and s is the total number of fundamental circuits in the network.

(7) *Eccentricity centrality*

Eccentricity centrality of a node is the inverse of the longest geodesic path from the node to other nodes. A node with the smaller longest geodesic path tends to be the important node in network (Khansari *et al.*, 2016)

$$y_i = \max(d(ij))^{-1}$$

where $d(i, j)$ is the weighted shortest path between nodes i and j.

5.2.8.3 *Core centrality*

The k-core of a network refers to the subgraph after repeatedly removing the nodes with node degree not greater than k (Kitsak *et al.*, 2010). The coreness of a node is k if the node exists in the k-core but not in $k+1$-core. The maximal k of nodes is the coreness of the network. It is possible that a node with great degree has a small coreness k.

5.2.8.4 *Spectral centralities*

Definition and calculation of spectral centralities depends upon the eigenvalues and the eigenvectors of adjacency matrix and Laplacian matrix of the graph (Khansari *et al.*, 2016).

(1) *Laplacian centrality*

Laplacian centrality is based on the Laplacian matrix, and the later contains information on dynamics and geometry of the network (Pauls and Remondini, 2012). Laplacian matrix is expressed as (Khansari *et al.*, 2016)

$$L = X - W$$

where X is the degree matrix of the network, W is the weighted adjacency matrix. Suppose the eigenvalues of L are $\lambda_1, \lambda_2, \ldots, \lambda_n$. The Laplacian energy of the graph is

$$E = \sum_{i=1}^{n} \lambda_i^2$$

Laplacian centrality of a node is the difference between Laplacian energy of network with and without the node

$$y_i = (E - E_{-i})/E$$

(2) *PageRank centrality*

PageRank centrality denotes a node's relative importance in the network (Sarma *et al.*, 2011). The node's importance depends on the importance of its neighbors (Khansari *et al.*, 2016)

$$y_i = (I - dW)^{-1} + (1 - d)P_i$$

where d is damping factor, usually set to 0.85. W is the normalized weighted adjacency matrix and P is the preference network.

More centrality measures can be found in Khansari *et al.* (2016).

The following Matlab codes are for calculation of degree centrality, closeness centrality, and betweenness centrality.

```
choice=input('Input the type of topological property of nodes (1: Degree
centrality; 2: Closeness centrality; 3: Betweenness centrality): ');
adj=input('Input the excel file name of adjacency matrix D (e.g., raw.xls,
    etc.
Adjacency matrix is D=(dij)v*v, j=1,2,..., v): ','s');
adj=xlsread(adj);
m=size(adj,1);
degr=zeros(1,m);
```

```
switch choice
   case 1
      degr=sum(adj);
   case 2
      [ss,pat,distances,paths]=Dijkstra(adj);
      for i=1:m
      degr(i)=1/sum(distances(i,:));
      end
   case 3
      [ss,pat,distances,paths]=Dijkstra(adj);
      for i=1:m
      degr(i)=pat(i)/ss;
      end
end
fprintf('Node centrality values\n');
degr
```

The function, Dijkstra.m, which is used to calculate the shortest path and corresponding distance between two nodes, can be found in Chapter 2 and Zhang (2016p).

5.2.8.5 *k-neighbor and k-core*

If the node i and node j are connected with an edge, then the two nodes are neighbors. If the node i connects to node j by going through k edges, then two nodes are called k-nearest neighbor (k-neighbor) for each other. For a directed graph, there are two types of k-nearest neighbors, k-out and -in nearest neighbors. Starting from the node i, if there are k positive-directed edges between nodes i and j, and j is called the k-out nearest neighbor of i and, i is called the k-in nearest neighbor of j. In a network, if any node has at least k neighbors that belong to the network, the network is called a k-core network. To find the core of a complex network means finding all k-core networks in the complex network. A large total k-core value means a compact network (Li and Zhang, 2013).

5.2.9 *Methods of network evolution and dynamics*

5.2.9.1 *Network evolution methods*

Network evolution modeling (Bond, 1989; Rossberg *et al.*, 2005; Zhang, 2012c, 2015b, 2016f) can be used to find crucial nodes.

The nodes that cause greater changes of network structure during network evolution are crucial nodes (Zhang, 2012d). Sensitivity analysis can be conducted in network evolution modeling to find crucial nodes (Zhang, 2016q).

5.2.9.2 *Network dynamics methods*

Zhou (2016a) held that the network dynamics (i.e., network kinetics)-based methods can be categorized as three indices as follows:

(1) **Path counting.** Path counting is the most used method. Proposition of many structural indices also use the similar principle. The most representative index is dynamical influence (Klemm *et al.*, 2012; Pei and Makse, 2013). In the dynamical influence, suppose the state of nodes is the vector x, and node dynamics is a Markov dynamics, i.e., it can be written as $x(t+1) = Mx(t)$. It can be proved that the influence $DI(i)$ (in terms of convergence state) of node i is the component of left eigenvector corresponding to the largest eigenvalue of M on the node i. Klemm *et al.* (2012) noted that for the SIR dynamic (kinetic) model, if the infection rate is just at the critical value, then the DI will degrade to the eigenvector centrality in the structural indices.

Using the idea of path counting, Ide *et al.* (2013) gave a new understanding of the famous alpha centrality (Bonacich and Lloyd, 2001): Alpha centrality can be used directly to characterize the influence of nodes in SIR dynamic model where alpha=beta/delta in alpha centrality, and beta is the infection rate and delta is the recovery rate. In a sense, the routing centrality can also be treated as path counting, provided that the routing table has been fixed (Dolev *et al.*, 2010).

(2) **Time-aware methods.** Ghanbarnejad and Klemm (2012) analyzed the Boolean dynamics. They abstracted an activity matrix A, where the matrix element a_{ij} is the probability with which the node j will change from the change of node i. Liu *et al.* (2016) reported a work to obtain the node influence index of the discrete SIR dynamic state. This index is actually a finite time truncation of alpha centrality.

(3) A lot of other innovative methods and indices have been proposed to measure node importance (Li *et al.*, 2012), e.g., how to define the influence of a node on the evolutionary game (Simko and Csermely, 2013), the influence of a node on seepage (Piraveenan *et al.*, 2013), etc.

5.3 Further Discussion

The indices and methods above fall into three categories, node perturbation, network analysis (structural indices), and network dynamics/evolution (Zhang, 2012d). Node perturbation methods, e.g., adjacency matrix index, flow change index, node perturbation index, etc., identify crucial nodes by comparing structural changes of the network resulted from changes of each node. These methods need a large amount of experiments. They are highly reliable methods. Network dynamics/evolution methods include network evolution modeling, e.g., community assembly modeling, etc. These methods need to have a deep insight into the mechanism of network dynamics and need to build a reliable model for network evolution. They are high reliable also, but a lot of work should be done before they can normally function. Network analysis (the analysis on structural indices) methods, like criticality index, centrality indices, branch flourishing index, etc., require the information of the network itself only, and thus cost much less than other methods. Nevertheless, they identify crucial nodes only by analyzing static connection structure of nodes and are thus, theoretically, less reliable than other methods. Connection strength–connection number method, e.g., node importance index, is mainly a static network analysis method. However, if the connection strength is measured by between-node correlation in the process of network evolution, it is a network dynamics method also.

Some of these indices or methods may also be used to identify crucial links (i.e., connections) in networks. In this case, the change of a link refers to the presence/absence of a link, or the change of flow in the link, etc., (Zhang, 2012d).

In practical application of indices and methods above, more professional knowledge and approach should be further considered.

For example, to improve the quality for identification of crucial genes/proteins/metabolites, the following aspects should be noted: (1) For the protein–protein networks, it is necessary to find the parameters and attributes that mostly relate to the crucial proteins (Zhang, 2016p). In addition, we may fully mine the information of the existing parameters and integrate various parameters to find in-depth information on crucial proteins. (2) For the metabolic networks, we can mainly investigate the paths relating the integrity of functionality of the metabolic network in order to analyze the cruciality of particular nodes or links in these paths. (3) For the gene regulation networks, we should analyze not only the static characteristics of nodes, such as the amplitudes of the main topological parameters, but also the dynamic characteristics (network dynamics) of nodes, such as the level of gene expression and the node dynamics of local structure.

5.4   Application Examples

5.4.1   *Application in metabolic pathway of nonalcoholic fatty liver disease*

Use the data of metabolic pathway of nonalcoholic fatty liver disease from KEGG-PATHWAY database (Kanehisa Laboratories, 2016), Zhang and Feng (2017) calculated three centralities of cytokines/metabolites in metabolic pathway of nonalcoholic fatty liver disease, as indicated in Table 1. Generally, FFAs, ROS, and JNK1, etc., are the crucial cytokines/metabolites (Table 2).

5.4.2   *Application in gene, regulatory, and protein–protein networks*

The study of crucial proteins is mostly focused on the use of degree centrality. Through the removal analysis of *Saccharomyces cerevisiae* and *Escherichia coli*, it has been confirmed that the importance of a protein is involved in a number of interactions with other proteins in the network. Crucial proteins have usually more interactions than others (Csermely *et al.*, 2005).

Table 1 Three centralities of cytokines/metabolites in metabolic pathway of nonalcoholic fatty liver disease (Zhang and Feng, 2017).

Cytokines/ metabolites	Degree	Closeness	Between- ness	Cytokines/ metabolites	Degree	Closeness	Between- ness
1	3	0.0038	0.0393	32	2	0.0035	0.0639
2	3	0.004	0.0486	33	3	0.0042	0.1164
3	3	0.0039	0.0945	34	3	0.004	0.0503
4	7	0.0049	0.1721	35	3	0.0039	0.0519
5	2	0.0039	0.0328	36	3	0.0037	0.0344
6	5	0.004	0.0448	37	1	0.0037	0.0328
7	3	0.0033	0.0896	38	8	0.0047	0.2322
8	4	0.0038	0.1137	39	3	0.0046	0.2339
9	4	0.0043	0.2224	40	1	0.0037	0.0328
10	2	0.0035	0.077	41	1	0.0037	0.0328
11	2	0.003	0.053	42	2	0.0041	0.1831
12	2	0.0028	0.0399	43	2	0.0034	0.1552
13	3	0.0037	0.0672	44	2	0.0029	0.1262
14	3	0.0041	0.177	45	2	0.0025	0.0962
15	3	0.0038	0.1694	46	2	0.0022	0.065
16	3	0.0027	0.0967	47	1	0.0019	0.0328
17	1	0.0023	0.0328	48	2	0.0038	0.1508
18	1	0.0023	0.0328	49	3	0.0034	0.1574
19	2	0.0026	0.065	50	2	0.0028	0.1262
20	2	0.0038	0.1235	51	2	0.004	0.0377
21	2	0.0033	0.1022	52	3	0.0025	0.0967
22	3	0.0031	0.1175	53	1	0.0022	0.0328
23	3	0.0046	0.2743	54	1	0.0022	0.0328
24	2	0.0042	0.0328	55	2	0.0033	0.0328
25	2	0.0042	0.0448	56	2	0.0041	0.0464
26	3	0.0041	0.0459	57	2	0.0031	0.1262
27	3	0.0042	0.0902	58	1	0.0023	0.0328
28	3	0.0043	0.0683	59	2	0.0034	0.065
29	6	0.0047	0.2426	60	1	0.0028	0.0328
30	4	0.0044	0.0546	61	8	0.0052	0.477
31	4	0.0045	0.0776				

Notes: 1: IL6, 2: IL6R, 3: SOCS3, 4: TNF-α, 5: TNFR1, 6: NF-κB, 7: INS, 8: INSR, 9: IRS1/2, 10: PI3K, 11: Akt, 12: GSK-3, 13: LXR-α, 14: RXR, 15: SREBP-1c, 16: ChREBP, 17: L-PK, 18: LEP, 19: ObR, 20: ACDC, 21: adipoR, 22: AMPK, 23: PPAR-α, 24: Cdc42, 25: Rac1, 26: MLK3, 27: JNK1/2, 28: ASK1, 29: JNK1, 30: AP-1, 31: c-Jun, 32: ITCH, 33: IRE1α, 34: TRAF2, 35: IKKβ, 36: IL1, 37: CYP2E1, 38: ROS, 39: FasL, 40: IL8, 41: TGF-β1, 42: PERK, 43: elF2α, 44: ATF4, 45: CHOP, 46: Bim, 47: Bax, 48: Fas, 49: CASP8, 50: Bid, 51: CxI/II/III/IV, 52: CxII, 53: CxIII, 54: CxIV, 52: Cytc, 53: CASP3, 54: CASP7, 55: Oxysterol, 56: Fatty Acyl-CoA, 57: Lipogenic enzymes, 58: Glucose, 59: XBP1, 60: C/EBPα, and 61: FFAs, respectively.

Table 2 Mean ranking of three centralities of cytokines/
metabolites (Zhang and Feng, 2017).

Cytokines/ metabolites	Mean ranking	Cytokines/ metabolites	Mean ranking
61	1.3333	25	31.6667
38	3	10	32
29	3.6667	21	33
4	4.6667	24	33
9	7.3333	44	34.6667
23	8.6667	5	35
39	12.3333	52	35
14	12.6667	36	35.3333
31	14.6667	56	36
8	17.3333	57	36.3333
30	17.3333	32	37
15	18	50	37
33	18	11	38.3333
27	19.3333	51	38.6667
3	20	19	39.6667
28	20.6667	45	40.3333
42	21.3333	59	41
6	22.6667	12	42
2	23.6667	46	45.3333
13	25.3333	37	46
49	25.6667	40	47
26	26	41	48
22	26.6667	55	48.3333
7	27	17	54.3333
20	27	18	55.3333
1	27.6667	60	55.3333
34	28	58	58.3333
35	29	53	58.6667
48	29.3333	47	59.6667
16	30	54	59.6667
43	30.6667		

According to the node degree, from greater to smaller, Igor and Ron (2005) chose 1,061 nodes as hubs from the network of 4,743 proteins and 23,294 interaction in *S. cerevisiae*. They found that 43% are crucial nodes, significantly higher than the expected 20% randomly chosen, and the degree of crucial nodes is about twice the degree of noncritical nodes.

Watts and Strogatz (1998) have shown that the lethality caused by the removal of a protein from the yeast proteome is associated with the amplitude of centrality measure of the protein, which is significantly superior to randomization method in identifying crucial proteins. Screening 1% of nodes from high to low values of centrality measure can find up to 60% of the crucial proteins.

Lemke *et al.* (2004) studied the *E. coli* metabolic network. They defined the destructivity d of an enzyme as the number of downstream metabolites that cannot be synthesized due to the absence of the enzyme in the network, and they found that large d values were rare, and only a few values contain more than 503 deadly enzymes.

Igor and Ron (2005) also studied crucial node problems from many small regulatory networks (including transcription factors and their regulated target genes). They found that the regulators (proteins) with more regulated targets tended to be crucial nodes. The more functionalities of a gene, the more likely it is to be a crucial gene in the network.

5.5   Identification of Subnetworks/Modules

5.5.1   *Adjacent nodes based method*

Assume the adjacency matrix of a network X with m nodes is $a = (a_{ij})_{m \times m}$ where $a_{ij} = 1$, if two nodes v_i and v_j are adjacent; $a_{ij} = 0$, if v_i and v_j are not adjacent, $a_{ij} = a_{ji}$, and $a_{ii} = 0$; $i, j = 1, 2, \ldots, m$. A subnetwork/module is a node set, and links within the subnetwork/module are much more than links towards outside the subnetwork/module.

Suppose we do not know subnetworks/modules in advance, and they need to be determined by cluster analysis. The method (Zhang, 2016c) is

(1) Define between-node similarity as follows:

$$r_{ij} = \sum_{k=1}^{m}(a_{ik}a_{jk})/m \quad a_{ij} = 1$$
$$r_{ij} = 0 \quad\quad\quad\quad\quad a_{ij} = 0$$
$$i, j = 1, 2, \ldots, m; \quad i \neq j$$

where $0 \leq r_{ij} \leq 1$. $r_{ij} = 1$, means that the sets of adjacent nodes of two connected nodes, i and j, are the same; $r_{ij} = 0$, means that two nodes, i and j, are unconnected. $r_{ij} = r_{ji}$, i, $j = 1, 2, \ldots, m$. Between-node distance is then defined as

$$d_{ij} = 1 - r_{ij}$$

where $0 \leq d_{ij} \leq 1$.

At the start, m nodes are naturally m subnetworks/modules respectively.

(2) Calculate between subnetwork/module distance. Suppose there are two subnetworks/modules, X and Y. The distance between X and Y, based on the longest distance between two nodes, is

$$d_{XY} = \max d_{ij}, \quad i \in X, \ j \in Y$$

(3) In the sets of clusters (subnetworks/modules), choose two clusters, A and B, with the minimum d_{XY} to combine into a new cluster (new subnetwork/module)

$$d_{AB} = \min d_{XY}$$

(4) Return (2) to repeat the cluster procedure, until all of m nodes are eventually clustered into a cluster (network). If the minimum d_{AB} for two or more cluster procedures are the same, only the last cluster procedure in these cluster procedures is chosen for use.

(5) For each cluster in the same cluster procedure, weight all links in the cluster with the same weight, r_{AB} (i.e., $1 - d_{AB}$), and mark different clusters with different IDs, until all clusters and cluster procedures are traversed. To avoid weight 0 for singular links, let $r_{AB} = 10^{-10}$, if $r_{AB} = 0$.

Finally, two matrices, linkWeightMat and linkClusterIDs, are achieved. Two links with both the same weight in linkWeightMat and the same cluster ID in linkClusterIDs belong to the same subnetwork/module. Two links with the same weight in linkWeightMat but different cluster IDs in linkClusterIDs belong to two subnetworks/modules at the same hierarchical level. However, a link with an unique cluster ID in linkClusterIDs does not belong to any subnetworks/modules. A subnetwork/module of the greater weight is the more connected subnetwork/module.

The following are Matlab codes of the algorithm (linkWeight.m):

```
clear
str=input('Input the file name of adjacency matrix a (e.g., raw.txt,
raw.xls, etc.
Adjacency matrix is D=(dij)m*m, where m is the number of nodes in the
network. dij=1, if vi and vj are adjacent, and dij=0, if vi and vj are not
adjacent; dii=0; i,
j=1,2,..., m): ','s');
a=load(str);
dim=size(a); m=dim(1);
for i=1:m-1
for j=i+1:m
if (a(i,j)~ =0)
r(i,j)=0;
for k=1:m
if ((a(i,k)==a(k,j)) & (a(i,k)~=0)) r(i,j)=r(i,j)+1; end
end
r(i,j)=r(i,j)/m; r(j,i)=r(i,j);
d(i,j)=1-r(i,j); d(j,i)=d(i,j);
else r(i,j)=0; r(j,i)=0; d(i,j)=1; d(j,i)=1;
end; end; end
d1=d; bb1=1;
u(bb1)=0; nu(bb1)=m;
for i=1:nu(bb1) x(bb1,i)=i; end
for i=1:nu(bb1) y(bb1,i)=1; end
while (nu(bb1)>1)
aa=1e+10;
for i=1:nu(bb1)-1
for j=i+1:nu(bb1)
if (d(i,j)<=aa) aa=d(i,j); end
end; end
aa1=0;
for i=1:nu(bb1)-1
for j=i+1:nu(bb1)
if (abs(d(i,j)-aa)<=1e-06)
aa1=aa1+1; v(aa1)=i; w(aa1)=j;
end; end; end
for i=1:nu(bb1) s(i)=0; end
nn1=0;
for i=1:aa1
if ((v(i)~=0) & (w(i)~=0))
nn1=nn1+1;
for j=1:aa1
if ((v(j)==v(i)) |(v(j)==w(i)) |(w(j)==w(i)) |(w(j)==v(i)))
s(v(j))=nn1; s(w(j))=nn1;
if (j~=i) v(j)=0; w(j)=0; end; end
```

```
end
v(i)=0; w(i)=0;
end; end
for i=1:nn1
y(bb1+1,i)=0;
for j=1:nu(bb1)
if (s(j)==i)
for k=1:m
if (x(bb1,k)==j) x(bb1+1,k)=i; end
end
y(bb1+1,i)=y(bb1+1,i)+y(bb1,j);
end; end; end
for i=1:nu(bb1)
if (s(i)==0)
nn1=nn1+1;
for k=1:m
if (x(bb1,k)==i) x(bb1+1,k)=nn1; end
end
y(bb1+1,nn1)=y(bb1,i); end
end;
bb1=bb1+1; u(bb1)=aa; nu(bb1)=nn1;
for i=1:nu(bb1)-1
for j=i+1:nu(bb1)
d(i,j)=-1e+10;
for k=1:m
if (x(bb1,k)==i)
for kk=1:m
if (x(bb1,kk)==j)
if (d1(k,kk)>d(i,j)) d(i,j)=d1(k,kk); end
end; end; end; end
d(j,i)=d(i,j);
end; end; end
for k=1:m
y(bb1,k)=1; end
for i=bb1-1:-1:1
rr=0;
for j=1:nu(i+1)
ww=0;
for k=1:m
if (y(i+1,k)==j) ww=ww+1; v(ww)=k; end
end
vv=0;
for ii=1:ww
ee=0;
for jj=ii-1:-1:1
if (x(i,v(ii))==x(i,v(jj))) y(i,v(ii))=y(i,v(jj)); break; end
ee=ee+1;
end
if (ee==ii-1) vv=vv+1; y(i,v(ii))=rr+vv; end
end
```

```
rr=rr+vv;
end; end
for k=1:bb1
rs(k)=1-u(k);
end;
s=1; i=0;
while (m>0)
ss=1;
for j=s+1:bb1
if (rs(j)==rs(s)) ss=ss+1; end;
end
s=s+ss; i=i+1; la(i)=s-1;
if (s>=bb1) break; end
end
bb1=i;
yy=zeros(m);
for k=1:bb1
for i=1:nu(la(k))
for j=1:m
if (y(la(k),j)==i) yy(k,j)=i; end;
end; end; end
for k=1:bb1
rss(k)=rs(la(k)); uu(k)=u(la(k)); nuu(k)=nu(la(k));
end
linkWeightMat=zeros(m);
linkClusterIDs=zeros(m);
id=0;
for k=1:bb1
for i=1:nuu(k)
numm=0;
for j=1:m
if (yy(k,j)==i) numm=numm+1; temp(numm)=j; end
end
sim=max(rss(k),1e-10);
id=id+1;
for ii=1:numm-1
for jj=ii+1:numm
if ((a(temp(ii),temp(jj))~=0) & (linkWeightMat(temp(ii),temp(jj))==0))
linkWeightMat(temp(ii),temp(jj))=sim; linkWeightMat(temp(jj),temp(ii))=sim;
linkClusterIDs(temp(ii),temp(jj))=id; linkClusterIDs(temp(jj),temp(ii))=id;
end
end; end; end; end
dif=zeros(1,m);
su=0;
for i=1:m-1
for j=i+1:m
sm=0;
for k=1:su
if ((linkClusterIDs(i,j)~=0) & (linkClusterIDs(i,j)~=dif(k)))
sm=sm+1;
```

```
end; end
if (sm==su) su=su+1; dif(su)=linkClusterIDs(i,j); end
end; end
ma=max(max(linkClusterIDs));
for k=1:su
for i=1:m-1
for j=i+1:m
if ((linkClusterIDs(i,j)~=0) & (linkClusterIDs(i,j)==dif(k)) &
(linkClusterIDs(i,j)~=ma))
linkClusterIDs(i,j)=k; linkClusterIDs(j,i)=k;
end; end; end; end
idnew=su;
for i=1:m-1
for j=i+1:m
if (linkClusterIDs(i,j)==ma)
idnew=idnew+1; linkClusterIDs(i,j)=idnew; linkClusterIDs(j,i)=idnew;
end; end; end
for k=1:idnew-1
s=0;
for i=1:m-1
for j=i+1:m
if (k~=linkClusterIDs(i,j)) s=s+1; end
end; end
if (s==(m*(m-1)/2))
for i=1:m-1
for j=i+1:m
if (linkClusterIDs(i,j)>k)
linkClusterIDs(i,j)=linkClusterIDs(i,j)-1;
linkClusterIDs(j,i)=linkClusterIDs(i,j); end;
end; end;
end; end
lab=zeros(1,idnew-1);
for k=1:idnew-1
s=0;
for i=1:m-1
for j=i+1:m
if (k==linkClusterIDs(i,j)) s=s+1; end
end; end;
lab(k)=s;
end
iss='';
for k=1:idnew-1
la=0;
for i=1:m-1
for j=i+1:m
if (linkClusterIDs(i,j)==k) weig=linkWeightMat(i,j); la=1; break; end;
end;
if (la==1) break; end
end;
if (lab(k)>1) iss=strcat(iss,'\nLinks in subnetwork/module (cluster)
```

```
ID:',num2str(k),'(Link weight=',num2str(weig),')\n'); end;
if (lab(k)==1) iss=strcat(iss,'\nStandalone link ID:',num2str(k),'(Link
weight=',num2str(weig),')\n'); end;
for i=1:m-1
for j=i+1:m
if (k==linkClusterIDs(i,j))
iss=strcat(iss,'(',num2str(i),',',num2str(j),') '); end
end; end;
end
fprintf(iss)
fprintf('\n\nLink weights\n')
linkWeightMat
fprintf('\nLink cluster IDs\n')
linkClusterIDs
```

5.5.2 *Connection strength based methods*

In connection strength based methods, between–node connection strength (weight, correlation, etc.) is used to determine if the two nodes belong to the same subnetwork or modules. The possibility of two nodes belonging to the same subnetwork/module increases with the connection strength between the two nodes. A variety of methods, including cluster analysis, are available to identify subnetworks/modules (Zhang, 2012e, 2016d, 2017j).

Chapter 6

Detection of Network Types

Suppose the portion of nodes with k-degree is p_k. The node degree is a random variable and its frequency distribution is called degree distribution. In the random network, degree distribution is binomial distribution, and its limit model is Poisson distribution. In a random network, the majority of nodes have the same degree on average. In the complex network, degree distribution is mostly a power-law distribution, and the network is called scale-free network (Barabasi and Albert, 1999; Barabasi, 2009; Zhang and Zhan, 2011). A property of the scale-free network is that the structure and the evolution of network are inseparable. Scale-free networks constantly change because of the arrival of nodes and links (Barabasi and Albert, 1999).

Actually, the frequency distribution of node degree in natural networks is a continuum, which traverses from the uniform distribution to highly aggregated distributions, e.g., power-law distribution. Correspondingly, the variance or variation of node degree changes continuously from simple networks to complex networks. That is, in the more relaxed statistical conditions, the node degree of a network may spontaneously follow several classical probabilistic distributions rather than a specific probabilistic distribution only. Therefore, various probabilistic distributions and aggregation indices can be jointly used to detect network types (Zhang and Zhan, 2011; Zhang, 2012e).

6.1　Methods

6.1.1　*Indices*

Different ranges of an index represent different network types, as indicated below:

(1) **Skewness.** It is used to measure the degree of skewness of a degree distribution relative to the symmetric distribution (Sokal and Rohlf, 1995; Zhang and Zhan, 2011)

$$S = v \sum (d_i - \bar{u})^2 / [(v-1)(v-2)s^3]$$

where $\bar{u}, s^2$: mean and variance of degree; v: number of nodes; d_i: degree of node i, $i = 1, 2, \ldots, v$. The smaller the skewness is, the more complex the network is.

(2) **Coefficient of variation.** In a random network, the majority of nodes have the same degree as the average. The coefficient of variation H is used to describe the type of a network (Zhang and Zhan, 2011)

$$H = s^2/\bar{u},$$
$$\bar{u} = \sum d_i/v, \quad s^2 = \sum (d_i - \bar{u})^2/(v-1)$$

where $\bar{u}, s^2$: mean and variance of node degree; v: number of nodes; d_i: the degree of node i, $i = 1, 2, \ldots, v$. The network is generally a random network, if $H \leq 1$. If $H \leq 1$, calculate $\chi^2 = (v-1)H$, and if $\chi^2_{1-\alpha}(v-1) < \chi^2 < \chi^2_\alpha(v-1)$ the network is a strict random network, otherwise it is a network with the uniform degree distribution. It is a complex network if $H > 1$, and to a certain extent, network complexity increases with H.

(3) **Network entropy.** The entropy of network $E = s^2 - \bar{u}$. A more complex network has a larger entropy. The network is generally a random network if $E \leq 0$, and it is a complex network if $E > 0$ (Zhang and Zhan, 2011).

(4) **Aggregation index.** Network type can be determined by using the following aggregation index also (Zhang and Zhan, 2011)

$$H = v^* \sum d_i(d_i - 1) \Big/ \Big[\sum d_i \Big(\sum d_i - 1\Big)\Big]$$

The network is generally a random network if $H \leq 1$. If $H \leq 1$, calculate $\chi^2 = H(\sum d_i - 1) + v - \sum d_i$, and the network is a complete random network if $\chi^2 < \chi_\alpha^2(v - 1)$, otherwise it is a network with uniform degree distribution. It is a complex network if $H > 1$. Network complexity increases with H.

Different from the following probabilistic distributions, in particular power-law distribution, the indices above provide a new category of methods to judge whether a network is a complex or simple network.

6.1.2 *Probabilistic distributions*

Some probabilistic distributions can be used to detect network types (Zhang, 2012e, 2016i). Each probabilistic distribution denotes a specific frequency distribution of node degree.

(1) Binomial distribution (uniform distribution)

$$p_r = q^n, \qquad\qquad r = 0$$
$$p_r = p(n - r + 1)p_{r-1}/(r\ q), \quad r = 1, 2, \ldots, n$$

where $p = \sum if_i/(n\sum f_i), q = 1 - p$, $i = 0, 1, 2, \ldots, n$; n is the maximum degree, f_i is the number of nodes with degree $i, i = 0, 1, 2, \ldots, n$. Strictly speaking, binomial distribution is the uniform distribution. Nevertheless, for convenience we may call it random distribution also.

(2) Poisson distribution (random distribution)

$$p_r = e^{-m}, \qquad r = 0$$
$$p_r = mp_{r-1}/r, \quad r > 0$$

where, m: mean degree.

(3) Exponential distribution (aggregated distribution)

$$F_x = 1 - e^{-\lambda x}, \quad x \geq 0$$

(4) Power-law distribution (aggregated distribution)

$$p_x = x^{-\alpha}, \quad x \geq x_{\min}$$

Power-law distribution is fitted following the method of Aaron Clauset (http://www.santafe.edu/~aaronc/powerlaws/). Node degrees are directly used in the fitting. For binomial distribution, Poisson distribution and exponential distribution, frequency distributions are firstly calculated from node degrees, and the frequencies of zero are omitted. This leads to more reasonable results different from Zhang and Zhan (2011).

The following are Matlab codes, revised from Zhang and Zhan (2011), netType.m, to calculate degree distribution and detect network type using various indices and frequency distributions

```
adj=input('Input the excel file name of adjacency matrix D (e.g., raw.xls,
etc. Adjacency matrix is D=(dij)v*v, where dij=dji=0 means no link between
nodes i and j; dij=-dji, and |dij|=1, means there is an link between nodes i
and j; dij=dji=2, means there are parallel links between nodes i and j;
dii=3 means there is a self-loop for node i; dii=4 means isolated node;
dii=5 means isolated node i with self-loop; i, j=1,2,..., v):','s');
d=xlsread(adj);
v=size(d,1);
rr=10;
deg=zeros(1,v);
frq=zeros(1,rr);
for i=1:v
deg(i)=0;
for j=1:v
if (abs(d(i,j))==1) deg(i)=deg(i)+1; end
if ((d(i,j)==2) | (d(i,j)==3) | (d(i,j)==5)) deg(i)=deg(i)+2; end
end; end
temp=sortrows(deg',-1);
deg=temp';
pp=sum(deg);
qq=sum(deg.*(deg-1));
fprintf('\n');
it=(deg(1)-deg(v))/rr;
for i=1:rr
frq(i)=0;
for j=1:v
```

```
if ((deg(j)>=(deg(v)+(i-1)*it)) & (deg(j)<(deg(v)+i*it))) frq(i)=frq(i)+1;
   end
end; end
fprintf('Frequency distribution of node degree:\n');
for i=1:rr
if (frq(i)~=0) fprintf([num2str(deg(v)+it/2+(i-1)*it) ' ']); end
end
fprintf('\n');
rexp=rr;
rr=sum(frq~=0);
p=zeros(1,rr);
pr=zeros(1,rr);
fr=frq(frq~=0);
for i=1:rr
fprintf([num2str(fr(i)/v) ' ']);
end
fprintf('\n\n');
meann=mean(deg);
varr=var(deg);
skew=v/((v-2)*sqrt(varr));
fprintf(['Skewness of degree distribution:' num2str(skew) '\n']);
h=v*qq/(pp*(pp-1));
fprintf(['Aggregation index of the network:' num2str(h) '\n']);
if (h<=1) fprintf('It is a random network.\n');
else fprintf('It is a complex network.\n');
end
fprintf('\n');
h=varr/meann;
fprintf(['Variation coefficient H of the network: ' num2str(h) '\n']);
fprintf(['Entropy E of the network: ' num2str(varr-meann) '\n']);
if (h<=1) fprintf('It is a random network.\n\n');
else fprintf('It is a complex network.\n\n');
end
%Binomial distri., pr= Crn pr qn-r, r=0,1,2,...,n;
ss=sum((0:rr-1).*fr);
pp=ss/(v*(rr-1));
qq=1-pp;
pr0=qq^(rr-1);
for i=1:rr-1
if (i==1) pr(i)=(rr-i)*pp*pr0/(i*qq); else pr(i)=(rr-i)*pp*pr(i-1)/(i*qq);
   end end
chi=0;
for i=0:rr-1
if (i==0) hk=pr0*v; else hk=pr(i)*v; end
if ((i==0) & (pr0==0)) hk=fr(i+1); end
if ((i>0) & (p(i)==0)) hk=fr(i+1); end
if (i==0) chi=chi+(pr0*v-fr(i+1))^2/hk; else
   chi=chi+(pr(i)*v-fr(i+1))^2/hk; end
end
fprintf(['Binomial distribution Chi-square=' num2str(chi) '\n']);
```

```
fprintf(['Binomial p=' num2str(pp) '\n']);
k1=20.09;
if (chi<=k1)
if (chi>=0)
fprintf('Node degree is binomially distributed. ');
fprintf('It is a random network.\n\n');
end;
else fprintf('It is likely not a random network\n\n');
end
%Poisson distri., pr = e-λλr/r! , r=0,1,2,...
pr0=exp(-meann);
for r=1:rr-1
if (r==1) pr(r)=meann/r*pr0;
else pr(r)=meann/r*pr(r-1); end
end
chi=0;
for i=0:rr-1
if (i==0) hk=pr0*v; else hk=pr(i)*v; end
if ((i==0) & (pr0==0)) hk=fr(i+1); end
if ((i>0) & (p(i)==0)) hk=fr(i+1); end
if (i==0) chi=chi+(pr0*v-fr(i+1))^2/hk; else
   chi=chi+(pr(i)*v-fr(i+1))^2/hk; end
end
fprintf(['Poisson distribution Chi-square=' num2str(chi) '\n']);
fprintf(['Poisson distribution m=' num2str(meann) '\n']);
k1=20.09;
if (chi<=k1)
if (chi>=0)
fprintf('Node degree is Poisson distributed.');
fprintf('It is a random network.\n\n');
end;
else fprintf('It is likely not a random network\n\n');
end
%Exponential distri., F(x) =1-e^{-λx}, x≥0
chi=0;
for i=1:rexp
if (frq(i)~=0)
k1=deg(v)+it/2+(i-1)*it;
k2=deg(v)+it/2+i*it;
pp=v*(exp(-k1/meann)-exp(-k2/meann));
chi=chi+(frq(i)-pp)^2/pp;
end; end
fprintf(['Exponential distribution Chi-square=' num2str(chi) '\n']);
fprintf(['Exponential distribution lamda=' num2str(1/meann) '\n']);
k1=20.09;
if (chi<=k1)
if (chi>=0) fprintf('Node degree is exponentially distributed.\n\n'); end;
else fprintf('It is not a exponentially distributed network\n\n');
end
x=deg;
%Power law distri., p(x)=x^{-α}, x≥x_{min}
%Source: http://www.santafe.edu/~aaronc/powerlaws/
```

```
%Copyright (C) 2008-2010 Aaron Clauset (Santa Fe Institute)
x=reshape(x,numel(x),1);
xmins=unique(x);
xmins=xmins(1:end-1);
dat=zeros(size(xmins));
z=sort(x);
for xm=1:length(xmins)
xmin=xmins(xm);
z=z(z>=xmin);
n=length(z);
a=n./sum(log(z./xmin));
cx=(0:n-1)'./n;
cf=1-(xmin./z).^a;
dat(xm)=max(abs(cf-cx));
end
D=min(dat);
xmin=xmins(find(dat<=D,1,'first'));
z=x(x>=xmin);
n=length(z);
alpha=1+n./sum(log(z./xmin));
L=n*log((alpha-1)/xmin)-alpha.*sum(log(z./xmin));
%L is the log-likelihood of the data x>=x_min under the fitted power law.
fprintf(['Power law distribution Kolmogorov-Smirnov goodness-of-fit
statistic D=' num2str(D) '\n']);
if (D<(1.63/sqrt(n)))
fprintf('Node degree is power-law distributed, it is a scale-free complex
network\n');
end
fprintf(['Power law alpha=' num2str(alpha) '\n']);
fprintf(['Power law xmin=' num2str(xmin) '\n']);
```

6.2 Application Example

Using the data of tumor-related pathways FAS, JAK-STAT, JNK, MARK, $p53$, etc., (ABCAM, 2012; Huang and Zhang, 2012; Li and Zhang, 2013; Zhang, 2016a, 2016e, 2016i), the network types are detected as indicated in Table 1.

The results demonstrate that the joint use of various indices and frequency distributions is more sensitive than single use of them, e.g., power-law distribution. The distribution of node degree follows power-law distribution in all cases, but significant differences exist between them in finer indices and distribution.

Once more, it is proved that the frequency distribution of node degree is a continuum rather than the absolute probabilistic distributions.

Table 1   Detection of network types of tumor-related pathways.

	VEGF	EGF	TGF	TNF	Fas	HGF	JNK	JAK-STAT	MARK	mTOR	p53	PPAR	Ras
Skewness	1.26	0.55	0.79	0.79	0.47	0.76	0.28	0.67	0.47	0.71	0.36	1.18	0.68
Aggregation index	0.73	1.19	0.92	0.89	1.47	0.97	1.95	0.98	1.31	0.99	1.95	0.73	1.06
Network type	Random	Complex	Random	Random	Complex	Random	Complex	Random	Complex	Random	Complex	Random	Complex
Variation coefficient	0.27	1.46	0.81	0.72	2.12	0.94	4.22	0.95	1.83	0.97	3.37	0.36	1.12
Entropy	-1.92	1.11	-0.42	-0.73	2.63	-0.12	10.72	-0.12	2.21	-0.06	5.83	-1.46	0.27
Network type	Random	Complex	Random	Random	Complex	Random	Complex	Random	Complex	Random	Complex	Random	Complex
Binomial distribution													
χ^2	0.17	6.92	7.52	7.23	52.38	1.46	16.55	12.93	17.69	6.45	24.81	3.11	0.5
p	0.26	0.14	0.27	0.31	0.08	0.27	0.12	0.24	0.18	0.27	0.1	0.43	0.3
Binomial distri.?	Yes	Yes	Yes	Yes	No	Yes	Yes	Yes	Yes	Yes	No	Yes	Yes
Network type	Random	Random	Random	Random	—	Random	Random	Random	Random	Random	—	Random	Random
Poisson distribution													
χ^2	121.27	358.75	38.73	24.97	464.72	56.29	671.95	70.26	124.92	92.93	402.68	8.22	48.82
m	2.63	2.44	2.21	2.58	2.35	2.06	3.33	2.6	2.66	2.26	2.46	2.29	2.17
Poisson distri.?	No	No	No	No	No	No	No	No	No	No	No	Yes	No
Network type	—	—	—	—	—	—	—	—	—	—	—	Random	—

Exponential distribution													
χ^2	216.36	63.09	65.52	83.87	100.66	61.59	36.94	80.92	30.18	51.39	72.08	149.89	48.88
λ	0.38	0.41	0.45	0.39	0.43	0.49	0.3	0.38	0.38	0.44	0.41	0.44	0.47
Exponential distri.?	No	No	No	No	No	No	No	No	No	No	No	No	No
Network type	—	—	—	—	—	—	—	—	—	—	—	—	—
Power-law distribution													
Kolmogorov–Smirnov goodness-of-fit statistic D	0.33	0.36	0.33	0.28	0.2	0.36	0.2	0.2	0.22	0.33	0.29	0.38	0.36
α	8.39	3.36	8.04	4.61	3.07	7.95	5.55	4	3.7	4.38	3.75	7.95	3.18
x_{min}	4	6	4	3	3	6	10	4	5	4	5	3	4
Power-law distri.?	Yes	Yes	Yes	Yes	Yes	Yes	Yes	Yes	Yes	Yes	Yes	Yes	Yes
Network type	Complex	Complex	Complex	Complex	Complex	Complex	Complex	Complex	Complex	Complex	Complex	Complex	Complex

Chapter 7

Comparison of Network Structure

7.1 Nonparametric Statistic Comparison of Network Structure

The structural indices of networks include the node degree, network connectance, aggregation strength, and many others (Dormann, 2011; Zhang, 2011c; Zhang and Zhan, 2011). Sometimes we need to compare the structural difference between networks. Nonparametric statistics may be used in the difference comparison (Solow, 1993; Manly, 1997; Zhang, 2007). In this section, the algorithm, based on previous studies, is introduced to statistically compare between-network structural difference (Zhang, 2011a).

7.1.1 *Algorithm*

The algorithm is used to compare the difference in structural composition between networks (Zhang, 2011a, 2012e). Suppose a_{ij} is the mass (or degree, etc.) of node j in network $i, i = 1, 2, \ldots, n$; $j = 1, 2, \ldots, s$. First, define between-network distance measures, i.e., Euclidean distances, Manhattan distances, Chebyshov distance, Pearson correlation (based distance), as follows

$$d_{ij} = \left(\sum_{k=1}^{s} (a_{ik} - a_{jk})^2 / s \right)^{0.5}$$

$$d_{ij} = \sum_{k=1}^{s} |a_{ik} - a_{jk}| / s$$

$$d_{ij} = \max_k |a_{ik} - a_{jk}|$$

$$d_{ij} = 1 - \sum_{k=1}^{s}((a_{ik} - a_{i\mathrm{bar}})(a_{jk} - a_{j\mathrm{bar}})) \Big/ \left(\sum_{k=1}^{s}(a_{ik} - a_{i\mathrm{bar}})^2\right)$$

$$\times \sum_{k=1}^{s}(a_{jk} - a_{j\mathrm{bar}})^2\bigg)^{0.5}$$

where $a_{i\mathrm{bar}}$ and $a_{j\mathrm{bar}}$ are means of a_{ik}'s and a_{jk}'s.

If $\min a_{ij} < 0$, let $a_{ij} = a_{ij} - \min a_{ij}$, $i = 1, 2, \ldots, n$; $j = 1, 2, \ldots, s$. Suppose z_{ij} is the decimal numbers of a_{ij} if network data contain the decimal value a_{ij}, and calculate $c_{ij} = 10^{z_{ij}}$. Let $a_{ij} = a_{ij}\max c_{kl}$, $i = 1, 2, \ldots, n$; $j = 1, 2, \ldots, s$. Through these transformations, all of the values in network data become integers that are equivalent to numbers of individuals. If there exists no difference, the distribution of individuals in networks i and j will be a result of allocating the mixed network values at random into two networks of size equaling those of the original network (Solow, 1993; Manly, 1997; Zhang, 2007). Assume the two networks to be tested are i and j, which contain $\sum_{k=1}^{s} a_{ik}$ and $\sum_{k=1}^{s} a_{jk}$ individuals, respectively. The $\sum_{k=1}^{s} a_{ik} + \sum_{k=1}^{s} a_{jk}$ individuals of the combined network are randomly reallocated into two randomized networks with $\sum_{k=1}^{s} a_{ik}$ and $\sum_{k=1}^{s} a_{jk}$ labeled individuals. Calculate the expected absolute distance between the two randomized networks and compare whether it is not less than the absolute distance between the true networks i and j. Repeat the simulation many times and calculate the number of the expected are not less than the absolute distance between i and j, and take the percentage as the p value. The p value is used to make statistical test (the threshold p value for test may be defined as 0.05, 0.01, etc.). If the calculated p value is less than p threshold, then the structure composition of networks i and j are statistically different.

The Matlab codes, NetStructComp.m, of the algorithm are listed below

```
raw=input('Input the excel file name of node-by-network data (e.g., raw.xls,
    etc.
raw data a=(aij)m*n, There are n networks, each network has maximum m
```

```
    nodes.):','s');
sim=input('Input the maximum number of simulations (e.g., 100):');
sel=input('Choose distance measure (1: Euclidean distance; 2: Manhattan
distance; 3: Chebyshov distance; 4: Pearson correlation):');
sig=input('Input the significance level (e.g., 0.05):');
a=xlsread(raw);
m=size(a,1); n=size(a,2);
f=zeros(n,n);
pva=zeros(n,n);
a1=zeros(1,m);
a2=zeros(1,m);
for k=1:n-1
for l=k+1:n
a1=a(:,k)';
a2=a(:,l)';
pvalue=randTest(a1,a2,sim,sel);
pva(k,l)=pvalue;
if (pvalue<=sig) f(k,l)=1; end
end; end
fprintf('Network pairs with significant statistic difference in structure
(with p values):\n');
for i=1:n-1
k=0;
for j=i+1:n
if (f(i,j)==1)
fprintf(['('num2str(i)','num2str(j)')' '('num2str(pva(i,j))')']);
k=k+1;
end
end
fprintf('\n');
end
```

The functions, randTest.m, euclideandis.m, manhattandis.m, chebyshovdis.m, and correcoeffdiff.m, used in the algorithm above are as follows

```
function pvalue=randTest(x,y,sim,sel)        %pvalue: calculated p value.
if ((min(size(x))~=1)|(min(size(y))~=1))     %sim: times of randomizations.
error('Both x and y are vectors');
end              %x and y: two vectors to be tested. x and y are row vectors.
m=max(size(x));
if (max(size(y))~=m)
error('Vector sizes do not match.');
end
if (sim<=1)
error('No. randomizations are too less.');
end
dum=min(min(x),min(y));
if (dum<0)
x=x-dum;
```

```
y=y-dum;
end
ma=-1e10;
for j=1:2
for i=1:m
in=1;
if (j==1) dum=x(i);
else dum=y(i);
end
while (m~=0)
if ((abs(dum-floor(dum))<1) & (~(abs(dum-floor(dum))<=1e-10)))
in=in*10;
dum=dum*10;
if ((floor(dum+1e-10))~=(floor(dum))) break; end
else break; end
end
if (in>ma) ma=in; end
end; end
x=x.*ma.*1.0;
y=y.*ma.*1.0;
switch sel
    case 1
        dxy=euclideandis(x,y);
    case 2
        dxy=manhattandis(x,y);
    case 3
        dxy=chebyshovdis(x,y);
    case 4
        dxy=correcoeffdiff(x,y);
end
nrx=sum(x);
nrxy=sum(sum(x+y));
fr=0;
for sm=1:sim
ar=floor(x+y);
col=sum(ar);
br(1)=ar(1);
for i=2:m
br(i)=br(i-1)+ar(i);
end
cols=randperm(nrxy);
p1(1:m)=0;
for j=1:m
if (ar(j)==0) continue; end
if (j==1) temp=0;
else temp=br(j-1);
end
for i=1:nrx
if ((cols(i)>temp) & (cols(i)<=br(j))) p1(j)=p1(j)+1; end
end; end
```

```
p2=ar'-p1;
switch sel
    case 1
        dum=euclideandis(p1,p2);
    case 2
        dum=manhattandis(p1,p2);
    case 3
        dum=chebyshovdis(p1,p2);
    case 4
        dum=correcoeffdiff(p1,p2);
end
if (abs(dum)>=abs(dxy))
fr=fr+1;
end
end
pvalue=fr/sim;

function distance =euclideandis(x,y) %x and y: two vectors to be tested.
if (max(size(x))~=max(size(y)))
    error('Array sizes do not match.');
end
if ((min(size(x))~=1)|(min(size(y))~=1))
    error('Both x and y are vectors');
end
distance =sqrt(sum((x-y).^2))/max(size(x));

function distance = manhattandis(x,y) %x and y: two vectors to be tested.
if (max(size(x))~=max(size(y)))
    error('Array sizes do not match.');
end
if ((min(size(x))~=1)|(min(size(y))~=1))
    error('Both x and y are vectors');
end
distance=sum(abs(x-y))/max(size(x));

function distance=chebyshovdis(x,y) %x and y: two vectors to be tested.
if (max(size(x))~=max(size(y)))
    error('Array sizes do not match.');
end
if ((min(size(x))~=1)|(min(size(y))~=1))
    error('Both x and y are vectors');
end
distance=max(abs(x-y));

function diff=correcoeffdiff(x,y) %x and y: two vectors to be tested.
m=max(size(x));
if (m~=max(size(y)))
    error('Array sizes do not match.');
end
if ((min(size(x))~=1)|(min(size(y))~=1))
```

```
    error('Both x and y are vectors');
end
xbar=mean(x);
ybar=mean(y);
aa=sum(x.*y)-ybar*sum(x)-xbar*sum(y)+m*xbar*ybar;
bb=sum(x.^2)-2*xbar*sum(x)+m*xbar^2;
cc=sum(y.^2)-2*ybar*sum(y)+m*ybar^2;
diff=1-aa/sqrt(bb*cc);
```

7.1.2 *Application example*

Choose the weed data of rice fields in four cities (networks) of Pearl River Delta, China. In total, 25 plant families (nodes) were found (Wei, 2010), as indicated in Table 1 (Zhang, 2011a).

Table 1 Abundance of plants around rice fields in four cities of China.

Plant family	Zhongshan	Zhuhai	Dongguan	Guangzhou
Gramineae	1056.9	184.6	439.3	193.6
Compositae	11.1	95	43.3	63.4
Amaranthaceae	31.1	56	93.4	49
Commelinaceae	0	52.2	14.4	1.4
Onagraceae	0	0.1	2.3	1
Urticaceae	0.3	0	11.9	3.4
Menispermaceae	0	0	0	0.1
Cyperaceae	0	0	0	26.1
Caryophyllaceae	0	0	5.3	6.7
Polygonaceae	0.4	4.2	6.9	3.8
Acanthaceae	0	0	0	0.3
Solanaceae	0.1	0	0.2	0.4
Umbelliferae	0	34.9	0	4.8
Lythraceae	0	0	0	1.6
Scrophulariaceae	0.7	3.9	1.7	2.4
Oxalidaceae	0	0	0	0.2
Chenopodiaceae	1.1	0.4	0.3	0.1
Haloragaceae	0	0	0	4.6
Campanulaceae	0	0	0	0.7
Plantaginaceae	0	0.3	0	0
Rubiaceae	0	0.1	0	0
Euphorbiaceae	0	0	0.1	0
Convolvulaceae	0	0.1	0	0
Pontederiaceae	0	0.1	0	0
Portulacaceae	24.6	0	8	0

Choose Euclidean distance measure, significance level $p = 0.01$, and 1,000 randomizations, the network pairs with significant statistic difference in structure ($p \approx 0$) are (1,2), (1,3), (1,4), (2,3), (2,4), (3,4), i.e., all network pairs have significant statistic difference.

Another data set is the arthropod data of nine rice fields of Pearl River Delta, China. In total, five arthropod groups (nodes) were found (Wei, 2010), as indicated in Table 2 (Zhang, 2011a).

Choose Euclidean distance measure, significance level $p = 0.01$, and 1,000 randomizations, the field (network) pairs with significant statistic difference in structure ($p \approx 0$, in exception of the pair (1,5), (2,7) and (7,8)) are as follows:

(1,2) (1,3) (1,5)($p = 0.0010$)

(2,3) (2,7)($p = 0.0020$) (2,8) (2,9)

(3,4) (3,5) (3,6) (3,7) (3,8) (3,9)

(4,8) (4,9)

(5,8) (5,9)

(6,8) (6,9)

(7,8)($p = 0.0020$) (7,9)

It is obvious that the rice fields 4, 5, 6, and 7 are significantly different from rice fields 8 and 9.

Table 2   Arthropod abundance in nine rice fields.

	Herbivorous insects	Neutral insects	Predatory insects	Parasitic insects	Spiders
1	42.0	0.0	5.3	3.0	8.3
2	66.4	0.0	7.9	4.3	5.7
3	298.8	0.0	10.5	3.2	10.8
4	58.1	0.0	8.9	3.1	6.9
5	50.2	0.0	6.5	3.5	4.5
6	90.6	0.0	19.6	5.6	8.0
7	53.0	0.0	10.0	3.0	7.0
8	36.1	0.1	6.9	3.2	8.2
9	40.3	0.0	8.4	2.4	11.9

7.2   Nonparametric Statistic Comparison of Community Structure

The structure of community refers to species composition, population size, etc., caused by environment/climate conditions, and dynamic inter- or intra-specific interactions (Damgaard, 2011; Lüi, 2011; Rai, 2011; Watts and Worner, 2011; Zhang, 2011c; Zhang and Chen, 2011). Nonparametric statistics may be used in the difference comparison between communities (Clarke, 1993; Schoenly and Zhang, 1999). This section describes Clarke (1993) algorithm that statistically compares between-community structure difference (Zhang, 2011e, 2012e).

7.2.1   *Algorithm*

The algorithm is used to compare the comprehensive difference in structure composition (i.e., taxa and individual number, randomness of taxa, taxa placement, etc.) between two communities (Clarke, 1993; Schoenly and Zhang, 1999; Zhang, 2011e, 2012e).

Assume there are s taxa in both community 1 and community 2. Community 1 has m samples and community 2 n samples. In total, there are $ts = m + n$ samples in the combined community of communities 1 and 2. Given a_{ij}, the individual number of taxons i in sample j, $i = 1, 2, \ldots, s$; $j = 1, 2, \ldots, ts$. Calculate the distance (similarity) between sample i and j, $i = 1, 2, \ldots, ts - 1$; $j = i, \ldots, ts$. The following distance (similarity) measures, Euclidean distance, Manhattan distance, Pearson correlation, Point correlation, quadratic correlation and Jaccard coefficient can be calculated:

$$d_{ij} = \left(\sum_{k=1}^{s} (a_{ki} - a_{kj})^2 / s \right)^{1/2}$$

$$d_{ij} = \sum_{k=1}^{s} |a_{ki} - a_{kj}| / s$$

$$d_{ij} = \sum_{k=1}^{s}((a_{ki} - a_{i\text{bar}})(a_{kj} - a_{j\text{bar}})) \Big/ \left(\sum_{k=1}^{s}(a_{ki} - a_{i\text{bar}})^2\right.$$

$$\times \left.\sum_{k=1}^{s}(a_{kj} - a_{j\text{bar}})^2\right)^{1/2}$$

$$d_{ij} = (ad - bc)/((a+b)(c+d)(a+c)(b+d))^{1/2}$$

$$d_{ij} = \sin((a+d-(b+c))/(a+b+c+d)*3.1415926/2)$$

$$d_{ij} = (b_i + b_j)/(c_i + c_j - e)$$

$$i = 1, 2, \ldots, ts - 1; j = i, \ldots, ts$$

In the last three measures, both sample i and sample j take values 0 or 1, $i, j = 1, 2, \ldots, s$. a is number of both sample i and sample j take value $0, b$ is number of sample i takes 0 and sample j takes 1, c is number of sample i takes 1 and sample j takes 0, and d is number of both sample i and sample j take value 1. b_i is the nonzero number present in sample i but not in sample j, b_j is the nonzero number present in sample j but not in sample i, c_i and c_j is the nonzero number in sample i and sample j respectively, and e is nonzero number shared by sample i and sample j.

Let $b_k = d_{ij}$, $i = 1, 2, \ldots, ts - 1$; $j = i, \ldots, ts$; $k = 1, 2, \ldots,$ $(ts^*ts - ts)/2 + ts - 1$. Rank b_k from small to large values, then reranked b_k and its ranking value g_k are thus given, $k = 1, 2, \ldots, (ts^*ts - ts)/2 + ts - 1$. For each of reranked b_k, $k = 1, 2, \ldots, (ts^*ts - ts)/2 + ts - 1$, if its corresponding two samples belong to the same community, then let $h_k = 1$, $g_k = k$, or else let $d_k = 1$, $f_k = k$. Given the number of $h_k = 1$, is k_p, the number of $d_k = 1$, is r_p, the sum of g_k is s_p, the sum of f_k is c_p. Calculate r measure

$$r = 4^*(c_p/r_p - s_p/k_p)/(ts^*(ts - 1))$$

and let $r_0 = r$, i.e., the observed r value. Using Monte Carlo technique, randomly divide all of d_{ij}, $i = 1, 2, \ldots, ts - 1$; $j = 1, 2, \ldots, ts$, into two communities with random number of samples in first community, the first community has m_1 samples and the second community has $ts - m_1$ samples. Let $b_k = d'_{ij}$, where d'_{ij} is d_{ij} after randomization, $i = 1, 2, \ldots, ts - 1$; $j = i, \ldots, ts$; $k = 1, 2, \ldots, (ts^*ts - ts)/2 + ts - 1$.

Repeat the above procedures from which the r for this randomization can be calculated. For v randomizations, record the total number of $r \geq r_0$ as w, and expected and standard deviation of r can be derived also. Finally, calculate the p value

$$p = (w + 1)/(v + 1)$$

If p is less than 0.05, or 0.01, then the difference of structure composition between community 1 and community 2 is statistically significant.

The following are the Matlab codes, CommStrucComp.m, of the algorithm

```
ra=input('Input the excel file name of sample-by-attribute data of two
communities (e.g., raw.xls, etc. The matrix is z=(zij)(m+n)*ss, where m and
n are total numbers of samples for two communities respectively, ss is the
total number of taxa):','s');
m=input('Input the number of samples of the first community: ');
n=input('Input the number of samples of the second community: ');
sele=input('Input distance measure (1: Euclidean Dist.; 2: Manhattan Dist.;
3: Pearson Correlation; 4: Point Correlation; 5: Quadratic Correlation; 6:
Jaccard Coefficient): ');
sim=input('Input the number of randomizations (e.g., 100, 500): ');
a=(xlsread(ra))';
ss=size(a,1);
ts=m+n;
cor=0;
for i=1:ts-1
for j=i:ts
ds(i,j)=0;
for k=1:ss
if (sele==1) ds(i,j)=ds(i,j)+(a(k,i)-a(k,j))^2; end
if (sele==2) ds(i,j)=ds(i,j)+abs(a(k,i)-a(k,j)); end
end
if (sele==1) ds(i,j)=sqrt(ds(i,j)/ss); end
if (sele==2) ds(i,j)=ds(i,j)/ss; end
if (sele==3)
xbar=0;ybar=0;
for k=1:ss
xbar=xbar+a(k,i);
ybar=ybar+a(k,j);
end
xbar=xbar/ss;
ybar=ybar/ss;
aa=0;bb=0;cc=0;dd=0;
for k=1:ss
dd=dd+(a(k,i)-xbar)*(a(k,j)-ybar);
```

```
bb=bb+(a(k,i)-xbar)^2;
cc=cc+(a(k,j)-ybar)^2;
end
ds(i,j)=dd/sqrt(bb*cc);
end
if ((sele==4)|(sele==5)|(sele==6))
nn1=0;rr1=0; aa=0;bb=0;cc=0;dd=0;
for k=1:ss
if (a(k,i)~=0) nn1=nn1+1; end
if (a(k,j)~=0) rr1=rr1+1; end
if ((a(k,i)==0) && (a(k,j)==0)) aa=aa+1; end
if ((a(k,i)==0) && (a(k,j)~=0)) bb=bb+1; end
if ((a(k,i)~=0) && (a(k,j)==0)) cc=cc+1; end
if ((a(k,i)~=0) && (a(k,j)~=0)) dd=dd+1; end
end;
switch (sele)
    case 4
        ds(i,j)=(aa*dd-bb*cc)/sqrt((aa+bb)*(cc+dd)* (aa+cc)*(bb+dd));
    case 5
        ds(i,j)=sin((aa+dd-(bb+cc))/(aa+bb+cc+dd)*3.1415926/2);
    case 6
        ds(i,j)=(cc+bb)/(nn1+rr1-dd);
end; end
ds(j,i)=ds(i,j);
end; end
for j=1:ts
h(j)=j;
end
r0=rvalue(m,ts,ds,h);
fprintf(['Observed r=' num2str(r0) '\n']);
hs=0;ha=0;
for sm=1:sim
zm=0;om=0;
for j=1:ts
cols(j)=round(rand()+0.5);
if (cols(j)==0) zm=zm+1;w(zm)=j; end
if (cols(j)==1) om=om+1;v(om)=j; end
end;
for j=1:ts
if (j<=zm) h(j)=w(j); end
if (j>zm) h(j)=v(j-zm); end
end
r=rvalue(zm,ts,ds,h);
if (r>=r0) hs=hs+1; end
ha=ha+r;
d(sm)=r;
end
ha=ha/sim;
dev=0;
for i=1:sim
```

```
dev=dev+(d(i)-ha)^2/(sim-1);
end
dev=sqrt(dev);
fprintf(['Expected r=' num2str(ha) '\n']);
fprintf(['Standard deviation of expected r=' num2str(dev) '\n']);
p=(hs+1)/(sim+1);
if (p<0.01) fprintf(['p=' num2str(p) '\n']);
fprintf(['Significant difference exists at 0.01 significance level' '\n']);
end
if (p<0.05) fprintf(['p=' num2str(p) '\n']);
fprintf(['Significant difference exists at 0.05 significance level' '\n']);
end
if (p<0.1) fprintf(['p=' num2str(p) '\n']);
fprintf(['Significant difference exists at 0.1 significance level' '\n']);
end
if (p>=0.1) fprintf(['p=' num2str(p) '\n']);
fprintf(['No significant difference' '\n']);
end
```

The following is the function, rvalue.m, used in the algorithm above

```
function r=rvalue(tp,ts,ds,h)
en=0;
for i=1:ts-1
for j=i+1:ts
en=en+1;
b(en)=ds(h(i),h(j));
end; end
for j=1:en
g(j)=j;
end
for i=1:en-1
k=i;
for j=i:en-1
if (b(j+1)<=b(k)) k=j+1; end
end
l=g(i); g(i)=g(k); g(k)=l;
vv=b(i); b(i)=b(k); b(k)=vv;
end
rw=0; rb=0;
rwn=0; rbn=0;
for k=1:en
em=0;
for i=1:ts-1
for j=i+1:ts
em=em+1;
if (em~=g(k)) continue; end
if (((i<=tp) & (j<=tp))|((i> tp) & (j>tp)))
rw=rw+k;rwn=rwn+1;
end
```

```
if ((i<=tp) & (j>tp))
rb=rb+k; rbn=rbn+1;
end
end; end; end
if (rwn==0) rw1=0; else rw1=rw/rwn; end
if (rbn==0) rb1=0;else rb1=rb/rbn; end
r=4*(rb1-rw1)/(ts*(ts-1));
```

7.2.2 Application example

We have obtained a set of data investigated in rice fields of four cities of Pearl River Delta, Guangzhou (23 samples), Zhongshan (17 samples), Zhuhai (23 samples), and Dongguan (17 samples) in September 2008 (Wei, 2010). In total, 58 arthropod families were found (Zhang, 2011e).

Choose different distance (similarity) measures and set 1,000 randomizations. The results, as indicated in Table 3, show that there is no significant difference between these cities in the arthropod composition. From p values in Table 3, the family composition of arthropods between Zhuhai and Dongguan is relatively more different.

Table 3 The p values for city pairs.

	Zhongshan	Zhuhai	Dongguan	Guangzhou
Euclidean				
Zhongshan		0.859	0.566	0.858
Zhuhai			0.217	0.584
Dongguan				0.887
Pearson				
Zhongshan		0.855	0.554	0.864
Zhuhai			0.207	0.565
Dongguan				0.863
Point				
Zhongshan		0.870	0.582	0.858
Zhuhai			0.221	0.570
Dongguan				0.846
Jaccard				
Zhongshan		0.849	0.577	0.872
Zhuhai			0.205	0.557
Dongguan				0.856

7.3   Network Matrix-based Methods

If we want to compare the similarity of two networks A and B, based on their network matrices $A = (a_{ij})$ and $B = (b_{ij})$, i, $j = 1, 2, \ldots, m$, where m is the number of nodes, a_{ij} and b_{ij} are the weights between two nodes i and j in two networks, respectively. Comparing between-node weight pairs (a_{ij}, b_{ij}), i, $j = 1, 2, \ldots, m$; $i \neq j$ (i.e., $(a_{12}, b_{12}), (a_{13}, b_{13}), \ldots, (a_{m\,m-1}, b_{m\,m-1})$) by using Pearson correlation, Euclidean distance, Jaccard coefficient, Point correlation, or the methods above, we can achieve the test results for network similarity. Generalized Matlab codes for Pearson correlation between two network matrices A and B are as follows:

```
AA=input('Input the excel file name of network matrix A: ','s');
A=xlsread(AA);
BB=input('Input the excel file name of network matrix B: ','s');
B=xlsread(BB);
m=size(A,1); mm=size(A,2); n=size(B,1); nn=size(B,2);
if ~((m==mm) & (n==nn) & (m==n))
error('Network matrices A and B should be the square matrices of the same
size.');
end
% Matrices A and B are the square matrices of the same size.
sig=input('Input significance level(e.g., 0.01): ');
A1=reshape(A, m*m, 1);   %Transform matrix A to vector A1
B1=reshape(B, m*m, 1);   %Transform matrix B to vector B1
r=corr(A1,B1);
fprintf(['Pearson correlation r=' num2str(r) '\n']);
tvalue=abs(r)/sqrt((1-r^2)/(m*m-2));
p=(1-tcdf(tvalue,m*m-2))*2;
sigma=p<sig;
if (sigma==1) fprintf(['Pearson correlation is statistically significant
(p=' num2str(p) ')\n']); end
if (sigma==0) fprintf(['Pearson correlation is not statistically
significant (p=' num2str(p) ')\n']); end
```

If both of network matrices A and B are Boolean type, point correlation and χ^2-test can be used (Zhang, 2017k).

Using the method, randTest.m, in nonparametric statistic comparison of network structure, we can compare the statistic difference between two networks, represented by two network matrices, A and B, respectively. The Matlab codes are as follows:

```
AA=input('Input the excel file name of network matrix A: ','s');
A=xlsread(AA);
```

```
BB=input('Input the excel file name of network matrix B: ','s');
B=xlsread(BB);
m=size(A,1); mm=size(A,2); n=size(B,1); nn=size(B,2);
if ~((m==mm) & (n==nn) & (m==n))
error('Network matrices A and B should be the square matrices of the same
size.');
end
% Matrices A and B are the square matrices of the same size.
sim=input('Input the maximum number of simulations (e.g., 100): ');
sel=input('Choose distance measure (1: Euclidean distance; 2: Manhattan
distance;
3: Chebyshov distance; 4: Pearson correlation): ');
sig=input('Input the significance level (e.g., 0.05): ');
A1=reshape(A, m*m, 1); %Transform matrix A to vector A1
B1=reshape(B, m*m, 1); %Transform matrix B to vector B1
pvalue=randTest(A1,B1,sim,sel);
if (pvalue<sig) fprintf(['Difference is statistically significant (p='
num2str(pvalue)
')\n']); end if (pvalue>=sig) fprintf(['Difference is not statistically
significant (p='
num2str(pvalue) ')\n']); end
```

Suppose that the difference $A_1 - B_1$ between two vectors A_1 and B_1, as generated from two network matrices A and B in the Matlab codes above, follows the normal distribution. If there is not statistic difference between A and B, the mean of $A_1 - B_1$ should be zero. The *t*-test can be used to test the statistic difference. The Matlab codes are as follows:

```
AA=input('Input the excel file name of network matrix A: ','s');
A=xlsread(AA);
BB=input('Input the excel file name of network matrix B: ','s');
B=xlsread(BB);
m=size(A,1); mm=size(A,2); n=size(B,1); nn=size(B,2);
if ~((m==mm) & (n==nn) & (m==n))
error('Network matrices A and B should be the square matrices of the same
size.');
end
% Matrices A and B are the square matrices of the same size.
alpha=input('Input significance level(e.g., 0.01)');
A1=reshape(A, m*m, 1); %Transform matrix A to vector A1
B1=reshape(B, m*m, 1); %Transform matrix B to vector B1
[h,p,ci,stats]=ttest(A1-B1,0,alpha,0);
if (h==1) fprintf(['Difference is statistically significant (p='
num2str(p) ')\n']);
else fprintf(['Difference is not statistically significant (p=' num2str(p)
')\n']); end
```

7.4   Other Methods

Luna *et al.* (2013) proposed a method to compare structural difference of biological networks based on dominant vertices. They propose a pseudo-distance between networks, built around the notions of determination and dominancy, the concepts introduced in the context of regulatory dynamics on networks. They use the proposed pseudo-distance to compare biological networks and equivalent homogeneous, scale-free, and geometric three-dimensional random networks. It is found that even when bacterial networks are characterized with different levels of details, the proposed pseudo-distance captures all these characteristics (Luna *et al.*, 2013).

Part 3
Network Dynamics, Evolution, Simulation, and Control

Chapter 8

Network Dynamics

In this chapter, network dynamics means the dynamics of the network with fixed nodes (variables). Discussion will focus on the dynamic models.

8.1 Differential Equations and Motion Stability

8.1.1 *Differential equations*

Given the ordinary differential equations of a dynamic network with n nodes (e.g., species, metabolites, etc., Zhang, 2010, 2016r)

$$dx/dt = f(x, t)$$

where

$$x(t) = (x_1(t), x_2(t), \ldots, x_n(t))^{\mathrm{T}}$$

$$f(x, t) = (f_1(x, t), f_2(x, t), \ldots, f_n(x, t))^{\mathrm{T}}$$

The network can be categorized by the form of $f(x, t)$: (1) $f(x, t) = f(x, t)$, nonlinear and nonstationary system; (2) $f(x, t) = f(x)$, nonlinear stationary system; (3) $f(x, t) = Ax$, linear stationary system; (4) $f(x, t) = A(t)x$, nonstationary linear system (Zhang, 2010, 2016r).

Let

$$dx/dt \approx (x(k+1) - x(k))/h$$

the network system becomes

$$x(k+1) = x(k) + h\ f(x(k), k), \quad k = 0, 1, 2, \ldots$$

where h is the time step. The numerical solution of the differential equations can thus be solved.

Furthermore, we can use Runge–Kutta method to obtain numerical solution of the differential equations. Runge–Kutta method is expressed as follows:

$$K_1 = h\ f(x_k, t_k)$$
$$K_2 = hf(x_k + K_1/2, t_k + h/2)$$
$$K_3 = hf(x_k + K_2/2, t_k + h/2)$$
$$K_4 = hf(x_k + K_3, t_k + h)$$
$$h = t_{k+1} - t_k$$

where h is the step length of iteration. The numerical solution is $x_{k+1} = x_k + (K_1 + 2K_2 + 2K_3 + K_4)/6$, $k = 0, 1, 2, \ldots$

8.1.2 *Constructing differential equations from sampling data*

Given the differential equation of ordinary and stationary linear system of a network with n nodes

$$dx/dt = Ax$$
$$x(t_0) = x_0$$

Suppose the systematic matrix A is unknown. Discretizing the differential equations above, we obtain difference equation $\Delta x/\Delta t = Ax$. Choose a variable, e.g., x_1, transform the equation

(Qi, 2007)

$$\Delta x_1/(x_1\Delta t) = a_{01} + a_{11}x_2/x_1 + a_{21}x_3/x_1 + \cdots + a_{n-1\,1}x_n/x_1$$

$$\Delta x_2/(x_1\Delta t) = a_{02} + a_{12}x_2/x_1 + a_{22}x_3/x_1 + \cdots + a_{n-1\,2}x_n/x_1$$

$$\cdots$$

$$\Delta x_n/(x_1\Delta t) = a_{0\,n} + a_{1\,n}x_2/x_1 + a_{2\,n}x_3/x_1 + \cdots + a_{n-1\,n}x_n/x_1$$

Let $y = (\Delta x_1/(x_1\Delta t), \ \Delta x_2/(x_1\Delta t), \ldots, \ \Delta x_n/(x_1\Delta t))$, $z_i = x_{i+1}/x_1$, $i = 1, 2, \ldots, n-1$. The linear regression equations are as follows

$$y_i = a_{0i} + a_{1i}z_1 + a_{2i}z_2 + \cdots + a_{n-1\,i}z_{n-1}, \quad i = 1, 2, \ldots, n$$

Given m sets of dynamic data. Based on the method above, we get the $(m-1) \times (n-1)$ data matrix of independent variables (Z_{ij}) and the $(m-1) \times n$ data matrix of dependent variables (y_{ij}).

Let $L = (l_{ij})$, $i, j = 0, 1, \ldots, n-1$. $D = (d_{ij})$, $i = 0, 1, \ldots, n-1$, $j = 1, \ldots, n$. Let $l_{00} = m - 1$, and

$$l_{0j} = \sum_{k=1,\ldots,m-1} z_{kj}, \quad l_{i0} = \sum_{k=1,\ldots,m-1} z_{ki}, \quad l_{ij} = \sum_{k=1,\ldots,m-1} z_{ki}\,z_{kj},$$

$$i, j = 1, 2, \ldots, n-1$$

$$d_{0j} = \sum_{k=1,\ldots,m-1} y_{kj}, \quad d_{ij} = \sum_{k=1,\ldots,m-1} z_{ki}\,y_{kj},$$

$$i = 1, 2, \ldots, n-1; \ j = 1, 2, \ldots, n$$

The systematic matrix $A = L^{-1} \times D$, where $A = (a_{ij})$, $i = 0, 1, \ldots,$ $n-1; j = 1, 2, \ldots, n$. The significance test on regression coefficients and correlation coefficients can be conducted.

Suppose a simple network has three nodes (variables), and 15 years of data on network dynamics are available as the follows:

	1	2	3
1	21.22	22.22	20.46
2	12.87	16.46	16.75
3	7.80	12.19	13.72
4	4.73	9.03	11.23
5	2.87	6.69	9.19
6	1.74	4.95	7.52
7	1.05	3.67	6.16
8	0.64	2.72	5.04
9	0.38	2.01	4.13
10	0.23	1.49	3.38
11	0.14	1.10	2.77
12	0.08	0.81	2.26
13	0.05	0.60	1.85
14	0.03	0.44	1.52
15	0.01	0.33	1.24

Using the method above, the systematic matrix A is achieved as

$$
\begin{array}{ccc}
-0.4593297 & 0.0556004 & -0.0188667 \\
0.048661 & -0.2976385 & 0.0116218 \\
-0.0342655 & 0.0267867 & -0.1898725
\end{array}
$$

8.1.3   *Motion stability and discrimination*

Let's suppose $x \in \Omega \subset R^n$, $t \in I = (t_1, t_2)$, the domain of definition of $f(x,t)$ is $\Omega \times I$ and $f(x,t)$ is continuous on $\Omega \times I$. If there is a constant K, such that (Zhang, 2010, 2016r)

$$|f(x,t) - f(y,t)| \leq K|x - y|, \quad \forall x, y \in \Omega, \ t \in I$$

$f(x,t)$ satisfies Lipschitz condition on $\Omega \times I$. $f(x,t)$ satisfies Lipschitz condition, if $\partial f_i(x,t)/\partial x_i, i = 1, 2, \ldots, n$, are finite. Given a system, $dx/dt = f(x,t)$, if $f(x,t)$ is continuous and satisfies Lipschitz condition on $\Omega \times I$, then there is a constant $c > 0$, and an unique solution $x = x(t)$ on $[t_0 - c, t_0 + c]$ for any $(x_0, t_0) \in \Omega \times I$, where $x(t)$ is continuous and $x(t_0) = x_0$.

Assume $z(t)$, $y(t)$, and $x(t)$ are the given motion, perturbation and observed motion, respectively, i.e., $z(t) = x(t) - y(t)$. We have

$$dy/dt = f(y(t) + z(t), t) - f(z(t), t) = g(y, t)$$

Given the perturbation equation of a known system, $dz/dt = f(z, t)$, is as follows:

$$dy/dt = g(y, t), \quad g(0, t) = 0$$

Inside a B-neighborhood of $y(t) = 0$, i.e.,

$$\{(y_1(t), y_2(t), \ldots, y_n(t)) | y_i(t) < B, \ i = 1, 2, \ldots, n\}$$

given a real value $0 < \varepsilon < B$; if there is a real value, $\delta = \delta(\varepsilon, t_0)$, such that the perturbed motion $y_i(t)$, satisfies

$$|y_i(t)| < \varepsilon, \quad i = 1, 2, \ldots, n; \ \forall t \geq t_0$$

when $|y_i(0)| \leq \delta$, $i = 1, 2, \ldots, n$; the given motion $z(t)$, is called to be stable. If the given motion $z(t)$ is stable and

$$\lim_{t \to \infty} y_i(t) = 0, \quad i = 1, 2, \ldots, n$$

it is asymptotic stable. If there is a real value $\varepsilon > 0$, such that a value, δ, that satisfies with the stability condition, is not existent, the given motion is unstable.

If $x(t) = x_e = (x_{1e}, x_{2e}, \ldots, x_{ne})^{\mathrm{T}}$, or $f(x, t) = 0$, the system is called to be in an equilibrium state, and x_e is the equilibrium state. For any real value $\varepsilon > 0$, if there is a value δ, such that

$$|x_i(t) - x_{ie}| < \varepsilon, \quad i = 1, 2, \ldots, n; \ \forall t \geq t_0$$

when $|x_i(0) - x_{ie}| \leq \delta$, $i = 1, 2, \ldots, n$; x_e is stable. If x_e is stable and

$$\lim_{t \to \infty} x_i(t) = x_{ie}, \quad i = 1, 2, \ldots, n$$

the state x_e is asymptotic stable. If there is a real value $\varepsilon > 0$, such that a value δ, that satisfies with the stability condition, is not existent, the state x_e is unstable (Yan and Zhang, 2000).

If the nonlinear function $f(x,t)$ is considerably smooth, the system can be linearized around the equilibrium state x_e

$$d\Delta x/dt = A(t)\Delta x(t)$$

by taking

$$x(t) = x_e + \Delta x(t)$$
$$f(x,t) = x_e + A(t)\Delta x(t)$$

where $A(t) = \partial f(x,t)/\partial x|_{x=x_e}$. Suppose $A(t) = A$, and $|A^{-1}| \neq 0$. If the eigenvalues of A are negative real values, x_e is a stable node; if the eigenvalues are conjugate complex numbers with negative real parts, x_e is a stable focal point.

Given a real function $V(x)$, defined in a neighborhood Ω of the origin, and $V(0) = 0$. V is single valued on Ω and is continuously derivative in respect to x_i, $i = 1, 2, \ldots, n$. $V(x)$ is positive definite (negative definite), if for any $x \in \Omega$, in exception of $x = 0$, $V(x) > 0$ ($V(x) < 0$); $V(x)$ is half positive definite (half negative definite), if for any $x \in \Omega$, $V(x) \geq 0$ ($V(x) \leq 0$); $V(x)$ is sign changing, if $V(x)$ will be positive, zero, or negative values for different $x \in \Omega$.

Based on the definition of $V(x)$, the Lyapunov theorem, used to discriminate the stability of nonlinear systems, is as follows (Zhang, 2010, 2016r):

(1) The system is stable on the origin if there is a positive (negative) definite function $V(x)$ on the neighborhood Ω of the origin and the derivative of $V(x)$ is half negative (positive) definite;

(2) The system is asymptotic stable on the origin if there is a positive (negative) definite function $V(x)$ on the neighborhood Ω of the origin and the derivative of $V(x)$ is negative (positive) definite;

(3) The system is unstable on the origin if there is a function $V(x)$ on the neighborhood Ω of the origin and the derivative of $V(x)$ is positive (negative) definite, but $V(x)$ itself is not half negative (positive) definite.

Known the differential equations of ordinary and stationary linear system of a network with n nodes, $dx/dt = Ax$, $x(t_0) = x_0$, where

$x = (x_1, x_2, \ldots, x_n)$, the solution of the system is $x(t) = e^{At}x(t_0)$. Let $h = t(k+1) - t(k)$, then the solution is

$$x(t(k+1)) = e^{Ah}x(t(k)), \quad k = 0, 1, 2, \ldots$$

where $e^{Ah} = E + Ah + (Ah)^2/2! + (Ah)^2/3! + \cdots$, and A is the systematic matrix, x_0 is initial state. Here, let $h = 1$.

If all eigenvalues of the equation

$$|A - \lambda E| = 0$$

have negative real part, the system is asymptotic stable; if their real parts are not greater than 0, and the eigenvalues with real part 0 are 1st-order, the system is stable, otherwise the system is unstable (Zhang, 2010, 2016r).

8.1.4 *Stochastic differential equation*

Once a continuously changing network (system) is perturbed by an external random variable, a chaotic term, $E(t)$, must be incorporated in its differential equation (Langevin equation; Zhang, 2010, 2016r)

$$dx/dt = f(x, t) + V(t)$$

The chaotic term, $E(t)$, is a discrete function of time. The solution of the differential equation is

$$x(t) = x(t_0) + \int_{t_0}^{t} f(x, \tau)d\tau + v(t) - v(t_0)$$

which is equivalent to

$$dx(t) = f(x, t)dt + dv(t)$$

The equation above is a stochastic differential equation.

A coefficient $w(t)$ may be included in chaotic term to generate a new equation

$$dx(t) = f(x,t)dt + w(t)dv(t)$$

and its solution is

$$x(t) = x(t_0) + \int_{t_0}^{t} f(x,\tau)d\tau + \int_{t_0}^{t} w(t)dv(t)$$

where the later integral term is defined by Wiener integral. If $w(t)$ depends on x also, the following Ito integral and Stratonovich integral should be used

Ito integral

$$\int_{t_0}^{t} w(x(t))dv(t) = \lim_{t_i \to 0} \sum w(x(t_{i-1}))(v(t_i) - v(t_{i-1}))$$

Stratonovich integral

$$\int_{t_0}^{t} w(x(t))dv(t) = \lim_{\tau_i \to 0} \sum w(x(\tau_i))(v(t_i) - v(t_{i-1}))$$

where $\tau_i = (t_i - t_{i-1})/2$. The Ito and Stratonovich stochastic differential equations and their solutions are represented by

$$dx_i = f_i(x(t),t)dt + \sum_j w_{ij}(x(t))dv_j(t)$$

$$x_i(t) = x_i(s) + \int_{s}^{t} f_i(x,\tau)d\tau + \sum_j \int_{s}^{t} w_{ij}(x(\tau))dv_j(\tau)$$

$$i = 1, 2, \ldots, n$$

8.2 Dynamics of Some Networks

8.2.1 *Epidemic systems*

8.2.1.1 *Viral epidemic system of the insect pest*

We have developed the mathematical model for insect pest — nucleopolyhedrovirus disease — crop system dynamics (Zhang *et al.*, 2011; Zhang, 2016r). The processes for temperature time course, crop growth, polyhedron inactivation, virus contamination by cadavers, virus contamination due to secretions by infected larvae, infection by horizontal transmission and polyhedron ingestion, infection by vertical transmission, insect population dynamics, and crop injury are included in this model. Immigration and emigration of adults were also incorporated in the model. The model includes a group of ordinary differential equations that describe insect population dynamics, as well as some models to describe temperature dynamics, crop growth, virus dynamics, etc. Given the crop, virus, insect-related parameters, and initial conditions, the model will produce the uninfected and infected insect populations, crop growth, virus accumulation dynamics, etc.

Insect population dynamics is simulated with a group of ordinary differential equations which are developed according to the principles of compartment modeling of Zhang *et al.* (1997).

Main state variables in the model include the densities of the healthy and infected stages of the insect, the densities of viral entities (polyhedron inclusion bodies, i.e., PIBs) on the leaves, and the leaf area of the crop.

The densities of m healthy larval stages, L1–Lm, are denoted by $x_1, \ldots, x_m$. The number of larval instars is flexible and can be input as a parameter (m). The densities of pupae, adults, and eggs are denoted by x_{m+1}, x_{m+2}, and x_{m+3}, respectively. The model takes only female individuals into account. The sex ratio is applied to reduce egg laying from total eggs to female eggs. Male adults are assumed to be sufficient in number to fertilize all female adults. The densities of the $m+3$ stages of infected individuals are written down as $y_1, \ldots, y_{m+3}$. Virus-killed individuals are indicated by the symbol z, indexed by stage. A portion of cadavers dying from infection is lost from leaves by falling to the ground and does not contribute to the horizontal transmission (Zhang, 2016r; Zhang *et al.*, 2011).

Some equations for the epidemic system are as follows.

Temperature over time is simulated with trigonometric functions

$$T = (T_0 - T_1)/2 \times \cos(\pi(1 + t - t_{\max})/(1 - t_{\max} + t_{\min}))$$

$$+ (T_0 + T_1)/2 \quad 0 \leq t \leq t_{\min}$$

$$T = (T_2 - T_1)/2 \times \sin(\pi(t - (t_{\max} + t_{\min})/2)/(t_{\max} - t_{\min}))$$
$$+ (T_2 + T_1)/2 \quad t_{\min} \leq t \leq t_{\max}$$
$$T = (T_2 - T_3)/2 \times \cos(\pi(t - t_{\max})/(1 - t_{\max} + t_{\min}))$$
$$+ (T_2 + T_3)/2 \quad t_{\max} \leq t \leq 1$$

where t is time (d) in a day, T is temperature (°C) at time t, $t_{\max}$ is the time (d) with the highest temperature during a day, $t_{\min}$ is the time (d) with the lowest temperature during a day, T_0 is yesterday's highest temperature, while T_2 and T_1 are today's highest temperature and lowest temperature, respectively. Finally, T_3 is tomorrow's lowest temperature.

In a situation without insect injury, the growth of leaf area is in logistic form

$$dL/dt = rL(1 - L/L_{\min})$$

where t is cumulative effective day degrees, L is leaf area (m² leaf per m² ground) at physiological time t, $L_{\max}$ is the maximum leaf area, and r is the relative growth rate of leaf area, $r > 0$.

Inactivation of sprayed polyhedra can be described by assuming a constant relative inactivation rate

$$V = V_0 e^{-rt}$$

where t is time (d), V is the density per m² ground of infective polyhedra on leaves (PIBs per m² ground) at time t, V_0 is the initial polyhedra on leaves, and r is the relative inactivation rate of polyhedra.

The dynamics of virus-killed larvae are described as

$$dz_i/dt = b_i'' \times y_i, \quad i = 1, 2, \ldots, m$$

where z_i is the cumulative number of dead virus-killed larvae, b_i'' is the dying rate of the ith instar infected larvae due to infection (1/d), while y_i is the number of infected larva per m² ground, $i = 1, 2, \ldots, m$.

Dynamics of contaminated leaf area are

$$dP/dt = \sum_{i=1}^{m} dz_i/dt \times (1 - F_i) \times G_i \times (1 - P/L_p)$$

where dP/dt is the rate at which contaminated leaf area (P) increases, L_p is the present leaf area per m^2 ground, P is the contaminated leaf area at time t, F_i is the fraction loss of bodies of virus-killed larva due to falling from leaves and bodies that didn't break down on leaves, and G_i is the leaf area (m^2) contaminated per dying larva of instar i, when it is naturally broken down, $i = 1, 2, \ldots, m$.

The rate of increase of the amount of polyhedra released by cadavers on contaminated leaves is

$$dV/dt = \sum_{i=1}^{m} dz_i/dt \times (1 - F_i) \times E_i$$

where E_i is the polyhedron production of ith instar per virus-killed larva (PIBs per virus-killed larvae), $i = 1, 2, \ldots, m$. V is the cumulative number of polyhedra on contaminated leaves by time t.

The cumulative leaf injury (per m^2 ground) over time up till time t is

$$L_{\text{loss}} = \sum_{i=1}^{m} \int_{t_0}^{t+dt} (C_i \times x_i(\tau) + C_i' \times y_i(\tau)) d\tau$$

Crop leaf area (per m^2 ground) at time $t + dt$ is calculated as the leaf area without insect injury, L, minus the cumulative leaf area loss, L_{loss}, i.e., $L - L_{\text{loss}}$.

Healthy L1s

$$dx_1/dt = e_{m+3} x_{m+3} - (b_1 + e_1) x_1 - d_1 x_1$$

The first term in the differential equation for healthy L1s denotes egg hatch, the second term attrition plus development into L2, and the third term infection.

Healthy L2s–Lms

$$dx_i/dt = e_{i-1} x_{i-1} - (b_i + e_i) x_i - d_i x_i, \quad i = 2, \ldots, m$$

The meaning of terms in the equations for L2–Lm is analogous to those for L1.

Healthy pupae

$$dx_{m+1}/dt = e_m x_m - (b_{m+1} + e_{m+1})x_{m+1}$$

The meaning of terms in the equation for pupae is analogous to those for the larvae. The third term is lacking because pupae do not feed and therefore do not contract infection.

Healthy adults

$$dx_{m+2}/dt = e_{m+1} x_{m+1} - (b_{m+2} + W_1)x_{m+2} + R_1$$

The first term in the equation for healthy adults indicates emergence of pupae. The second term includes attrition and emigration. The third term is immigration.

Healthy eggs

$$dx_{m+3}/dt = \beta_1 e_{m+2} x_{m+2} - (b_{m+3} + e_{m+3})x_{m+3} + \beta_1'(1-h)e_{m+2}' y_{m+2}$$

The first term in the equation for eggs is the production of female eggs. The second term includes attrition and development into L1. The third term includes the production of uninfected eggs by infected female adults.

Infected L1s

$$dy_1/dt = e_{m+3}' y_{m+3} - (b_1' + e_1')y_1 - b_1'' y_1 + d_1 x_1$$

The first term in the differential equation for infected L1s indicates the hatching of infected eggs into infected L1s. The second term includes attrition and development into infected L2. The third term is the death rate of infected L1s due to virus infection. The final term is the recruitment of new infected L1s due to horizontal or spray-induced infection.

Infected L2s–Lms

$$dy_i/dt = e_{i-1}' y_{i-1} - (b_i' + e_i')y_i - b_i'' y_i + d_i x_i, \quad i = 2, \ldots, m$$

The terms in the equation for infected L2–Lm are analogous those in the equation for infected L1.

Infected pupae

$$dy_{m+1}/dt = e'_m y_m - (b'_{m+1} + e'_{m+1})y_{m+1} - b_{m+1''} y_{m+1}$$

The equation for infected pupae is analogous to those for larvae, without the last term, which quantifies horizontal and spray-induced infection in larvae.

Infected adults

$$dy_{m+2}/dt = e'_{m+1} y_{m+1} - (b'_{m+2} + W'_1)y_{m+2} - b_{m+2''} y_{m+2} + R'_1$$

The equation for infected adults has four terms. The first indicates emergence of the infected pupae. The second term includes attrition and emigration of infected adults. The third term is virus-induced death. The fourth term is immigration.

Infected eggs

$$dy_{m+3}/dt = h\beta'_1 e'_{m+2} y_{m+2} - (b'_{m+3} + e'_{m+3})y_{m+3}$$

The equation for infected eggs has two terms. The first term indicates laying of infected eggs (vertical transmission) by infected females, taking into account the probability of passing the virus on to the egg (h) and the sex ratio (β'_1). The second term covers attrition and development into L1.

8.2.1.2 *Ecoepidemic model with two-strains-diseased prey*

Ecoepidemic models describe general systems in which two or more populations interact, but are subject to some diseases (Elena *et al.*, 2013). Elena *et al.* (2013) developed an ecoepidemic model in which predators and prey interact and two diseases affect the very same species. In their model, Elena *et al.* (2013) consider strains that cannot both simultaneously affect the same individual. Further, any individual infected by any one of them cannot contract the other strain, nor can the latter replace the former. Therefore, neither coinfection nor super-infection are possible in the model. Moreover, they assume that only sound prey are captured. The infected are thus identifiable, and avoided by the predators that do not feed on them. Let S denote the healthy prey, I the diseased prey of type A and Y those of type

B, and P be their predators. Consider the situation in which available for the predators are also other food sources (the last equation); the model is

$$dS/dt = r(1 - (S + I + Y)/k)S - \lambda IS - \beta YS - aSP + \gamma I + \varphi Y$$
$$dI/dt = \lambda IS - \mu I - \gamma I$$
$$dY/dt = \beta YS - \nu Y - \varphi Y$$
$$dP/dt = n(1 - P/H)P + eSP$$

The first three equations represent the prey dynamics. In the first equation, when the total prey population experiences intra-specific pressure, sound prey reproduce logistically and can be infected with one of the two strains and are hunted by the predators. The final two terms denote the fact that both diseases are assumed to be recoverable (Elena *et al.*, 2013). The second and third equations contain the epidemics dynamics. Individuals enter these classes via successful contacts among susceptibles and infected of the "right" strain, and leave them by natural plus disease-related mortalities, respectively μ and ν, or by recovering, at respective rates γ and ϕ. The last term of the fourth equation gives the reward predators have from hunting.

Details of the model and its application can be found in Elena *et al.* (2013).

8.2.2   *Network modeling framework for predicting climate change impacts on biodiversity*

Ferrarini (2013c) suggested that loop analysis (Puccia and Levins, 1985) would provide predictions on the possible direction of change in species abundances and would be suitable only if the direction of the effects of climate perturbations is required, not their magnitude.

Loop analysis uses signed digraphs to represent networks of interacting variables (nodes). Each connection between two nodes represents a nonzero coefficient of the community matrix. Press perturbations may act by changing one or more parameters in the growth rate of the variables. Taking the inverse of the community matrix provides an estimate of the direction of change in the equilibrium level of variables in response to these parameter changes (Ferrarini, 2013c).

The element a_{ij} of the matrix represents the effect of variable j on the growth variable i. The following equation is solved for a moving equilibrium

$$dx_i/dt = f_i(x_1, x_2, \ldots, x_n; c_1, c_2, \ldots, c_h)$$

where $x_1, x_2, \ldots, x_n$ represent the variables (nodes) and $c_1, c_2, \ldots, c_h$ the parameters. Responses of abundances or biomass are arranged in a table of predictions whose signs show the predicted direction of change (Ferrarini, 2013c). The entries in a table denote variations expected in all the column variables when parameter inputs affect each row variable. In the paper of Ferrarini (2013c), interacting variables (nodes) may represent species, while sign $(+, 0, -)$ of induced connections (links, or edges) among species could emerge from the following rules. Suppose that climate change favors species i and disfavors species j, this causes species i to likely expand its niche-hindering species j, in case they are in contact. Since each species can undergo three types of climate-induced effects (positive, negative, or null), a couple of species can undergo: $(+1, -1)$ interaction when one species is favored and the other one disfavored; $(-1, -1)$ interaction when they are favorite or disfavored, and $(0, 0)$ represents no interaction (when the two species are both indifferent to climate change). Moreover, there might be self-damped terms associated with density-dependent control (i.e., organisms with spatial limitations) or continuous supply of the species from outside the system. Details can be found in Ferrarini (2013c).

8.2.3 Boolean network model for evaluating the effect of a silenced gene

Gene regulation networks are one of the most attractive topics in biology. Gene regulation is implemented through indirect interaction between protein coding genes that configure the expression level of one another. Prevention of expression of any genes in gene regulation at the levels of transcription or translation indicates the gene silencing event. Gene silencing can be considered a type of loss-of-function mutation in which the altered gene product lacks the molecular

function of the silenced gene (Nowak, 2006), which can be simulated
by Boolean network models (Haliki and Kazanci, 2017).

In the Boolean network model, the state of each gene, i.e., Boolean
variables and the possible phenotypic transitions, is determined by
the states of the other genes in the network with the Boolean functions that govern each gene (Albert, 2004). For a Boolean network
model, microarray experiments should first be processed to binary
in the experimental data from time series (Hakamada *et al.*, 2001).
As an example, Boolean GRN models have been successfully used to
describe genetic regulatory systems (Hickman and Hodgman, 2009).

In the Boolean GRN models, genes and their corresponding proteins are nodes of the network, and each node is assigned to a binary
value $g_i(t) \in \{0,1\}$ with 1 for active and 0 for inactive (Hickman and
Hodgman, 2009; Haliki and Kazanci, 2017). Any cellular phenotypes
are represented by their expression patterns

$$\Phi(t) = \{g_1(t), g_2(t), \ldots, g_N(t)\}$$

where N is the number of genes. Gene interactions are directed links
(edges). All interactions are represented by the adjacency matrix
$W = (w_{ij})$, which is just the GRN, where $w_{ij} \in \{-1,0,1\}$ represents
an interaction from gene j to gene i (Lau *et al.*, 2007). If there exists
an interaction between genes i and j, the interaction must be acti-
vating (1) or inhibiting (−1). The rule for change of the expression
state of each gene $g_i(t)$ of the phenotype $\Phi(t)$ as time t in discrete
timesteps, reflecting the regulation of gene i's expression by other
genes, is expressed below (Haliki and Kazanci, 2017)

$$\begin{aligned}
g_i(t+1) &= 1, & \text{if } \Sigma_j w_{ji} g_j(t) > 0 \\
g_i(t+1) &= 0, & \text{if } \Sigma_j w_{ji} g_j(t) < 0 \\
g_i(t+1) &= g_i(t), & \text{if } \Sigma_j w_{ji} g_j(t) = 0
\end{aligned}$$

Implementing the rule for every gene and thereafter cellular phenotype goes to the next timestep. In a series of updated phenotypes,
the phenotype repeating itself in the phenotype trajectory is the
stable state once it has been reached (Haliki and Kazanci, 2017).
Furthermore, some limit cycles may exist (Glass and Kauffman, 1973;
Kauffman, 1969).

Using their Boolean network model, Haliki and Kazanci (2017) examined what types of results in gene silencing would bring about the dynamics of gene regulatory mechanisms. First, they define the event (Jablonka and Lamb, 2005) in Boolean manner. In their model, one particular active gene is chosen and its state is fixed at zero.

By the silencing of active kth gene

$$\Phi(t) = \{g_1(t), g_2(t), \ldots, g_k(t), \ldots, g_N(t)\}$$
$$g_k(t) = 0$$

they define a new threshold function for the ith gene, which is counted as a target of g_k, by subtracting the contribution of silenced gene from the sum of $w_{ji} \, g_j(t)$

$$I_i(t) = \sum_j w_{ji} g_j(t) - w_{ki} \, g_k(t)$$

And the gene regulation rule becomes as below

$$\begin{aligned} g_i(t+1) &= 1, & \text{if } I_i(t) > 0 \\ g_i(t+1) &= 0, & \text{if } I_i(t) < 0 \\ g_i(t+1) &= g_i(t), & \text{if } I_i(t) = 0 \end{aligned}$$

The application details of the Boolean network model can be found in Haliki and Kazanci (2017).

There are numerous network dynamic models, from simple to complex, e.g., those developed by Nedorezov (2011), Nedorezov and Neklyudova (2014), Shakil *et al.* (2015a, 2015b). More details can be found in the literature.

Chapter 9

Network Robustness and Sensitivity Analysis

9.1 Network Robustness

Network (or system; they are used in this book with the same or similar meaning unless special note is made) robustness is a relatively new concept based on stability. In this chapter, the discussion of Zhang (2016n) on network robustness is introduced.

9.1.1 *Concepts*

9.1.1.1 *Definition and implication*

Perturbation of a network's structure/parameters occurs occasionally. Network perturbation occurs due to (1) external disturbances, and (2) internal factors such as unknown variables or mechanisms, structural catastrophe, slow drift of parameters/properties, etc. Robustness refers to a network's capacity for maintaining some performance (stability, topological structure, functionality, flux, etc.) if the network's internal structure (or parameters) is perturbed. It denotes the insensitivity of network's performance to perturbation of its structure/parameters (Zhang, 2016d). According to the specific types of network's performances, robustness is divided into stability robustness and performance robustness.

Robustness is the key for the survival of networks in the case of abnormal and dangerous situations. Robust control refers to control theory and methodology that maintains satisfied performance when network model encounters perturbation or other uncertain disturbances (Ferrarini, 2011a, 2013b, 2014, 2015).

In terms of robustness of biological systems, Alon (2006) defined robustness as that a biological network can almost make its basic functions irrelevant to original biochemical parameters, while nonrobustness was defined as fine-turned, i.e., system properties greatly change when biochemical parameters are perturbed (Alon *et al.*, 1999). Robustness is related to the survival of organisms, which is a response to internal perturbation and which reflects the capacity of an organism's internal organization (Zhang and Zhang, 2008).

Exploiting the mechanism of robustness is very important for understanding biological networks, through which we can understand how biological networks maintain their own features under various disturbances, such as changes in the environment (lack of nutrition level, chemical induction, temperature), internal faults (DNA damage, genetic failure of metabolic pathways), etc., (Zhang and Wang, 2009; Gao and Guo, 2011).

9.1.1.2   *Robustness and stability*

Stability is divided into state stability and structure stability. State stability is a network's capacity to maintain its operation state after it is disturbed by external factors, while the network's structure/parameters is/are maintained. State stability includes uniform stability, asymptotic stability, etc. Structure stability denotes a system's capacity to maintain its structure/parameters after it/they is/are disturbed by external factors.

Robustness is a mapping from network's structure/parameters to network's performance, driven by the network's perturbation, while stability is driven by external disturbance.

Both robustness and stability are determined by a network's structure/parameters.

9.1.2   **Network robustness**

Complex networks are characterized by some or all of the properties of self-organization, self-similarity, attractors, small-world, and scale-free. Some of these properties are occasionally disturbed or even destroyed.

Network structure robustness is the capacity of the network for maintaining its functionality when network nodes or links are damaged (by random failures, malicious attacks, etc.) (Newman *et al.*, 2006; Barabasi and Albert, 1999; Zhu and Liu, 2012). It includes both resistance capacity (connection robustness) of network structure to external damage and restoration capacity (restoration robustness) of network structure if it is damaged (Du *et al.*, 2010). Research on network robustness have been widely reported (Kwon and Cho, 2007; Ash and Newt, 2007; Gao *et al.*, 2006; Wang *et al.*, 2006a, 2006b). Kwon and Cho (2007) studied the relationship between feedback structure and network structure robustness and found that scale-free network model may evolve more feedback structures than random graph model and its network structure robustness enhanced considerably. Ash and Newt (2007) optimized network structure using evolutionary algorithm and found that clustering, modularity, and length of long paths all have important effects on the network structure robustness. Gao *et al.* (2006) demonstrated that for most food webs the attacks based on betweenness centrality are more effective than that based on node degree. Wang *et al.* (2006a, 2006b) argued that network structure robustness to random failures can be improved by optimizing the efficiency of the network (average inverse length of paths). So far, network structure robustness is usually represented by network connectivity (connection robustness). Restoration robustness is seldom studied.

Besides network structure robustness, Zhang (2016n) defined two more categories of robustness, network parameter robustness and comprehensive robustness. Network parameter robustness refers to a network's capacity, without any structural changes, for maintaining between-node flows (fluxes)/link weights if it is perturbed. Comprehensive robustness refers to a network's capacity that not only topological structure of the network, e.g., nodes and links, are not or less changed, but also between-node flows, link weights, nodes' state values are maintained even if the network is perturbed. Comprehensive robustness considers both structure and parameter changes of a network.

In addition to conventional indices for network robustness, Zhang (2016n) proposed to use the inverse of some indices of global sensitivity analysis (GSA) such as Sobol index (Sobol, 1993), Extended Fourier Amplitude Sensitivity Test, etc., (Tarantola *et al.*, 2002; Xu *et al.*, 2004; Zhang, 2012d, 2016q) as indices of network robustness.

As a basic formula, for instance, we may define network robustness as (Zhang, 2016n)

$$R = 1/(dN/dp/N)$$

or

$$R = 1/(dN/dp)$$

where N: network measure, dN: final variation of network measure due to perturbation; p: strength of perturbation, which is usually expressed by variation of network measure itself, e.g., number of removed nodes/links, or amount of reduced flux, etc. The greater the R value, the stronger the robustness.

There are many network measures for use, i.e., total number of nodes or links, network fluxes, etc.

9.1.2.1 *Structure robustness*

(1) Connection robustness

Connection robustness refers to the capacity of a network to maintain connectivity among remaining nodes if some nodes are attacked and destroyed. Suppose N_r nodes with maximal degree along with their links are simultaneously removed from a network X. Thus, connection robustness is (Dodds *et al.*, 2003)

$$R = c/(n - n_r)$$

where n is the total number of nodes in the original network, n_r is the number of nodes removed, and c is the number of nodes in the maximal connected subgraph (i.e., components; Zhang, 2012e) once n_r nodes are removed.

(2) Restoration robustness

The full information of a node with unknown or incomplete information can be achieved by consulting the information of its adjacent nodes, as used in detection of key terrorist in the terrorist organization (Bohannon, 2009). Restoration robustness refers to the capacity of a network to restore missed nodes/links if they are destroyed.

Restoration robustness in terms of nodes (D) and links (L) are

$$D = 1 - (n_r - n_s)/n$$
$$L = 1 - (l_r - l_s)/l$$

respectively, where n: total number of nodes in the original network, l: total number of links in the original network, n_s: number of nodes restored, n_r: number of nodes removed, l_s: number of links restored, and l_r: number of links removed.

9.1.2.2 *Parameter robustness*

Suppose the topological structure of a network, e.g., nodes and links, is not changed if the network is perturbed. Parameter robustness refers to a network's capacity, without any structural changes, for maintaining between-node flows (fluxes)/link weights if it is perturbed. The following index, revised from adjacency matrix index, was proposed (Zhang, 2012d) to represent parameter robustness. Following the definition of Zhang (2012d), suppose flows (fluxes)/link weights matrix of a network with n nodes is $w = (w_{ij})_{n \times n}$, where w_{ij} is the flow (flux) or the weight of the link between nodes v_i and v_j; $w_{ij} = 0$ if there is not a link between nodes v_i and v_j; $i, j = 1, 2, \ldots, n$. The index is (Zhang, 2016n)

$$S = \sum_i \sum_j |w_{ij0} - w_{ijt}|$$

where w_{ijt}, w_{ij0}: flow (flux) or link weight between nodes v_i and v_j after and before a network is perturbed. The lower the S value, the stronger but robustness.

9.1.2.3   *Comprehensive robustness*

Following the definition of Zhang (2012d), suppose adjacency matrix of a network with n nodes is $d = (d_{ij})_{n \times n}$, If $d_{ij} = d_{ji} = 0$, then there is no link between nodes v_i and v_j; if $d_{ij} = d_{ji} = 1$, then there is a link between nodes v_i and v_j. Suppose w_{ij} is the flow (flux) or the weight of the link between nodes v_i and v_j, $i, j = 1, 2, \ldots, n$. The index for comprehensive robustness is (Zhang, 2016n)

$$S = \sum_i \sum_j |w_{ij0}d_{ij0} - w_{ijt}d_{ijt}|$$

In addition to the measures on network robustness, other robustness measures can be defined according to various performance indices of a network, e.g., network type, variation coefficient, network entropy, etc (Zhang and Zhan, 2011). For example, the robustness of network entropy is defined as a network's capacity to maintain network entropy if the network's structure (or parameters) is perturbed.

For some biological networks, e.g., ecosystems, network performance indices may be additionally defined as total productivity, number of functional groups, etc.

9.1.3   *Structure robustness of typical networks*

In the random network of n nodes, two nodes are connected at a certain probability p (Zhang, 2012e). In the regular network, nodes are connected following certain rules. For example, the nearest neighbors coupling network (Wang *et al.*, 2006a, 2006b), namely, for a given k (k is an even number), n nodes in the network are linked to generate a ring in which each node is only connected with its $k/2$ neighborhood nodes. Scale-free network is the most popular network (Barabasi and Albert, 1999; Zhang, 2012e). Small-world network was proposed by Watts and Strogatz (1998). The methods for network generation/evolution are from Barabasi and Albert (1999) (scale-free network), Watts and Strogatz (1998) (small-world network), Wang *et al.* (2006a, 2006b) (regular network), and Zhang (2012e) (random network).

9.1.3.1 *Connection robustness*

For the random network, increasing connection probability p enhances connection robustness. Network connectivity will not be destroyed if $p \geq 0.3$ (the threshold density for connection robustness; Zhang, 2016n).

With the increase of the nodes removed from scale-free network, the decrease of network connection capacity produces "emergence" phenomena. Connection robustness will enhance with the increase of network density.

Generally, the connection robustness of regular network and random network is stronger than that of scale-free network. In terms of connection robustness, small-world network is the interim type between regular network and random network. Connection robustness of all networks produces "emergence" with the change of number of the nodes removed and of network density.

9.1.3.2 *Restoration robustness*

(1) Node restoration

Suppose the node i is adjacent to node j. If node i is removed from node j, we may try to restore node i and the link to node j based on the information of node j.

For a random network, if its network density is less than 0.3, the restored nodes decline with the increase in the number of the nodes removed. Removed nodes can be restored if network density is not less than 0.3 (threshold density for node restoration robustness). Similar to threshold density for connection robustness, the threshold density for node restoration robustness declines with the increase of network size (Zhang, 2016n).

A scale-free network can be thoroughly restored if only a few of nodes are removed. After the removed nodes have reached a threshold, the number of restored nodes will decline quickly. However, node restoration robustness increases with network density.

In the small-world network, for a fixed reconnection probability p, the decline of node restoration rate shows "emergence" with the increase of nodes removed. The increase of node degree or reconnection probability p will enhance node restoration robustness.

Generally, the node restoration robustness of all kinds of networks produces "emergence" with the change in the number of nodes removed and in network density.

(2) Link restoration

For the random networks, link restoration robustness increases with network density. Nevertheless, link restoration robustness will be almost irrelevant to network density once network density reaches a threshold (Zhang, 2016n).

The link restoration robustness of scale-free networks decreases linearly with the increase of number of the nodes removed, and has no significant relation with network density.

For the small-world networks, with a fixed reconnection probability p, the decline of link restoration rate shows "emergence" as the increase of the nodes removed. The increase of node degree or reconnection probability p will enhance link restoration robustness.

In summary, both link and node restoration robustness of the random network are the best among four types of networks, and the regular network is the worst. The small-world network is the interim type between the random network and the regular network. Node restoration robustness of scale-free network is better, but its link restoration robustness is as worse as the regular network.

In addition to network evolution methods and resultant networks above, a more complex network evolution model has demonstrated that network density (connectance; Zhang, 2011c, 2012b, 2012e) increases with time (Zhang, 2016f), which means that the network evolves to the stronger robustness according to the conclusions above. As theoretical models, all of the network evolution methods and resultant networks above do not limit the number of links connected to a node, i.e., node degree is limitless in these methods and networks.

9.1.4 *Facilitation of network structure robustness*

Different from the theoretical networks above, practical networks sometimes demonstrate different mechanisms (e.g., the food webs with random degree distributions were highly fragile to the removals of species, see Montoya and Sole, 2003) for robustness maintaining due to the limitation of node degree (e.g., many species have only one or two links in food webs), etc. Based on both theoretical analysis and practical observations, Zhang (2016n) summarized the following methods to enhance structure robustness of a given network

(1) Maintaining a certain network density/connectance/connectivity (Dunne *et al.*, 2002; Allesina *et al.*, 2005; Zhang, 2011c, 2012e). It has been reported that a food web with the higher connectance has more numerous reassembly pathways and can thus restore faster from perturbation (MacArthur, 1955; Law and Blackford, 1992; Zhang, 2012e). However, some models suggested that food webs with the lower connectance restored faster after a disturbance (May, 1973; Pimm, 1991; Chen and Cohen, 2001; Cohen *et al.*, 1990; Zhang, 2012e). Therefore, for a network with a fixed number of nodes, maintaining (increasing or reducing) a certain number of links may improve network structure robustness.

(2) Deploying network circuits (Alon, 2006; Zhang, 2012e, 2016d). Deploying some circuits in network means the existence of some feedback controls, which help to improve network robustness. Feedback control, in particular negative feedback control, is the basis of robustness and stability of bio-systems and bio-processes, e.g., the chemotaxis and heat shock response of *Escherichia coli*, biological rhythms, cell cycles, etc (Oleksiuk *et al.*, 2011).

(3) Constructing hierarchical subnetworks/modules/connected components (Zhang, 2012e, 2016c). A large network is always organized from various small mosaics (modules). Organizing mosaics to a large network will probably influence the robustness of the entire network (May, 1973). Pinnegar *et al.* (2005) used a detailed Ecopath with Ecosim model to examine the impacts of food web aggregation and the removal of weak linkages. They found that aggregation of a 41-compartment food web to 27 and 16 compartment systems greatly affected system properties (e.g., connectance, system omnivory, and ascendancy) and influenced dynamic stability. Highly aggregated webs restored more quickly

following disturbances compared to the original disaggregated model.

Existence of hierarchical subnetworks/modules/connected components may prevent local failures from diffusing across over a network, simplify the evolution and update of nodes and links, and thus help to enhance network structure robustness. In addition, utilization of several hierarchical subnetworks/modules/connected components may avoid malfunction of the major components in a network (Zhang, 2016c). A biological cell is a typical example. In a cell, mitochondria, ribosomes, chloroplasts, etc., are subnetworks/modules/connected components.

(4) Incorporating or enhancing redundancy. Besides useful nodes and links, adding redundant (i.e., temporarily not useful, or candidate) nodes, links, and circuits in a network will help to enhance structure robustness. Redundant nodes/links/circuits are expected to play key roles if some original nodes, links, and circuits are destroyed. In biological networks, repeated genes, homeotic genes (McAdams and Arkin, 1999), redundant metabolites, redundant multisignaling pathways, similar metabolic circuits (Edwards *et al.*, 2001), etc., are all examples of network redundancy.

Some methods above can be used to facilitate parameter robustness and comprehensive robustness. Actually, comprehensive robustness is more reasonable than structure robustness, because network structure is usually determined by the strengths of between-node interactions (i.e., link weights, between-node fluxes, etc.). As an example, past studies have demonstrated that under the condition of constant structure robustness, the increase in nodes and links in a network must be at the cost of weakening the added links (weak interactions, i.e., weak weights, weak fluxes, etc.; McCann, *et al.*, 1998; Paine, 1992; Zhang, 2011c, 2012e).

Computational simulation, e.g., the network evolution model (Zhang, 2015b), can be used to exploit the relationship between network structure and network robustness.

9.1.5   *Confused use of robustness and stability*

So far, most research used stability, etc., to describe the robustness of various biological systems. This includes such topics as "relationship between biodiversity and stability", "stability of ecosystems", "stability of metabolic networks", etc. It is unsurprising because the terminology "robustness" was defined as late as about 30 years ago, while "stability" had been used by ecologists for more than 40 years (May, 1973). Obviously, "stability" in these topics was in essence robustness in most situations. Definition and implication of robustness and stability, as discussed above, are extremely distinctive. We consider both external disturbance and internal perturbation in biodiversity and ecosystems. According to the substantial implication of so-called stability in these topics, I argue to use robustness to replace the misused or inaccurately used stability in these situations with the exception of fewer cases for stability-specific topics (Zhang, 2016n). In most of the situations, the exact and correct term should be "relationship between biodiversity and robustness", "robustness of ecosystems", "robustness of metabolic networks", etc.

9.2   Sensitivity Analysis

Sensitivity analysis explores the relationships of, e.g., structure *vs.* stability, input *vs.* output, of various systems. It helps us understand the structural properties, dynamic mechanisms, key components (or nodes, links, parameters, etc.), and other aspects of systems. A variety of methods for sensitivity analysis of system models have been proposed (McKay *et al.*, 1979; Downing *et al.*, 1985; Morris, 1991; Saltelli *et al.*, 1999, 2000; Sobol, 1993; Xu *et al.*, 2004). Networks are systems also. Like general systems, some networks can be described with mathematical models (Ferrarini, 2013b, 2014, 2015). Many networks evolve naturally, following some rules (Zhang, 2012e, 2015b, 2016f, 2016r). To find the most important nodes, links, or other parameters that determine network structure or performance is of significant in network analysis. In this section, I will introduce the earlier discussion on methods of sensitivity analysis (Zhang, 2016q). These methods were originally proposed for systems. As pointed out earlier, however, a system is in a sense a network. Thus they can be used in network analysis also. In network analysis, the model output in traditional sensitivity analysis represents network output, network stability, network flow, network structure, or other indices we used, and model input represents network nodes, network links, network parameters, etc., (Zhang, 2016q).

There are two types of sensitivity analyses in systems analysis, local sensitivity analysis and GSA (Saltelli, 2000). In the sensitivity analysis, a system (network) is represented by a mathematical model.

9.2.1 *Local sensitivity analysis*

Local sensitivity analysis (LSA) is mainly used to analyze local influence of parameters on model output. Gradients of parameters *vs.* model output can be achieved by using LSA. LSA is valuable to systems with simple mathematical representations, fewer undeterministic parameters, and easily derived sensitivity equation. LSA methods include directed derivation, finite difference, and Green function (Zhang, 2016q).

9.2.1.1 *Directed derivation*

For models with less independent variables and simple structure, directed derivation is a simple and rapid method for sensitivity analysis. Suppose the initial-value problem of a system is (Han *et al.*, 2008)

$$dy/dt = f(y, x), y(0) = y_0 \qquad (1)$$

where $y = (y_1, y_2, \ldots, y_n)$ is output vector, $x = (x_1, x_2, \ldots, x_m)$ is the input vector, and t is time. The sensitivity equation is

$$d(\partial y/\partial x_i)/dt = A \, \partial y/\partial x_i + \partial f/\partial x_i \qquad (2)$$

or

$$dS/dt = AS + C$$

where S is the sensitivity matrix, $A = (\partial f_j/\partial y_l)$ is the Jacobi matrix, and $C = (\partial f_j/\partial x_i)$ is the parametrical Jacobi matrix. Assume the network is time-invariant, i.e., $f(y, x) = 0$. y can be obtained by solving the equation, $f(y, x) = 0$. The static sensitivity matrix S with respect to x_i is achieved as

$$S = -A^{-1}C$$

9.2.1.2 *Finite difference*

In finite difference method, a perturbation of an input x_j, Δx_j, is made to obtain derivative of output to x_j. The forward difference scheme is usually used (Han *et al.*, 2008)

$$\partial y / \partial x_i \approx (y(x^j) - y(x)) / \Delta x_i, \quad j = 1, 2, \ldots, m$$

where $x^j = (x_1, x_2, \ldots, x_{j-1}, x_j + \Delta x_j, x_{j+1}, \ldots, x_m)$. The more precise scheme, central-difference scheme, is used also

$$\partial y / \partial x_i \approx (y(x^{j+}) - y(x^{j-})) / (2\Delta x_i), \quad j = 1, 2, \ldots, m \quad (3)$$

where $x^{j+} = (x_1, x_2, \ldots, x_{j-1}, x_j + \Delta x_j, x_{j+1}, \ldots, x_m), x^{j-} = (x_1, x_2, \ldots, x_{j-1}, x_j - \Delta x_j, x_{j+1}, \ldots, x_m)$.

9.2.1.3 *Green function*

The differential equation of Eq. (1) with respect to the initial value y_0 is (Han *et al.*, 2008)

$$dS(t, t_1)/dt = A(t)S(t, t_1) \quad (4)$$

where t, t_1: perturbation time and observation time; $x(t, t_1)$: sensitivity matrix. $S(t, t_1) = (\partial c_i(t)/\partial c_j^0(t_1))$, $S(t_1, t_1) = 1$, $t \geq t_1$. The solution of sensitivity matrix is made of two parts, the general solution of homogeneous equation of eq. (2) and the special solution of nonhomogeneous eq. (2). The general solution of homogeneous equation of equation is obtained by using eq. (3). And the special solution of nonhomogeneous equation is

$$T(t_1, t_2) = \int_{t_1}^{t_2} S(t_2, s)F(s)ds$$

LSA is used to test model sensitivity to the change of a single specified parameter only, and the remaining parameters are fixed. LSA does not consider the influence of interactions between model parameters on model output.

9.2.2 *Global sensitivity analysis*

GSA analyzes the joint influence of multiple parameters on model output and analyzes model sensitivity of both a single parameter and between-parameter interactions. It can be used in nonlinear, nonoverlapping, or nonmonotonous models (Zhang, 2016q).

9.2.2.1 *Qualitative GSA*

With less calculation, Qualitative GSA is used to rank parameters according to their sensitivities.

(1) Multivariable regression method

In this method, Latin hypercube sampling is used (McKay *et al.*, 1979; Downing *et al.*, 1985). It divides the cumulative probability distribution (i.e., the interval $(0, 1)$) of a parameter into multiple nonoverlapping intervals of equal length along y-axis, and thus is more effective than random sampling method. Each interval corresponds to an interval in x-axis. Randomly sampling a point in an interval of y-axis, we can obtain a corresponding parametrical value in an interval of x-axis. For the model with n parameters, each parameter has a cumulative probability distribution and m intervals are generated. In total, n^m sampling combinations are thus produced.

The procedure is that to arrange values of n parameters into a matrix of $n \times m$ and randomize elements of each column, then the m input of n parameters are thus obtained. Each row of the matrix can be taken as the input values of a parameter can be used as model input to get model output. Finally, build multivariable regression between model outputs and inputs and use regression coefficients or partial correlation coefficients as the sensitivity values of corresponding parameters. Stepwise regression is more reliable in this method (Qi *et al.*, 2016; Zhang, 2016p).

(2) Morris method

Morris method (Morris, 1991; Xu *et al.*, 2004) maps the domain of each parameter into $[0, 1]$ and discretizes the interval, such that each parameter takes values in $\{0, 1/(p-1), 2/(p-1), \ldots, 1\}$, where p is the number of sampling points of the parameter. Each parameter takes values randomly from p sampling points and obtains a vector $X = (x_1, x_2, \ldots, x_k)$, where k is the number of parameters.

Consider a matrix, $B = (b_{ij})_{(k+1) \times k}$, ($b_{ij} = 1$, if $i < j$; $b_{kk} = 0$, $b_{kj} = 1$, if $j < k$; $b_{k+1 \times j} = 1$; otherwise, $b_{ij} = 0$) and the change $\Delta = s/(p-1)$, where s is the constant. Use the adjacent two rows in the matrix ΔB as the model input and obtain two outputs y_1 and y_2. The sensitivity of parameter i can be calculated by $\Delta_i(X) = (y_1 - y_2)/\Delta$. Taking k pairs of adjacent rows as mode inputs, we may obtain sensitivities of k parameters (Xu *et al.*, 2004).

Values of elements in ΔB are not stochastic. Thus, in practical application, we can use a stochastic process to ensure the stochasticity of elements' values (Xu *et al.*, 2004) as follows: suppose $D^*_{k \times k}$ is a diagonal matrix in which any element has the same chance as 1 and -1. Suppose $J_{m \times k}$ is the unit matrix. Each column of elements of matrix $C_{m \times k} = (1/2)[(2B - J_{m \times k})D^* + J_{m \times k}]$ is equal to corresponding elements in B (or 1 change to 0, or 0 change to 1 in B); suppose X^* is the base vector of X. Each parameter in X randomly takes values in $\{0, 1/(p-1), 2/(p-1), \ldots, 1\}$, and suppose $P^*_{k \times k}$ is the random permutation matrix in which each row and each column contains one 1 only and others are zeros. Let $B^* = (J_{m \times 1}X^* + (\Delta/2)[(2B - J_{m \times k})D^* + J_{m \times k}])P^*$, B^* is the stochastic matrix of B. Use the adjacent two rows in the matrix B^* as the model input and obtain two outputs y_1 and y_2. The sensitivity of parameter i can be calculated by $\Delta_i(X) = (y_1 - y_2)/\Delta$. Replications can be made to obtain the mean sensitivity and standard deviation of each parameter. A low standard deviation means little interaction between the parameter and other parameters.

Morris method can be used to freeze less sensitive parameters and further make quantitative GSA.

9.2.2.2 *Quantitative GSA*

Quantitative GSA can quantitatively calculate the contribution proportion of uncertainty of model parameters to uncertainty of model output. It holds that variance of model output represents the uncertainty of model output. Before making quantitative GSA, qualitative GSA is always used to filter the less sensitive parameters from the model (Zhang, 2016q).

(1) Sobol method

A representative method of quantitative GSA is Sobol method (Sobol, 1993; Xu *et al.*, 2004; Han *et al.*, 2008). Sobol method is a Mont Carlo method based on variance.

Suppose the model is $y = f(x), x = (x_1, x_2, \ldots, x_k)$. x_i follows uniform distribution of [0,1], and $(f(x))^2$ is integrable. First, define a k-dimensional unit cube as the space domain of parameters, $\Omega = \{x|0 \leq x_i \leq 1; i = 1, 2, \ldots, k\}$. The core procedure of Sobol method is to decompose the function $f(x)$ into the sum of subterms

$$f(x) = f_0 + \sum_{i=1}^{k} f_i(x_i) + \sum_{i<j} f_{ij}(x_i, x_j) + \cdots + f_{1,2,\ldots,k}(x_1, x_2, \ldots, x_k)$$

(5)

The decomposition above is not unique. If

$$\int_0^1 f_i(x_i)dx_i = 0, \qquad\qquad \forall x_i, i = 1, 2, \ldots, k$$

$$\int_0^1 \int_0^1 f_{ij}(x_i, x_j)dx_idx_j = 0, \qquad\qquad \forall x_i, x_j, i < j$$

$$\int_\Omega f_{1,2,\ldots,k}(x_1, x_2, \ldots, x_k)dx_1dx_2 \ldots dx_k = 0$$

the decomposition (5) is unique, and

$$\int_\Omega f_{i1,i2,\ldots,is} f_{j1,j2,\ldots,jl}dx = 0$$

$$(i_1, i_2, \ldots, i_s) \neq (j_1, j_2, \ldots, j_l), \quad k = s + l$$

Calculate

$$f_0 = \int_\Omega f(x)dx$$

$$f_i(x_i) = -f_0 + \int_0^1 \ldots \int_0^1 f(x)dx_{-i}, \quad i = 1, 2, \ldots, k$$

$$f_{ij}(x_i, x_j) = -f_0 - f_i(x_i) - f_j(x_j)$$

$$+ \int_0^1 \ldots \int_0^1 f(x)dx_{-ij}, \quad i < j$$

where x_{-i} and x_{-ij} represent the parameters in exception of x_i and x_i together with x_j, respectively. Similarly, we can get $f_{1,2,\ldots,k}(x_1, x_2, \ldots, x_k)$. The total variance of model output, i.e., the effect of all parameters on model output is

$$V = \int_\Omega (f(x))^2 dx - (f_0)^2$$

The effect of a single parameter on model output is represented by its partial variance

$$V_i = \int (f_i)^2 dx_i$$

And the joint effect of several parameters on model output is represented by the partial variance

$$V_{i1,i2,\ldots,is} = \int_0^1 \ldots \int_0^1 (f_{i1,i2,\ldots,is})^2 dx_{i1} dx_{i2} \ldots dx_{is}$$

Square eq. (5) and thereafter make integration, we have

$$V = \sum_{i=1}^k V_i + \sum_{i<j} V_{ij} + \cdots + V_{1,2,\ldots,k}$$

and

$$S_{i1,i2,\ldots,is} = V_{i1,i2,\ldots,is}/V \quad (6)$$

where S_i is the sensitivity of parameter i, known as 1st sensitivity; $S_{i1,i2,\ldots,is}$ is the joint sensitivity of interactive parameters $i_1, i_2, \ldots$, and i_s, known as sth sensitivity.

The integrals above can be solved using Monte Carlo method

$$f_0 = \sum_{i=1}^{n} f(x_i)/n$$

$$V = \sum_{i=1}^{n} (f(x_i))^2/n - (f_0)^2$$

$$V_i = \sum_{j=1}^{n} f(x_{ij}^{(1)}, x_{-ij}^{(1)}) f(x_{ij}^{(1)}, x_{-ij}^{(2)})/n - (f_0)^2$$

$$\cdots$$

(2) Extended Fourier Amplitude Sensitivity Test

Extended Fourier Amplitude Sensitivity Test (EFAST) was proposed by Saltelli *et al.* (1999). In this method, the spectrum of Fourier series is obtained by Fourier transformation. The variances of model output arose from a single parameter and parameter interactions are calculated from frequency spectrum curve (Tarantola *et al.*, 2002; Xu *et al.*, 2004).

The model $y = f(x), x = (x_1, x_2, \ldots, x_k)$ can be transformed to $y = f(s)$ using a certain search function. The Fourier transformation is

$$y = f(s) = \sum_{j=-\infty}^{\infty} (A_j \cos(js) + B_j \sin(js))$$

where s is the independent parameter of all parameters, and

$$A_j = (1/(2\pi)) \int_{-\pi}^{\pi} f(s) \cos(js) ds$$

$$B_j = (1/(2\pi)) \int_{-\pi}^{\pi} f(s) \sin(js) ds$$

$$j \in Z = \{-\infty, \ldots, -1, 0, 1, \ldots, \infty\}$$

Frequency spectrum curve of Fourier series is $\wedge_j = A_j + B_j$, where $A_{-j} = A_j$, $B_{-j} = B_j$, $\wedge_{-j} = \wedge_j$. The variance of model output aroused from uncertainty of x_i is

$$V_i = 2 \sum_{p=1}^{\infty} \wedge_p w_i \qquad (7)$$

where w_i is a specified frequency. The total variance is

$$V = 2 \sum_{p=1}^{\infty} \wedge_p \qquad (8)$$

Take s in the $[-\pi, \pi]$ with the same interval, and input each sampled parameter to the model. Run the model many times, and A_j and B_j can be obtained from

$$A_j = \sum_{S_{k}=1}^{Ns} f(s_k) \cos(js_k)$$

$$B_j = \sum_{S_{k}=1}^{Ns} f(s_k) \sin(js_k)$$

$$j \in Z' = \{-(N_s - 1)/2, \ldots, -1, 0., 1, \ldots, (N_s - 1)/2\}$$

where N_s is the number of samples, $N_s = 2Mw_{\max}+1$. Finally, based on eq. (6)–(8), we can get sensitivity of each parameter.

EFAST costs much lower than Sobol. Both EFAST and Sobol require parameters that are irrelevant to each other.

In addition to the methods above, Zhang (2012d, 2012e) proposed a series of methods for determination of crucial nodes, which can be used in sensitivity analysis also, for example, Node Perturbation index (NP) (Zhang, 2012d). More methods that can be used in sensitivity analysis of networks, e.g., adjacency matrix index, flow change index, etc., have been discussed in Zhang (2012d) and other references.

Chapter 10

Network Control

A network is controlled to achieve an optimal or desired structure, state, dynamics, etc. The network is optimized to: (1) search for an optimal search plan, and (2) achieve a topological structure so that the network possesses relative stability (Zhang, 2012e). Nevertheless, the goal function of complex networks is always hard to be analytically expressed. Thus, we choose to achieve an approximation of goal function by nonparametric test or linearization methods. If the goal function is simple, parameter optimization can be realized through conventional optimization techniques. Otherwise, parameters can be optimized by using sensitivity analysis, statistical experiment, and heuristic methods.

This chapter focuses on the dynamic control of networks, as described in detail in Zhang (2010, 2016r). The dynamic control of a network aims to change topological structure and parameters of the network stage by stage so that the optimal or desired property (e.g., stability) of the network is achieved.

10.1 Conventional Control

In conventional control, we assume that as a system, network nodes are fixed, and dynamics of the network to be controlled can be described with ordinary differential equations. Moreover, network stability specifically means the state stability of network nodes if stability is the goal of network control.

10.1.1 *Linear feedback control*

A linear ordinary system (network) is not certainly stable. It will become stable by adding a negative feedback in the network. The linear feedback control of the network with n nodes (state variables) is (Qian and Song, 1983)

$$dx/dt = Ax + Bu, \quad x(t_0) = x_0$$
$$u = -Dx$$
$$y = Cx$$

where $x = (x_1, x_2, \ldots, x_n)$, $u = (u_1, u_2, \ldots, u_m)$, $y = (y_1, y_2, \ldots, y_p)$; A: system matrix, B: control matrix, C: output matrix, D: feedback matrix, and x_0: initial state. The closed system for the linear feedback control is

$$dx/dt = (A - BD)x$$

The solution for closed system is

$$x(t) = e^{(A-BD)t}x(t_0)$$

As a general case, if A, B, and C are time dependent, the solution will be

$$x(t) = \Phi(t, t_0)x(t_0) + \int_{t_0}^{t} \Phi(t, \tau)B(\tau)u(\tau)d\tau$$

and

$$y(t) = C(t)\Phi(t, t_0)x(t_0) + \int_{t_0}^{t} c(t)\Phi(t, \tau)B(\tau)u(\tau)d\tau$$

Let $h = t(k+1) - t(k)$, the output is

$$x(t_{k+1}) = e^{(A-BD)h}x(t_k)$$
$$y(t_{k+1}) = C\,x(t_{k+1})$$
$$k = 0, 1, 2, \ldots$$

and the control dynamics is

$$u(t_{k+1}) = -D\,x(t_{k+1}), \quad k = 0, 1, 2, \ldots$$

where

$$e^{(A-BD)h} = E + (A-BD)h + ((A-BD)h)^2/2! + ((A-BD)h)^3/3! + \cdots$$

here $h = 1$.

For the closed system, if all the eigenvalues for the equation

$$|A - BD - \lambda * E| = 0$$

have negative real parts, the system (i.e., the nodes' states) will be asymptotic stable; if all of them are not greater than 0 and the eigenvalues equivalent to 0 are 1st-order, then the system is stable; otherwise, the system is not stable.

10.1.2 Linear optimization control

The linear optimal control is (Qian and Song, 1983)

$$dx/dt = Ax + Bu$$

where $x = (x_1, x_2, \ldots, x_n)$, $u = (u_1, u_2, \ldots, u_m)$; A: system matrix, B: control matrix. The objective general function is

$$\min J = \int_0^\infty (x(t)Qx(t) + u(t)Ru(t))\,dt$$
$$u(t) = -Dx(t)$$

where D: feedback matrix, Q, R: general function matrices for state x and control u, respectively.

The closed loop system for the network is

$$dx/dt = (A - BD)x$$

First, solve Riccati equation

$$Az + zA + Q - zFz = 0$$

where $F = B * R^{-1} * B$, $z = (z_1, z_2, \ldots, z_n)$, the iterative method is

$$z(0) = \left(\int_0^h (e^{-At}Fe^{-At}\,dt) \right)^{-1}$$

where h is a constant, and

$$(A - Fz(n-1))z(n) + z(n)(A - Fz(n-1)) + Q + z(n-1)Fz(n-1) = 0$$

Solve $z(n)$, $n = 1, 2, \ldots$, until

$$\max |(z(n-1)_{ij} - z(n)_{ij}| <= E$$

where $1 \le i, j \le n$, then the Riccati solution is $z = z(n)$, and optimal control matrix is $D = R^{-1} * B * z$, and the optimal closed loop matrix is $A - Fz$, where

$$e^{At} = E + At + (At)^2/2! + (At)^3/3! + \cdots$$

If all the eigenvalues for

$$|A - BD - \lambda * E| = 0 \text{ (closed system)}$$

or

$$|A - \lambda * E| = 0 \text{ (open system)}$$

have negative real parts, the system will be asymptotic stable; if all of them are not greater than 0 and the eigenvalues equivalent to 0 are 1st-order, then the system is stable; otherwise, the system is not stable.

10.1.3   *Number of controllable states of linear ordinary systems*

It is necessary to determine the number of controllable states (nodes) of a system in order to effectively design the control system (Zhang, 2010, 2016r). Consider a linear ordinary system

$$dx/dt = Ax + Bu$$

where $x = (x_1, x_2, \ldots, x_n)$, $u = (u_1, u_2, \ldots, u_m)$; A: system matrix, B: control matrix. The number of controllable states is the rank of the matrix

$$(BAB \ldots A^{n-1}B)$$

If the rank is n, then the system is completely controllable.

10.1.4 Number of observable states of linear ordinary systems

Suppose the linear ordinary system is

$$dx/dt = Ax$$
$$y = Cx$$

where $x = (x_1, x_2, \ldots, x_n)$, $y = (y_1, y_2, \ldots, y_m)$; A: system matrix, C: output matrix. The number of observable states is the rank of the matrix

$$(C\ CA \ldots CA^{n-1})$$

If the rank is n, then the system is completely observable.

10.1.5 Discrete control systems

Consider a control system

$$dx/dt = f(x, u, t)$$
$$y(t) = g(x, t)$$

where $x = (x_1, x_2, \ldots, x_n)$, $u = (u_1, u_2, \ldots, u_m)$, $y = (y_1, y_2, \ldots, y_p)$, $f = (f_1, f_2, \ldots, f_n)$. Let

$$dx/dt \approx (x(k+1) - x(k))/T$$

the system becomes

$$x(k+1) = x(k) + T\ f(x(k), k), \quad k = 0, 1, 2, \ldots$$

where T is the time step. And the discrete control system is

$$x(k+1) = x(k) + T\ f(x(k),\ u(k), k)$$
$$y(k) = g(x(k), k)$$
$$k = 0, 1, 2, \ldots$$

For the linear ordinary system above, the discrete control system is

$$x(k+1) = x(k) + T\,(Ax(k) + Bu(k))$$
$$y(k) = Cx(k)$$
$$k = 0, 1, 2, \ldots$$

The general form of discrete linear system is

$$x(k+1) = A(k)x(k) + B(k)u(k))$$
$$y(k) = C(k)x(k)$$
$$k = 0, 1, 2, \ldots$$

The solution is

$$x(k) = \Phi(k,0)x(0) + \sum_{j=0}^{k-1} \Phi(k, j+1)B(j)u(j)$$

and the output is

$$y(k) = C(k)\Phi(k,0)x(0) + \sum_{j=0}^{k-1} C(k)\Phi(k, j+1)B(j)u(j)$$

For the linear ordinary system, the solution of discrete control system is

$$x(k) = A^k x(0) + \sum_{j=0}^{k-1} A^{k-(j+1)} Bu(j)$$

and the output is

$$y(k) = CA^k x(0) + \sum_{j=0}^{k-1} A^j Bu(k - (j+1))$$

Given the discrete system

$$x(k+1) = f(x(k),\ u(k), k)$$
$$x(k_0) = x_0$$

The optimization problem is to solve the control series

$$u(k_0), u(k_0 + 1), u(k_0 + 2), \ldots, u(k_f - 1)$$

such that the objective function

$$J = \sum_{i=k_0}^{k_f-1} L(x(i), u(i), i) + \theta(x(k_f), k_f)$$

be minimized or maximized.

10.2 New Perspectives of Network Control

In addition to the conventional methods given above, Ferrarini (2011c) proposed five unconventional thoughts on networks control as below:

(1) Conventional control, e.g., the control proposed by Liu *et al.* (2011), focuses on nodes, neglecting to enlarge the approach to edges as well. Although an n-node network could bear up to $n(n-1)$ links (or $n * n$ if we also consider self-links) among nodes, this could be conceived as a better chance of network control, not just like a computational difficulty. The higher the number of switches on which one can act, the higher the chance to make the actual network into the desired one. Several tools, like genetic and particle swarm optimization, can reasonably lower the seemingly intractable problem of edge control. Furthermore, the control of one edge leads to directly influence on both the nodes connected by the edge itself; hence, network control based on edge ascendancy could be more parsimonious than expected.

(2) Ferrarini observed that, while nodes are time-dependent by defi- nition, edges are not time-dependent by necessity in many types of network. It's well known that many stock-and-flows networks in ecology, geography, and socio-economic dynamics have almost constant links, at least on short and mid-time intervals, and are hence expressed on a yearly basis. The controllability of complex networks through discrete-time edge control could be more parsi- monious than a continuous-time node control in many situations.

(3) He argued that the concept of driver nodes used by Liu *et al.* (2011) could be translated to edges too. One could think of the control of edges which connect driver nodes as complementary to the control of just driver nodes. Because of the distributed nature

of complex networks, whose dynamics are mainly decentralized, it is feasible to control them by acting locally on permanently selected driver links, which could play the role of network pacemakers, and by exploiting the coupling effects between these and the rest of the network.

(4) Ferrarini observed that while the control of driver nodes has just one time-dependent solution to get the desired final state, the control of edges makes many solutions available to the controller due to the indirect effects that can derive higher-level properties of the system. The wider the network, the higher the number of solutions to the problem of controlling it by mastering its edges is. Because one solution is enough for the desired goal, the controller is not requested to find all the solutions available to get the desired final configuration, which would be truly prohibitive.

(5) Liu *et al.* (2011) do not mention the chance to employ a node exogenous to the network (Almendral *et al.*, 2009). Nodes in the network might behave as slave systems of the external node, and could be coupled to it through n links. In this way, the problem of network controllability could be translated into the control of just n links, instead of $n(n-1)$. If one external node is not enough, further nodes can be added up to satisfaction.

Various methods on network control have been proposed. More details can be found in the literature, e.g., Ferrarini (2011a, 2011c, 2013a, 2013b, 2014, 2015).

Chapter 11

Network Evolution

In a sense, network evolution is network dynamics also. However, different from the previous chapter on network dynamics, which focuses on the network dynamics with the fixed number of nodes, network evolution in this chapter treats network dynamics with changing number of nodes and links. The present chapter introduces the network evolution models proposed by Zhang (2015b, 2016f), and a method for phase recognition (classification) in network evolution (Zhang, 2017m).

11.1 A Generalized Network Evolution Model for Community Assembly Dynamics

Community assembly is the process by which species grow and interact to establish a community (Zhang, 2014c). It stresses the change of community over a single phase (Warren *et al.*, 2003). Community assembly may likely produce a diverse and stable community. Community assembly might be constrained by local (i.e., abiotic characteristics of habitat) or regional (i.e., composition of species pool, habitat isolation) factors (Bossuyt *et al.*, 2005). The theory and knowledge on community assembly provides basic rules for species assembly (Chase, 2003; Zhang, 2011c, 2012b, 2012c, 2012e, 2014c). Community assembly rule was firstly proposed by Diamond in 1975. He argued that community assembly is the process that species in the regional species bank join the local community through the multiple-layer filtering of the environment and biological interactions

(Diamond, 1975). Wilson and Roxburgh (1994) thought that the rules for plant community assembly are a series of potential rules restricting the presence or increase of species. So far, a variety of rules have been proposed, among which the most accepted rule is species cooccurrence hypothesis proposed by Diamond (1975). Fukami (2010) defined the mechanism of community assembly as a construction and conservation process of local community through sequential arrival of species from the external species bank and increase/extinction of species in the community. Community assembly theory acknowledges that the community tends to be stable over the time, and acknowledges the role of inter-specific interactions, in particular competition. So far, the ecological niche theory and neutral theory are two well-known interpretations of the mechanism for community assembly. Methods used in community assembly include establishing research plots, indoor simulation, etc. Some methods for community succession can be used in community assembly research.

Some researchers have built the predation structures of some communities (Dunne *et al.*, 2002). These models include cascade model (Cohen and Newman, 1985; Zhang *et al.*, 2014a), habitat model (Williams and Martinez, 2000), and the model for energy flows and functional groups (Fath *et al.*, 2007). In addition, Zhang (2012c) presented a simple probabilistic network model. Furthermore, a generalized, rule-based network evolution model for community assembly was proposed (Zhang, 2015b). The model was based on difference equations with different number of species in different stages of evolution. It consists of pioneer rule, invasion and growth rule, extinction rule, connection (flow) rule, termination rule, etc. Here, the model and the corresponding self-organization theory on community assembly are described.

11.1.1   *Model: CommAssembly*

Suppose there are in total of m species (i.e., m nodes in the network) in the species pool of a community being assembled. In general, the nonlinear differential equation of dynamics of the n-species ($n \leq m$)

community is

$$dx/dt = f(x,t)$$

where $x = (x_1, x_2, \ldots, x_n), f(x,t) = (f_1(x,t), f_2(x,t), \ldots, f_n(x,t))$. Suppose $f(x,t)$ is 2nd-order differentiable, then in a short time interval, the nonlinear equation can be approximated with a linear differential equation

$$dx/dt = A(t)x$$

where $x = (x_1, x_2, \ldots, x_n), A(t) = (a_{ij}(t))_{n \times n}, a_{ij}(t) = \partial f_i(x,t)/ \partial x_j(t), i, j = 1, 2, \ldots, n$. The linear differential equation can be transformed to a difference equation

$$\Delta x = A(t)x\Delta t$$

i.e.,

$$\Delta x_i = (a_{i1}(t)x_1 + a_{i2}(t)x_2 + \cdots + a_{in}(t)x_n)\Delta t, \quad i = 1, 2, \ldots, n$$

Without the loss of generality, let $\Delta t = 1$, and we have

$$\Delta x_i = a_{i1}(t)x_1 + a_{i2}(t)x_2 + \cdots + a_{in}(t)x_n, \quad i = 1, 2, \ldots, n$$

The rule for population dynamics is thus

$$x_i(t+1) = a_{i1}(t)x_1(t) + a_{i2}(t)x_2(t) + \cdots + a_{ii-1}(t)x_{i-1}(t)$$
$$+ (1 + a_{ii}(t))x_i(t) + a_{ii+1}(t)x_{i+1}(t)$$
$$+ \cdots + a_{in}(t)x_n(t), \quad i = 1, 2, \ldots, n \tag{1}$$

The coefficients, $a_{ij}(t), i, j = 1, 2, \ldots, n; i \neq j$, are correlated with various ecological interactions between species (e.g., competition, mutualism, predation, etc.). If both $a_{ij}(t) = 0$ and $a_{ji}(t) = 0$, the species i and j do not have niche overlap. The coefficients, $a_{ii}(t)$, $i = 1, 2, \ldots, n$, are correlated with intrinsic growth and resource availability of each species, etc.

Assign each species with an invasion probability $p_i(t), i = 1, 2, \ldots, m$, and an invasion strength (population size) $c_i(t), i = 1, 2, \ldots, m$.

(1) Pioneer rule. Let $t = 1$; randomly choose a species, i, with the invasion strength $c_i(t)$, and let it establish in the community. Let $x_i(t) = c_i(t)$. As described in rule (1), the species will grow to the next time step, following the rule $x_i(t+1) = (1+a_{ii}(t))x_i(t)$.

(2) Invasion and growth rule. $t = 2$; For each species in the species pool, randomly choose a species at its invasion probability from the species pool, in exception of i. If no species is chosen, $x_i(t + 1) = (1 + a_{ii}(t))x_i(t)$; otherwise, if the species j is chosen and established, let $x_j(t) = c_j(t)$, and we have the rule

$$x_i(t + 1) = (1 + a_{ii}(t))x_i(t) + a_{ij}(t)x_j(t)$$
$$x_j(t + 1) = (1 + a_{jj}(t))x_j(t) + a_{ji}(t)x_i(t) \tag{2}$$

The same species, for example, species j, established in earlier time, can invade the community again. In this case, let

$$x_j(t) \Leftarrow x_j(t) + c_j(t)$$

Once an invasion occurs, the population changes according to the equation group (2). If the population tends to zero, as described in the following extinction rule, the establishment of the species is not successful. Thus, species establishment is naturally included in the model.

Two or more species can be chosen and established simultaneously. In this case, several species and equations should be added in equation group (2), following the form of equation group (1).

(3) Extinction rule. Following the rule (1), repeatedly use invasion and growth rule. Suppose until certain time steps are conducted, the rule is the equation group (1), and a species, k (without loss of generality, suppose $k < i$), is removed from community because its population size becomes zero. Then the rule (1) becomes

$$x_i(t + 1) = a_{i1}(t)x_1(t) + a_{i2}(t)x_2(t) + \cdots a_{ik-1}(t)x_{k-1}(t)$$
$$+ a_{ik+1}(t)x_{k+1}(t) + \cdots + a_{ii-1}(t)x_{i-1}(t)$$
$$+ (1 + a_{ii}(t))x_i(t) + a_{ii+1}(t)x_{i+1}(t)$$
$$+ \cdots + a_{in}(t)x_n(t), \quad i = 1, 2, \ldots, k - 1, k + 1, \ldots, n \tag{3}$$

Two or more species can be removed simultaneously. In this case, several species and equations should be removed from equation group (1).

(4) Connection rule (flow rule). At each time step, for species i and $j(i \neq j)$

if $a_{ij}(t) \neq 0$, or $a_{ji}(t) \neq 0$, there is a connection (interaction) between species i and j

if $a_{ij}(t) \neq 0$ and $a_{ji}(t) = 0$, there is a connection from species j to i, and $a_{ij}(t)$ is the flow coefficient (connection weight)

if $a_{ij}(t) = 0$ and $a_{ji}(t) \neq 0$, there is a connection from species i to j; $a_{ji}(t)$ is the flow coefficient (connection weight)

if $a_{ij}(t) \neq 0$ and $a_{ji}(t) \neq 0$, there is a loop between species i and j; $a_{ij}(t)$ and $a_{ji}(t)$ are the flow coefficients (connection weights)

if $a_{ij}(t) = 0$ and $a_{ji}(t) = 0$, there is no connection between species i and j

In addition

if $a_{ij}(t) = 0$ and $a_{ji}(t) = 0$, $j = 1, 2, \ldots, n$; $i \neq j$, species i is an isolated (redundant) species

(5) Termination rule. Repeat the steps (2) to (4) until the community tends to be stable, i.e.,

$$x_i(t+1) = x_i(t) \quad i \in S \tag{4}$$

where S is the set of species occurred in the community, or until a certain number of iterations is achieved.

11.1.2 *Simplified model*

11.1.2.1 *Model*

The above model, CommAssembly, can be simplified for convenient demonstration as below:

In the equation group (1), if $x_i(t+1) \geq K_i(t+1)$, let $x_i(t+1) = K_i(t+1)$, $i = 1, 2, \ldots, n$, where $K_i(t)$ is environmental capacity of species i at time t, and further, let $K_i(t) = K_i, i = 1, 2, \ldots, m$;

Let $A(t) = A$, where A is a constant matrix;

Let $p_i(t) = p_i, i = 1, 2, \ldots, m$;

Let $c_i(t) = c_i, i = 1, 2, \ldots, m$.

K_i, A, p_i, and c_i are generated with random numbers. The simplified model and data generation method are included in the following Matlab codes

```
%Zhang WJ. 2015. A generalized network evolution model and self-organization
%theory on community assembly. Selforganizology, 2(3): 55-64
m=30;      %Set the number of species in species pool
tmax=1000;    %Maximum number of iterations
%Generate the coefficient matrix A
a=rand(m);
growfac=0.05;
interactfac=0.1;
nointeractfac=0.3;
neginteractfac=0.8;
for i=1:m; for j=1:m;
if (i==j) a(i,j)=a(i,j)*growfac; continue; end
if (a(i,j)<nointeractfac) a(i,j)=0;
else if (a(i,j)<neginteractfac) a(i,j)=-a(i,j)*interactfac; end;
end;
end; end
K=500+rand(1,m)*500; %Set the environmental capacity, Ki(t), for each
species; for simplified model only
p=ones(1,m)*rand()*0.1;   %Set the invasion probability, pi(t), for each
   species
c=ones(1,m)*rand()*5;   %Set the invasion strength, ci(t), for each species
x=zeros(1,m);
id=zeros(1,m);
t=1;
idx=round(rand()*m+0.5);
id(idx)=1;
x0=zeros(1,m);
x(idx)=c(idx);
while (t<tmax)
x0=x;
for i=1:m;
if (id(i)==0) x(i)=0; continue; end
s(i)=0;
for j=1:m;
if (id(j)==0) continue; end
s(i)=s(i)+a(i,j)*x(j);
```

```
end;
x(i)=s(i)+x(i);
end;
w=zeros(1,m); v=zeros(1,m);
nn=0;
fprintf(['t=' num2str(t) '\nSpecies list and population size\n'])
for i=1:m
if (id(i)==0) continue; end
if (x(i)>=K(i)) x(i)=K(i); end     % For simplified model only
if (x(i)>=0) id(i)=0; continue; else id(i)=1; end
nn=nn+1;
v(nn)=i;
w(nn)=x(i);
fprintf([num2str(i) '(x('num2str(i)')='num2str(round(x(i)*100)/100)')']);
end
fprintf(['\nTotal number of species=' num2str(sum(id)) '\nConnection and
connection weight (flow coefficient)\n']);
for i=1:nn
for j=1:nn
if (i==j) continue; end
if (a(v(i),v(j))~=0) fprintf([num2str(v(i)) ' ' num2str(v(j)) ' '
num2str(round(a(v(i),v(j))*1000)/1000) '\n']); end;
end; end
iso=zeros(1,m);
for k=1:nn
temprow=0;
for j=1:nn
if (j==k) continue; end
if (a(v(k),v(j))==0) temprow=temprow+1; end
end;
tempcol=0;
for j=1:nn
if (j==k) continue; end
if (a(v(j),v(k))==0) tempcol=tempcol+1; end
end;
if ((temprow==nn-1) & (tempcol==nn-1)) iso(k)=1; end
end
for i=1:nn
if (iso(i)==1) fprintf([num2str(iso(i)) '(isolated species)\n']); end
end
fprintf('\n\n');
if (sum(x==x0)==m) break; end
if (x==K) break; end                 % For simplified model only
t=t+1;
for i=1:m
if (rand()<p(i)) id(i)=1; x(i)=x(i)+c(i); end
end
end
```

11.1.2.2 *Result analysis*

Running the simplified model, a result set was obtained as follows:

```
t=1
Species list and population size
17(×(17)=2.34)
Total number of species=1
Connection and connection weight (flow coefficient)
1(isolated species)

t=2
Species list and population size
17(×(17)=2.45)
Total number of species=1
Connection and connection weight (flow coefficient)
1(isolated species)

t=3
Species list and population size
17(×(17)=2.44) 26(×(26)=4.29)
Total number of species=2
Connection and connection weight (flow coefficient)
17    26    −0.059
26    17     0.828

t=4
Species list and population size
17(×(17)=2.31) 26(×(26)=6.29)
Total number of species=2
Connection and connection weight (flow coefficient)
17    26    −0.059
26    17     0.828

t=5
Species list and population size
5(×(5)=1.77) 16(×(16)=7.64) 17(×(17)=1.51) 26(×(26)=9.41)
Total number of species=4
Connection and connection weight (flow coefficient)
 5    16    −0.05
 5    17    −0.066
 5    26    −0.034
16     5    −0.067
16    17    −0.078
16    26     0.892
17     5    −0.062
17    16    −0.058
17    26    −0.059
26     5     0.987
26    17     0.828
```

t=6
Species list and population size
5($\times$(5)=0.99) 16($\times$(16)=18.07) 21($\times$(21)=2.16) 26($\times$(26)=10.27)
Total number of species=4
Connection and connection weight (flow coefficient)
```
 5    16    -0.05
 5    26    -0.034
16     5    -0.067
16    21     0.842
16    26     0.892
26     5     0.987
26    21    -0.067
```

t=7
Species list and population size
16($\times$(16)=29.89) 21($\times$(21)=2.26) 26($\times$(26)=10.07)
Total number of species=3
Connection and connection weight (flow coefficient)
```
16    21     0.842
16    26     0.892
26    21    -0.067
```

t=8
Species list and population size
16($\times$(16)=42.14) 21($\times$(21)=2.37) 26($\times$(26)=10.12)
Total number of species=3
Connection and connection weight (flow coefficient)
```
16    21     0.842
16    26     0.892
26    21    -0.067
```
. . . .

t=26
Species list and population size
```
 1(×(1)=873.51)     2(×(2)=654.96)     3(×(3)=958.25)     4(×(4)=891.13)
 6(×(6)=640.55)     7(×(7)=953.75)     8(×(8)=555.54)     9(×(9)=894.91)
11(×(11)=883.97)   12(×(12)=831.03)   13(×(13)=886.55)   15(×(15)=588.44)
16(×(16)=510.21)   17(×(17)=916.82)   18(×(18)=736.6)    20(×(20)=544.53)
21(×(21)=557.56)   22(×(22)=583.99)   23(×(23)=986.26)   25(×(25)=799.51)
27(×(27)=622.39)   30(×(30)=655.3)
```
Total number of species=22
Connection and connection weight (flow coefficient)
(318 connections. They are omitted for page limitation)

In the case example, there are 30 species in the species pool. In exception of species IDs 5, 10, 14, 19, 24, 26, 28, and 29, a total of 22 species finally establish in the community. The number of species does not necessarily increase monotonously with the time The changes of No. species and No. interactions are shown in Fig. 1.

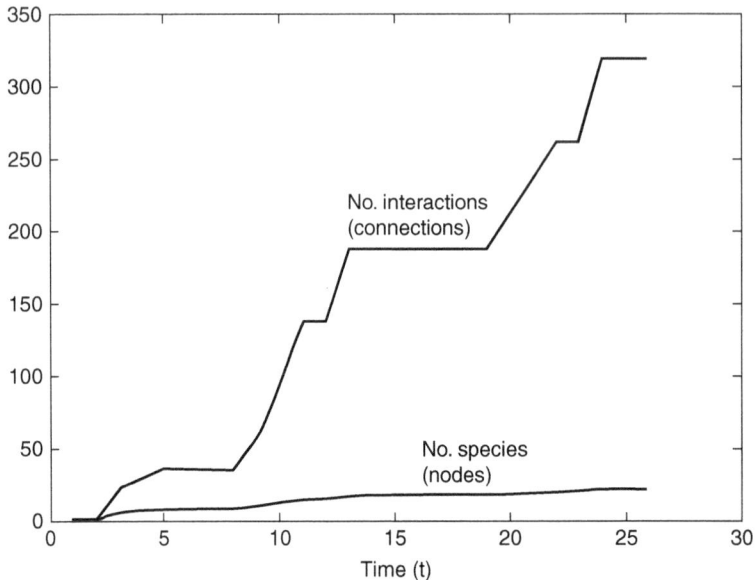

Fig. 1   The changes of No. species and No. interactions with the time (Zhang, 2015b).

In the network view, the nodes, connections, and topological structure change during the network evolution. In some conditions, the topological structure of the network will gradually stabilize with time. The mechanism of the present network evolution model is thoroughly different from previous ones (Barabasi and Albert, 1999; Zhang, 2012a, 2012c). In addition, the networks generated by my model are directed graphs rather than previous undirected graphs.

11.1.2.3   *Conclusions*

Some conclusions have been drawn from a number of running of the simplified model (Zhang, 2015b). The following findings are only from various settings of constant A, K_i, p_i, and c_i. With the changes of $A(t), K_i(t), p_i(t)$, and $c_i(t)$, community dynamics and species composition are expected to yield more diverse patterns.

(1) Effect of A. Set different sets of coefficients, $a_{ij}(t), i, j = 1, 2, \ldots, n$; $i \neq j$, community assembly dynamics will change accordingly, depending on signs and absolute values of $a_{ij}(t)$, and if $a_{ij}(t) = 0$ or not. If $a_{ij}(t) = 0$ and $a_{ji}(t) = 0$, $i, j = 1, 2, \ldots, n$; $i \neq j$, all species have no niche overlap, and the coefficients, $a_{ii}(t), i = 1, 2, \ldots, n$, are thus pure growth rates. With different growth rates, community assembly dynamics are different. The greater growth rates lead to faster growth of species population.

(2) Effect of p_i. The speed of community assembly increases with p_i. The species with small p_i are harder to invade the community.

(3) Effect of c_i. The greater c_i facilitate the species to reach their environmental capacity fastly.

(4) Effect of K_i. The smaller K_i lead to the community reaching its final state earlier.

11.1.2.4 *The self-organization theory on community assembly*

Using the present model to describe community assembly, the expected and natural dynamics may occur in certain conditions. In the field, however, community assembly is a natural process in which the community always evolves in natural ways (e.g., it gradually evolves to a diverse and stable climax). Thus, I argue that community assembly is a self-organization process. In terms of the model above, the natural community can spontaneously adjust its $A(t)$ and guarantee itself to evolve in a natural way. Temporal dynamics and species composition of community are dependent on species composition in species pool, between-species interactions (in which both niche differentiation and ecological interactions are included), intrinsic growth of species (which includes both the reproduction potentiality and the adaptability of the species to the environment), environmental capacity (resource availability), and the probability, strength and history of invasions, etc. The invasion, establishment, growth, and extinction of species follow a series of rules. Community assembly can be well modeled. It can be best described and modeled with self-organization approaches (Zhang, 2012e). The present model provides the basis for building advanced self-organization models of community assembly.

11.2 A Model for Perturbed Food Web Dynamics

Revised from the methods above (Zhang, 2015b), I assume the equilibrium state (time $t = 0$) for a species i is $x_i(0)$ and equilibrium weight of the link between species i and j is $e(j, i, 0)$. In addition, suppose $\Gamma(t)$ is the set of species at time t, and $\Omega(t)$ is the set of links at time t. The state at time $t+1$ for the species i, $x_i(t+1)$, is determined by the state of every other species j at time t and the weight of the link between them, $e(j, i, t)$. $e(j, i, t) = 0$ by definition if no edge exists between species i and j. $e(i, i, t)$ is the intrinsic growth rate of species i. We have the following rules for state transition of the network (nodes and links)

$$x_i(t + 1) = x_i(t) + \Delta x_i(t)$$

$$e(j, i, t + 1) = e(j, i, t) + \Delta e(j, i, t)$$

$$x_i(t) \in \Gamma(t), \quad e(j, i, t) \in \Omega(t)$$

where $\Delta x_i(t)$ and $\Delta e(j, i, t)$ are variation of the species state and the weight of the link between species i and species j, respectively, resulting from perturbation. As a self-organization process, the network will evolve to reach an equilibrium state, therefore

$$\Delta x_i(t) = f(x_j(t), \quad e(j, i, t)|_{x_j(t) \in \Gamma(t), e(j, i, t) \in \Omega(t)})$$

$$\Delta e(j, i, t) = g(x_j(t), \quad x_i(t)), x_j(t) \in \Gamma(t), \quad e(j, i, t) \in \Omega(t)$$

Take a linear approximation, we have

$$f(x_j(t), e(j, i, t)) = \sum_{x_j(t) \in \Gamma(t), e(j,i,t) \in \Omega(t)} \Delta x_j(t) e(j, i, t)$$

$$g(x_j(t), x_i(t)) = \alpha |\Delta x_j(t)| + \beta |\Delta x_i(t)|)$$

Thus,

$$x_i(t + 1) = x_i(t) + \sum_{x_j(t) \in \Gamma(t), e(j,i,t) \in \Omega(t)} \Delta x_j(t) e(j, i, t)$$

$$e(j, i, t + 1) = e(j, i, t) + \alpha |\Delta x_j(t)| + \beta |\Delta x_i(t)|)$$

If $x_i(t+1) \leq 0$, let $x_i(t+1) = 0$, i.e., the species i disappears at time $t + 1$. If $e(j, i, t + 1) \leq 0$, let $e(j, i, t + 1) = 0$, i.e., the link between species i and j disappears at time $t + 1$.

11.3 A Network Evolution Algorithm Based on Node Attraction

The following network evolution algorithm, proposed by Zhang (2016f), is based on the random network, and facilitated by node attraction.

11.3.1 *Algorithm*

Assume the initial network is a random network (this is popular in nature, e.g., the distribution of particles with different sizes with which the water vapor attaches to generate droplets, the random distribution of matter/nucleus of crystallization for the generation of stars/crystals, etc.), or a given initial network. When a node is ready to connect, it tends to connect to the node already having the most connections (i.e., the largest node), which accords with the general rule of node connecting (Barabasi and Albert, 1999). In addition, a node may randomly disconnect a connection, i.e., the addition of connections in the network is accompanied by the pruning of some connections.

Assume there are v nodes in the network. The expected initial connectance (connectance = practical connections/potential maximum connections) is c (i.e., initial condition) if the initial network is a random network generated by the algorithm, the expected final connectance is c_e (termination condition), the attraction factor of nodes is $\lambda(t, a)$ (it is a driving variable; $\lambda(t, a) > 0$, where a is the node degree), the probability of node connection is $p(t, a)$ (driving variable), the probability of node disconnection is $q(t, a, b)$ (it is a driving variable, where b is the connected node's degree), maximum number of iterations is *iter* (i.e., termination condition), and the confidence degree for detecting the statistical significance of network type is α (an auxiliary constant). $\lambda(t, a), p(t, a)$, and $q(t, a, b)$ are time and node (in particular node degree) changing. The procedures are as below (Zhang, 2016f):

(1) Generate the initial network. In the situation of random initial network, assume the adjacency matrix of the random network is $d = (d_{ij}), i, j = 1, 2, \ldots, v$, where $d_{ij} = d_{ji}, d_{ii} = 0$, and if $d_{ij} = 1$ or $d_{ji} = 1$, there is a connection between nodes i and j. For each pair of $i, j(i = 1, 2, \ldots, v - 1; j > i)$, generate a random value r, if $r < c, d_{ij} = 1$ and $d_{ji} = 1$. Otherwise, the initial network is a given network.

(2) Let $t = 1$. Calculate the degree of node, $a_i(t), i = 1, 2, \ldots, v$. The cumulative attraction strength of node 1 to node i is

$$p_i(t) = \sum_{j=1}^{i} a_j(t)^{\lambda(t, a_j)} \bigg/ \sum_{j=1}^{v} a_j(t)^{\lambda(t, a_j)}$$

(3) Generate/disconnect connections. For the node $i, i = 1, 2, \ldots, v$, generate a random value s, if $s < p(t, a_i)$, the node i is ready to connect to one of the remaining nodes. Let $p(t) = 0$. For the node $j, j = 1, 2, \ldots, v, j \neq i$, generate a random value w. $d_{ij} = 1$ and $d_{ji} = 1$, if $p_{i-1}(t) \leq w < p_i(t)$. For practical use, the interval $[p_{i-1}(t), p_i(t))$ represents the mass or volume of the particle i, the gravity of the celestial body i, the personality charm of the person i, the academic impact of the scientist i, etc.

For the node $i, i = 1, 2, \ldots, v$, generate a random value g; if $g < q(t, a_i, b)$, one of the connections of the node i, e.g., d_{ij}, is randomly disconnected, and let $d_{ij} = 0$ and $d_{ji} = 0$.

By doing so, a network at time t is generated. Various indices and methods, e.g., coefficient of variation (CV), aggregation index (AI), and entropy (Zhang and Zhan, 2011; Zhang, 2012e) can be used to detect the types and properties of the network.

(4) Calculate the connectance C of the network. Let $t = t + 1$ and return to (2), if C is less than the expected final connectance c_e; otherwise, the algorithm terminates, if C is not less than c_e, or the maximum iterations *iter* are achieved.

For convenience and simplicity, assume $\lambda(t,a) = \lambda, p(t,a) = p,$ $q(t,a,b) = q,$ i.e., the attraction factor of nodes, the probability of node connection, and the probability of node disconnection are constants for any degree of nodes at any time. Thus, we obtain a simplified version of the algorithm.

The following are Matlab codes for simplified version of the algorithm (Zhang, 2016f):

```
%Zhang WJ. 2016. A random network based, node attraction facilitated
    network
%evolution method. Selforganizology, 3(1): 1-9
v=input('Total number of nodes in the network=');
choice=input('Input the type (1: a random network generated by the
    algorithm; 2: a given network) of generating initial network:');
if (choice==1) ci=input('Expected initial connectance (=practical
connections/potential maximum connections; e.g., 0.05, etc)='); end
if (choice==2) adjstr=input('Input the file name of adjacency matrix of the
given initial network (e.g., raw.txt, raw.xls, etc. Adjacency matrix is
d=(dij)v*v, where v is the number of nodes in the network. dij=1, if vi and
vj are adjacent, and dij=0, if vi and vj are not adjacent; i, j=1,2,..., v:
',' s'); end
ce=input('Expected final connectance (=practical connections/potential
    maximum connections; e.g., 0.1, 0.15, etc)=');
lamda=input('Attraction factor of nodes (lamda; e.g., 2, 4, etc.
    lamda>0)=');
p=input('Probability of node connection (e.g., 0.1, 0.2)=');
q=input('Probability of node disconnection (e.g., 0, 0.01)=');
alpha=input('Confidential degree for detecting network type (e.g., 0.05,
    0.01)=');
iter=input('Permitted maximum iterations (e.g., 5000)=');
adj=zeros(v);
degr=zeros(1,v);
prop=zeros(1,v);
z=zeros(v);
if (choice==1)
for i=1:v-1
for j=i+1:v
if (rand()<ci) adj(i,j)=1; adj(j,i)=1; end
end; end; end
if (choice==2) adj=load(adjstr); end;
degr=sum(adj);
fprintf('Initial adjacency matrix\n')
disp([adj])
fprintf('Initial degree distribution\n')
disp([degr])
t=1;
while (v>0)
propdegr=degr.^lamda;
```

```
prop(1)=propdegr(1)/sum(propdegr);
for i=2:v;
prop(i)=prop(i-1)+propdegr(i)/sum(propdegr);
end
node=zeros(1,v);
nu=0;
for i=1:v
if (rand()<p) nu=nu+1; node(nu)=i; end
end
for k=1:nu;
for i=1:v
if (node(k)~=i) continue; end
lab=0;
ran=rand();
for j=1:v
if (j==i) continue; end
if (j==1) st=0; end
if (j>=2) st=prop(j-1); end
if ((ran>=st) & (ran<prop(j))) lab=1; adj(i,j)=1; adj(j,i)=1; break; end
end
if (lab==1) break; end
end; end
nodes=zeros(1,v);
for i=1:v
if (rand()<q) nodes(i)=1; end
end
for i=1:v-1
if (nodes(i)~=1) break; end
nuu=0;
for j=i+1:v
if (adj(i,j)==1) nuu=nuu+1; z(i,j)=nuu; end
end
np=round(rand()*nuu+0.5);
for j=i+1:v
if (z(i,j)==np) adj(i,j)=0; adj(j,i)=0; end
end; end
fprintf(['\n\nTime' num2str(t)])
fprintf(['\n\nAdjacency matrix\n'])
disp([adj])
degr=sum(adj);
fprintf('\nDegree distribution\n')
disp([degr])
cnow=(sum(degr)/2)/((v^2-v)/2);
fprintf(['\nConnectance=' num2str(cnow) '\n'])
meann=mean(degr);
varr=(std(degr))^2;
fprintf(['\nEntropy=' num2str(varr-meann) '\n'])
num=0;
cv=varr/meann;
fprintf(['\nCoefficient of variation (CV)=' num2str(cv) '.'])
```

```
x2=cv*(v-1);
sig=chi2cdf(x2,v-1);
if (sig<=alpha) fprintf('The network is a random network according to
   CV.\n'); end
if ((sig>alpha) & (cv>1)) fprintf('The network is a complex network
   according to CV.\n'); num=num+1; end;
summ=sum(degr);
summa=sum(degr.*(degr-1));
h=v*summa/(summ*(summ-1));
fprintf(['\nAggregation index (AI)=' num2str(h) '.'])
x2=h*(summ-1)+v-summ;
sig=chi2cdf(x2,v-1);
if (sig<=alpha) fprintf('The network is a random network according to
   AI.\n'); end;
if ((sig>alpha) & (h>1)) fprintf('The network is a complex network according
   to AI.\n'); num=num+1; end;
if (num>=2) fprintf('\nThe network is a complex network according to all
indices.\n'); end;
if (cnow>=ce) break; end
if (t>=iter) break; end;
t=t+1;
end
```

11.3.2   *Analysis*

Assume the initial network is a random network, generated by the algorithm.

11.3.2.1   *Dynamics of network evolution*

Running the simplified algorithm, the parameters set are $v = 30, c = 0.05, c_e = 0.2, \lambda = 3, p = 0.2, q = 0, \alpha = 0.05$. In this case, the connected nodes increase quickly until all nodes are connected. Connectance increases with time. Coefficient of variation (CV) increases quickly to an approximate upper asymptote. However, aggregation index increases quickly to a climax and declines thereafter. Overall, the three indices increase most quickly until all nodes are connected (Zhang, 2016f).

11.3.2.2   *Parametrical analysis*

(1) Effects of node attraction factor

With $v = 30, c = 0.05, c_e = 0.2, p = 0.2, q = 0, \alpha = 0.05$, run the simplified algorithm to analyze the effects of node attraction factor on network evolution. For $\lambda = 0.05$, 1, 2, 3, and 4, network connectance increases more quickly when λ is smaller. Coefficient of variation and aggregation index increases more quickly when λ is greater.

It is found that the required iterations increase with λ. With the same expected final connectance, the coefficient of variation, aggregation index, and entropy increase with λ, which means the complexity of the final network increases with λ.

(2) Effects of the probability of node connection p

Assume $v = 30, c = 0.05, c_e = 0.2, \lambda = 3, q = 0$, and analyze the effects of the probability of node connection p on network evolution. It is indicated that for $p = 0.01$, 0.05, 0.1, and 0.2, the three indices increase more quickly when p is greater.

(3) Effects of the probability of node disconnection q

Let $v = 30$, $c = 0.05$, $p = 0.2$, $c_e = 0.2, \lambda = 3$, and analyze the effects of the probability of node disconnection q on network evolution. It is demonstrated that for $q = 0$, 0.01, 0.02, and 0.03, network connectance increases more quickly when q is smaller. Coefficient of variation and aggregation index increases more quickly when q is greater.

(4) Effects of expected initial connectance c

Let $v = 30, c_e = 0.2, \lambda = 3, p = 0.2, q = 0$, and analyze the effects of the expected initial connectance c on network evolution. The results show that for $c = 0.01$, 0.03, 0.05, and 0.08, network connectance increases more quickly when c is greater. Coefficient of variation and aggregation index increases more quickly when c is smaller.

11.3.3   *Model's universality*

The changes of attraction factor λ can reflect various effects of the node degree on connection mechanism. The larger λ leads to generate complex networks, e.g., exponential law, power-law networks, etc. $\lambda \to 0$ means the trend to generate the random network. Even the changes of λ only may generate various networks from the random to the complex. Therefore, the above algorithm can be treated as a general model for network evolution.

Modeling results show that to generate power-law-distributed node degrees (i.e., to generate a power-law-distributed network), the likelihood of a node attracting connections is dependent upon the power function of the node's degree with a higher-order power (Zhang, 2016f).

The dynamics of network evolution is determined of $\lambda(t, a)$, $p(t, a), q(t, a, b)$, and c. Here, all parameters are simplified as constants. However, a more complex mechanism for network evolution can be achieved by setting reasonable forms of $\lambda(t, a), p(t, a)$, $q(t, a, b)$, depending on the networks being studied. In the algorithm, the addition of connections coincides with the general rule of node connecting (Barabasi and Albert, 1999). However, the mechanism for pruning of connections is still unknown (i.e., unknown $q(t, a, b)$); thus in the simplified algorithm, $q = 0$ is a better choice (Zhang, 2016f).

11.4 Phase Recognition of Network Evolution

Network evolution is a time series (Zhang, 2017m). During the process of evolution, network properties (topological structure, etc.) change with time (Zhang, 2012c, 2015b, 2016f), and the change is always not homogeneous. We want to classify the time series into several phases according to their homogeneity in network properties. The following method uses one-dimensional ordered cluster method to recognize different phases in network evolution (Zhang, 2017m).

11.4.1 *Algorithm*

One-dimensional ordered cluster method (Zhang and Fang, 1982; Qi, 2005) is used to recognize different phases in network evolution by clustering time points in a time series. Suppose a network has m properties and it evolves across n time points in the time series. The raw data for network evolution is $(x_{ij})_{m*n}$. If we choose to standardize the raw data into the matrix (a_{ij}), calculate (Zhang, 2017m)

$$a_{ij} = (x_{ij} - x_{bi})/s_i$$

where

$$x_{bi} = \sum_{j=1}^{n} x_{ij}/n$$

$$s_i = \left(\sum_{j=1}^{n} (x_{ij} - x_{bi})^2/(n-1) \right)^{1/2}, \quad i = 1, 2, \ldots, m$$

If data standardization is not needed, let $a_{ij} = x_{ij}$, $i = 1, 2, \ldots, m$; $j = 1, 2, \ldots, n$.

Different properties have different importance in the network. Thus each property can be differently weighted. Suppose the weights of m properties are w_i, $i = 1, 2, \ldots, m$, where $\sum w_i = 1$. The weighted data matrix is $a_{ij} = w_i \times a_{ij}$, $i = 1, 2, \ldots, m$; $j = 1, 2, \ldots, n$.

Calculate the distance between the adjacent two time points in network evolution, i.e., Euclidean distance, Manhattan distance, Pearson correlation-based distance, or Jaccard distance

$$e_{ii+1} = \left(\sum_{k=1,\ldots,m} (a_{ki} - a_{ki+1})^2/m \right)^{1/2}$$

$$m_{ii+1} = \sum_{k=1,\ldots,m} |a_{ki} - a_{ki+1}|/m$$

$$c_{ii+1} = \sum_{k=1,\ldots,m} (a_{ki} \times a_{ki+1}) \bigg/ \left(\sum_{k=1,\ldots,m} a_{ki}^2 \times \sum_{k=1,\ldots,m} a_{ki+1}^2 \right)^{1/2}$$

$$d_{ii+1} = (b_i + b_{i+1})/(c_i + c_{i+1} - e) \quad i = 1, 2, \ldots, n-1$$

where b_i is the number of properties of value 1 for time point i but value 0 for time point $i+1$; b_{i+1} is the number of properties of value 1 for time point $i+1$ but value 0 for time point i; c_i and c_{i+1} are the numbers of properties of value 1 for time point i and $i+1$ respectively. e is the number of properties of value 1 for both time point i and time point $i+1$.

According to different distance measures chosen, let $r_{ii+1} = e_{ii+1}$, or $r_{ii+1} = m_{ii+1}$, or $r_{ii+1} = 1 - c_{ii+1}$, or $r_{ii+1} = d_{ii+1}$. Find the shortest distance r_{ii+1}, $i = 1, 2, \ldots, n-1$, and combine the two corresponding time points i and $i+1$ into the same cluster. Similarly, take the minimum between-time point distance as the between-cluster distance. Find the two clusters with minimum between-cluster distance and combine into a single cluster. Finally, all time points are combined into a single cluster. The minimum between-cluster distance in a clustering hierarchical level is defined as the between-cluster distance of this hierarchical level.

The following are the Matlab codes for the algorithm (Zhang, 2017m):

```
raw=input('Input the excel file name of property-by-time point data (e.g.,
    raw.xls, etc.)((xij)m*n, m: number of properties; n: number of time
    points):','s');
sel=input('Data standardization or not (1: Data standardization; 2: No
    standardization):');
selw=input('Weighting properties or not (1: No weighting; 2: Weighting
    properties):');
sele=input('Choose distance measures (1: Euclidean distance; 2: Manhattan
distance; 3: Pearson correlation; 4: Jaccard coefficient):');
araw=xlsread(raw);
m=size(araw,1); n=size(araw,2);
a=araw';
if (selw==2)
weights=input('Input the excel file name of weights of properties (e.g.,
    weights.xls, etc.)(a row of m weights of properties):','s');
w=xlsread(weights);
end
q=zeros(1,m);
p=zeros(1,m);
r=zeros(1,n-1);
r1=zeros(1,n-1);
pp=zeros(1,n-1);
am=zeros(1,n-1);
x=zeros(n,n);
nu=zeros(1,n);
if (sel==1)
for i=1:m
p(i)=sum(a(:,i))/n;
q(i)=sum((a(:,i)-p(i)).^2);
q(i)=sqrt(q(i)/(n-1));
for j=1:n
a(j,i)=(a(j,i)-p(i))/q(i);
end; end; end
if (selw==2)
dum=sum(w);
if (dum~=1) w=w/dum; end
```

```
for i=1:m
a(:,i)=a(:,i)*w(i);
end; end
disp('Distance Between Two Adjacent Time Points');
for i=1:n-1
switch sele
  case 1
    aa=sum((a(i,:)-a(i+1,:)).^2);
    r1(i)=sqrt(aa/m);
  case 2
    aa=sum(abs(a(i,:)-a(i+1,:)));
    r1(i)=aa/m;
  case 3
    aa=0; bb=0; cc=0;
    aa=sum(a(i,:).*a(i+1,:));
    bb=sum(a(i,:).^2);
    cc=sum(a(i+1,:).^2);
    r1(i)=aa/sqrt(bb*cc);
  case 4
    aa1=0; bb1=0; cc1=0; rr1=0; nn1=0;
    for k=1:m
    if (abs(a(i,k))>=1e-08) aa1=aa1+1; end
    if (abs(a(i+1,k))>=1e-08) bb1=bb1+1; end
    if ((abs(a(i,k))>1e-08) & (abs(a(i+1,k))>=1e-08)) cc1=cc1+1; end
    if ((abs(a(i,k))<=1e-08) & (abs(a(i+1,k))>=1e-08)) rr1=rr1+1; end
    if ((abs(a(i,k))>=1e-08) & (abs(a(i+1,k))<=1e-08)) nn1=nn1+1; end
    end
    if ((aa1+bb1-cc1)==0) r1(i)=1;
    else r1(i)=(nn1+rr1)/(aa1+bb1-cc1);
    end
    end;
if ((sele==1) | (sele==2) | (sele==4)) r(i)=r1(i);
else if (sele==3) r(i)=1-r1(i); end
end; end
for i=1:n-1
fprintf(['r('num2str(i)' 'num2str(i+1)')']);
end
fprintf('\n');
for i=1:n-1
fprintf([num2str(r(i)) ' ']);
end
fprintf('\n\n');
pp=r;
for i=1:n-2
k=i;
for j=i:n-2
if (pp(j+1)<=pp(k)) k=j+1; end
end
aa=pp(i);
pp(i)=pp(k);
```

```
pp(k)=aa;
end
bb1=1;
for k=0:n-1
if (k==0) u=0; else u=pp(k); end
lab1=0;
for i=0:k-1
if (i==0) v=0; else v=pp(i); end
if (v==u) lab1=1; break; end
end
if (lab1==1) continue; end
for i=1:n-1
if ((r(i)-u)<1e-06) am(i)=1;
else am(i)=0; end
end
i=1;
nu(bb1)=1;
lab3=0;
while (i<=n-1)
lab2=0;
if (am(i)==0) x(bb1,i)=nu(bb1); x(bb1,i+1)=nu(bb1)+1; end
if (am(i)==1)
for j=i:n-1
if (am(j)==0) i=j; lab2=1; break;
else x(bb1,j)=nu(bb1); x(bb1,j+1)=nu(bb1);
if (j==n-1) lab3=1; break; end
end; end;
if (lab2==1) continue; end
if (lab3==1) break; end
end
i=i+1;
nu(bb1)=nu(bb1)+1;
end
fprintf('\n');
fprintf(['Distance=' num2str(u) '\n']);
%u(bb1)=u;
for i=1:nu(bb1)
fprintf('(');
for j=1:n
if (x(bb1,j)==i)
fprintf([num2str(j) ' ']);
end; end
fprintf(')');
end
bb1=bb1+1;
end
fprintf('\n');
```

Table 1 A series of data on network evolution, derived from the model of Zhang (2016f).

	Time points									
	1	2	3	4	5	6	7	8	9	10
Connectance	0.0555	0.0588	0.0612	0.0637	0.0669	0.0686	0.0718	0.0751	0.08	0.0833
Entropy	0.8735	1.5951	1.9388	1.8857	2.0686	2.0588	2.3265	2.7869	3.5429	4.0359
Coefficient of variation	1.3211	1.5539	1.6463	1.6044	1.6307	1.6127	1.6609	1.7573	1.9038	1.9892
Aggregation index	1.1166	1.1898	1.2125	1.1911	1.1896	1.1798	1.1851	1.2028	1.2271	1.2388

	Time points									
	11	12	13	14	15	16	17	18	19	20
Connectance	0.0882	0.0906	0.0939	0.0971	0.1004	0.1037	0.1078	0.1135	0.1184	0.12
Entropy	4.6367	4.9543	6.0531	6.2833	6.8694	7.4441	8.191	9.3037	9.4653	9.9012
Coefficient of variation	2.0733	2.1158	2.3159	2.32	2.3962	2.4654	2.5513	2.6733	2.6319	2.6839
Aggregation index	1.2446	1.2474	1.2816	1.2729	1.2792	1.2838	1.289	1.296	1.2767	1.2816

11.4.2 *An example*

Based on a series of data on network evolution, which are derived from the model of Zhang (2016f) (Table 1), using the algorithm, with data standardization, no weighting properties (four properties) and Euclidean distance, the resultant recognition of phases of network evolution is indicated in Table 2. In Table 2, for the same hierarchical level, the time points with the same background color belong to the same phase, and the time points without background color are independent time points (Zhang, 2017m).

Table 2 Recognition of phases of network evolution (Zhang, 2017m).

Hierar. Level	Distance	1	2	3	4	5	6	7	8	9	10	11	12	13	14	15	16	17	18	19	20
1	0.091	1	2	3	4	5	6	7	8	9	10	11	12	13	14	15	16	17	18	19	20
2	0.098	1	2	3	4	5	6	7	8	9	10	11	12	13	14	15	16	17	18	19	20
3	0.109	1	2	3	4	5	6	7	8	9	10	11	12	13	14	15	16	17	18	19	20
4	0.114	1	2	3	4	5	6	7	8	9	10	11	12	13	14	15	16	17	18	19	20
5	0.119	1	2	3	4	5	6	7	8	9	10	11	12	13	14	15	16	17	18	19	20
6	0.124	1	2	3	4	5	6	7	8	9	10	11	12	13	14	15	16	17	18	19	20
7	0.157	1	2	3	4	5	6	7	8	9	10	11	12	13	14	15	16	17	18	19	20
8	0.168	1	2	3	4	5	6	7	8	9	10	11	12	13	14	15	16	17	18	19	20
9	0.193	1	2	3	4	5	6	7	8	9	10	11	12	13	14	15	16	17	18	19	20
10	0.194	1	2	3	4	5	6	7	8	9	10	11	12	13	14	15	16	17	18	19	20
11	0.197	1	2	3	4	5	6	7	8	9	10	11	12	13	14	15	16	17	18	19	20
12	0.231	1	2	3	4	5	6	7	8	9	10	11	12	13	14	15	16	17	18	19	20
13	0.239	1	2	3	4	5	6	7	8	9	10	11	12	13	14	15	16	17	18	19	20
14	0.267	1	2	3	4	5	6	7	8	9	10	11	12	13	14	15	16	17	18	19	20
15	0.283	1	2	3	4	5	6	7	8	9	10	11	12	13	14	15	16	17	18	19	20
16	0.346	1	2	3	4	5	6	7	8	9	10	11	12	13	14	15	16	17	18	19	20
17	0.464	1	2	3	4	5	6	7	8	9	10	11	12	13	14	15	16	17	18	19	20
18	0.805	1	2	3	4	5	6	7	8	9	10	11	12	13	14	15	16	17	18	19	20

Chapter 12

Cellular Automata

Cellular automata are a kind of dynamic models with discrete time, discrete space, and discrete states (Zhang, 2016r). They are finite-state machines, i.e., each cell in the model has the finite number of states. They are mathematical models that have simple rules governing their replication and destruction and are used to model complex systems composed of simple units. As pointed out earlier, community assembly, along with many other network evolution processes, are self-organizing systems (Zhang, 2012e, 2016r). Cellular automata are self-organizing models with a series of simple rules. Therefore, they are suitable for modeling some kinds of network evolution.

Cellular automata were first proposed by von Neumann in 1950s The model was used to simulate the self-reproduction of biological cells. Wolfram (2002) analyzed the effects of 256 rules on the evolution outcomes of cellular automata and classified cellular automata based on their evolution behaviors.

The nature of parallel computation, locality, and conformity of cellular automata makes them an effective dynamic model to understand the global behavior and complex phenomena of various systems (networks). They have been widely used in a variety of areas. They have been used to simulate traffic breakdown, highway capacity, and synchronized flow (Kerner *et al.*, 2011); real-world urban processes (Sante *et al.*, 2010); and evacuation process with obstacles (Alizadeh, 2011). In addition, they are used to simulate reaction process of atom and molecular. For example, in the Belousov–Zhabotinsky reaction, the feedbacks and interactions between various moleculars may yield an oscillation in which two colors occur alternatively. The universe is considered as a multidimensional cellular automaton (Ilachinski, 2001). High-quality random numbers can be generated by using two dimensional cellular automata (Tomassini *et al.*, 2000).

Biology is the science for the origination of the concept of cellular automata and the most widely used area for cellular automata (Zhang, 2016r). Natural cellular automata are sometimes found in organisms and biology. For example, the shell pattern of a sea shell, *Conus textile*, is similar to a one-dimensional Wolfram cellular automaton (Coombes, 2009). Stomata on the plant leaves show complex and collective dynamics and perform emergent and distributed computation through communications and interactions between them (Peak, 2004). They are similar to a two-dimensional cellular automaton. Jiao and Torquato (2011) used the emergent behaviors from a cellular automaton model to simulate invasive tumor growth in heterogeneous microenvironments. Moreover, cellular automata have been used to simulate dynamics of HIV infection under antiretroviral therapy (González *et al.*, 2013). Based on the protein sequence only, various properties of proteins can be predicted using cellular automata (Xiao *et al.*, 2011).

12.1   Classification of Cellular Automata

12.1.1   *Evolutionary behavior-based classification*

Cellular automata are divided into the following four categories according to their behavioral patterns: homogeneous, periodic, chaos, and edge of chaos (Wolfram, 2002; Wang *et al.*, 2007; Zhang, 2016r):

(1) **Homogeneous.** Almost all patterns evolve to a stable and homogeneous state. The stable state does not change as time, and the initial random pattern disappears gradually. At the stable state, some cells are alive or all cells die.

(2) **Periodic.** Almost all patterns evolve to a relatively stable oscillation structure. The states of some cells change periodically with time. Some stochasticity in initial states can be filtered and

some are remain. A local change of the initial pattern may only produce a local effect.

(3) **Chaos.** Almost all of the initial patterns evolve to chaotic states. Any temporarily stable structure that occurred during evolution will be terminated by the environmental chaos. The changes in the initial pattern may be unlimitedly propagated to entire cellular automaton. Chaotic cellular automata are completely stochastic and unordered, without producing valuable structures.

(4) **Edge of chaos.** Almost all of the initial patterns evolve to the complex but not chaotic structures.

Overall, the final outcome of evolution of cellular automata is similar to that of chaos. However, the stable states or periodic states may occur in some local evolution. Such cellular automata are the interim model from periodic to chaotic ones.

To better describe the four categories of cellular automata, Wolfram defined a parameter $\lambda(0 < \lambda < 1)$, i.e., the proportion of a set of rules that transformed to nonzero states, for classifying cellular automata. In his definition, $\lambda = (m^{2r+1} - n_q)/m^{2r+1}$, where m: the number of states in the set of states $S = \{s_1, \ldots, s_m\}, r$: radius of neighbors, n_q: the number of outputs with 0. $\lambda = 0 \sim 0.1$: homogeneous cellular automata; $\lambda \approx 0.2$: periodic cellular automata; $\lambda = 0.3 \sim 0.6$: the cellular automata at the edge of chaos; $\lambda \geq 0.6$: cellular automata of chaos (Wang *et al.*, 2007).

12.1.2 *Space dimensionality-based classification*

Cellular automata are divided into three types according to their spatial dimensionality (Wolfram, 2002; Wang *et al.*, 2007; Zhang, 2016r):

(1) **One-dimensional.** In this model, cells distribute along a line, which is similar to the Turing machine.

(2) **Two-dimensional.** It is the most commonly used cellular automata. According to the different definitions of states of adjacent cells, such cellular automata include von Neumann type, Moore type, Margolus type, etc.

(3) **Three or higher-dimensional** cellular automata are seldom used owing to their complexity (Tomoyoshi and Tsutomu, 2002).

In addition, cellular automata can be classified based on their boundary conditions into periodic and random types.

12.2   Algorithms of Cellular Automata

Following some common rules, each time all cells in a cellular automaton update their states synchronously during the evolution, given the initial states of cells (Zhang, 2016r).

A cellular automaton contains some rules constructed by a series of models. It can be expressed as a conceptual structure as below

$$A = (L, d, S, N, F)$$

where L: space of cells, d: space dimensionality, S: the set of finite and discrete states of cells, N: the set of cells in the adjacent area, and f: local rules (Amoroso and Patt, 1972).

For example, in the elementary cellular automaton (one-dimensional) of Wolfram, $S = \{0, 1\}$, the radius is defined as that there is only one adjacent cell, and the local mapping rules are

$$S_i^{t+1} = f(S_{i-1}^t, S_i^t, S_{i+1}^t)$$

where i: the present cell, $i + 1$ and $i - 1$: the adjacent cells, t and $t + 1$: the present time and the next time. It means that the state of the present cell in the next time depends on its current state and the current states of its adjacent cells.

In the two-dimensional cellular automata, different neighborhood patterns may lead to different outcomes of evolution. Local rules of a two-dimensional cellular automaton can be

$$S_i^{t+1} = f(S_i^t, S_N^t)$$

where S_N^t is the combination of states of neighborhood cells.

12.3   Some Cases of Cellular Automata

12.3.1   *Diffusion in heterogeneous environment*

The cellular automata, proposed and discussed by Qi and Zhang (2002), Qi (2003), and Zhang (2012e, 2016r), can be used in the simulation of population diffusion in heterogeneous environment (Fig. 1).

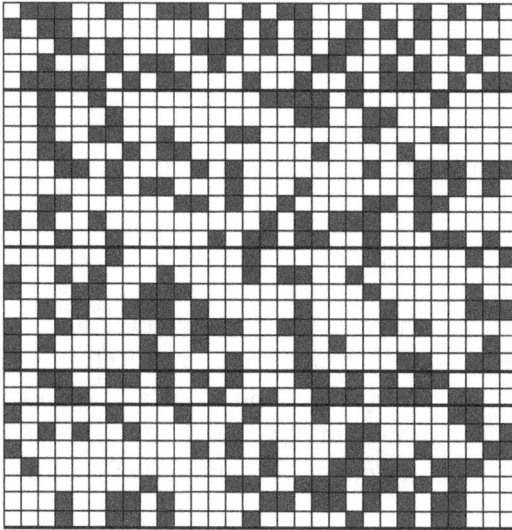

Fig. 1 Population diffusion in a heterogeneous environment (Zhang, 2016a).

Suppose an area is divided into $c \times f$ cells, each with the same area, $a_{ij}, i = 1, 2, \ldots, c; \; j = 1, 2, \ldots, f$. Population diffusion occurs in this region (Qi, 2003; Qi and Zhang, 2002; Zhang, 2012e, 2016r).

Given the sensitivity level of each cell against population diffusion, $q_k (q_k \in R, k = 1, 2, \ldots, m)$, the corresponding probability of being diffused $p_k (p_k \in [0, 1], k = 1, 2, \ldots, m)$, and the initial distribution of population diffusion (i.e., whether or not the population occurs in a cell before the simulation, $t = 0$), the initial distribution is

$$v_{ij} = 1, \quad \text{if cell } a_{ij} \text{ has been diffused}$$

$$v_{ij} = 0, \quad \text{if cell } a_{ij} \text{ is not yet diffused}$$

$$i = 1, 2, \ldots, c; \quad j = 1, 2, \ldots, f$$

In addition,

$$v_{0i} = v_{f+1_i} = 0, \quad i = 1, 2, \ldots, c + 1$$

$$v_{i0} = v_{ic+1} = 0, \quad i = 1, 2, \ldots, f + 1$$

Sensitivity distribution is also known

$$u_{ij} = q_k, \quad \text{if the sensitivity of cell } a_{ij} \text{ is } q_k(k = 1, 2, \ldots, m),$$
$$i = 1, 2, \ldots, c; \; j = 1, 2, \ldots, f$$

The probability distribution of population is

$$p_{ij} = p_k, \quad \text{if } u_{ij} = q_k, \quad k = 1, 2, \ldots, m,$$
$$i = 1, 2, \ldots, c; \; j = 1, 2, \ldots, f$$

Suppose each diffusion process occurs at unit time, and population diffuse to those cells with which the present cell has adjacent edges. For an undiffused cell x, the population will diffuse to the cell x with the diffusion probability p_i from the adjacent diffused cells. For the sensitivity k, calculate the number of cells available for the diffusion N_k

$$N_k = \sum_i \sum_j b_{ij}$$

where

$$b_{ij} = 1, \quad \text{if } (v_{ij} = 0) \wedge (p_{ij} = p_k) \wedge ((v_{i-1j} = 1) \vee$$
$$(v_{i+1j} = 1) \vee (v_{ij-1} = 1) \vee (v_{ij+1} = 1)) \text{ is true}$$
$$b_{ij} = 0, \quad \text{if } (v_{ij} = 0) \wedge (p_{ij} = p_k) \wedge ((v_{i-1j} = 1) \vee$$
$$(v_{i+1j} = 1) \vee (v_{ij-1} = 1) \vee (v_{ij+1} = 1)) \text{ is false}$$

The number of actually diffused cells with sensitivity k is

$$n_k = \text{INT}((100 p_k N_k + 0.5)/100), \quad k = 1, 2, \ldots, m$$

A diffused cell will restore to an undiffused one at the restoration rate r, i.e., the proportion of cells that can restore from diffused state during unit time

$$v_{ij}, \quad \text{if } 1 - r + R \geq 1$$
$$v_{ij} = 0, \quad \text{if } 1 - r + R < 1$$
$$i = 1, 2, \ldots, c; \; j = 1, 2, \ldots, f$$

where R is a random value, $0 \leq R \leq 1$.

Population dynamics shows different patterns with the change of restoration rate, diffusion probability, sensitivity levels, etc. Various equilibrium states may be reached with different combinations of parameters.

12.3.2 *Population diffusion in homogeneous environment*

Cellular automata have been used to describe population diffusion in a homogeneous rectangular area (Qi, 2003; Qi *et al.*, 2004; Zhang *et al.*, 2015).

Suppose an area has $c \times f$ cells. Given a cell, individuals in the cell diffuse to adjacent four cells, or, individuals in the adjacent four cells diffuse to the cell. Population diffuses until all cells have the same number of individuals. At this time, the number of individuals for immigration and migration are the same in each cell. The time step for diffusion is ∂t and the diffusion coefficient is a. The diffusion rules are thus (Zhang, 2016r)

$$\partial u(1,1,t)/\partial t = a^*(u(1,2,t) + u(2,1,t) - 2^*u(1,1,t))$$

$$\partial u(1,c,t)/\partial t = a^{**}(u(1,c-1,t) + u(2,c,t) - 2^*u(1,c,t))$$

$$\partial u(1,j,t)/\partial t = a^*(u(1,j-1,t) + u(2,j,t)$$
$$+ u(1,j+1,t) - 3^*u(1,j,t))$$

$$\partial u(f,1,t)/\partial t = a^*(u(f-1,1,t) + u(f,2,t) - 2^*u(f,1,t))$$

$$\partial u(f,c,t)/\partial t = a^*(u(f-1,c,t) + u(f,c-1,t) - 2^*u(f,c,t))$$

$$\partial u(f,j,t)/\partial t = a^*(u(f,j-1,t) + u(f-1,j,t)$$
$$+ u(f,j+1,t) - 3^*u(f,j,t))$$

$$\partial u(i,1,t)/\partial t = a^*(u(i-1,1,t) + u(i,2,t)$$
$$+ u(i+1,1,t) - 3^*u(i,1,t))$$

$$\partial u(i,c,t)/\partial t = a^*(u(i-1,c,t) + u(i,c-1,t)$$
$$+ u(i+1,c,t) - 3^*u(i,c,t))$$

$$\partial u(i,j,t)/\partial t = a^*(u(i-1,j,t) + u(i,j-1,t)$$
$$+ u(i+1,j,t) + u(i,j+1,t)$$
$$- 4^*u(i,j,t)) \quad i = 1,2,\ldots,f; \quad j = 1,2,\ldots,c$$

if $u(i,j,t) < 0$, then let $u(i,j,t) = 0$.

Regardless of diffusion process, the natural population dynamics in a cell can be described by Malthus model, Verhulst model, or oscillation model

$$\partial u(i,j,t)/\partial t = ru$$

$$\partial u(i,j,t)/\partial t = ru(K-u)/K$$

$$\partial u(i,j,t)/\partial t = -2\pi/T^*A^* \sin(2\pi t/T + \varphi)$$

where, $u = u(x,y,t)$: the population size of the cell (x,y) at time t; r: intrinsic rate, $-\infty < r < \infty$; K: environmental capacity of a cell; φ: phase, $0 \le \varphi \le 2\pi$; $A = (u-b)/\cos(\varphi)$; b: the mean number of individuals per cell; and T: oscillation cycle, $0 < T < \infty$.

The results show that overall the population grows in the logistic pattern (Qi, 2003; Qi *et al.*, 2004; Zhang *et al.*, 2015; Zhang, 2016r). A greater intrinsic rate will lead to a larger population size. On the other hand, population size of each cell will not exceed the environmental capacity. Due to the effect of diffusion, the population dynamics in a given cell is not certainly a Verhulst growth.

Population will reach a stable state with the same number of individuals (environmental capacity) in each cell, and population will change if a cell is disturbed, or the environmental capacity is changed, until the stable state is reached. It is a self-organizing process.

For a population dynamics with oscillation pattern (generally found in the situation of year by year changes of biological population), the oscillation amplitude decreases with the increase of diffusion coefficient in a given cell.

Chapter 13

Self-Organization

In previous chapters, I have stressed that most network evolution processes in nature are self-organizing systems. Actually, almost all biological systems (networks) are self-organizing systems (Hess and Mikhailov, 1994; Misteli, 2001; Clyde *et al.*, 2003; Motegi *et al.*, 2011). Therefore, the theory and methods on self-organization (i.e., self-organizology) are useful to handle the problems of network evolution in nature.

13.1 Basic Concept of Self-Organization

The organization with organizational instructions/forces from inside the system is called self-organization (Zhao and Zhang, 2013; Zhang, 2016r). Self-organizing systems are such systems which can evolve and improve the organization's behaviors or structure by themselves. In a self-organizing system, the system evolves spontaneously to produce an ordered structure according to some compatible rules. A system is called self-organizing system if there is no external intervention when the system is experiencing evolution. The stronger a system's self-organization capacity is, the stronger the system's ability to generate and maintain new functions (Zhang, 2016r).

Self-organization is a process by which some form of global order or coordination arises out of the local interactions between the components of an initially disordered system. The process is spontaneous, i.e., it is not directed or controlled by any agent or subsystem inside or outside the system; however, the rules followed by the process and its initial conditions may have been chosen or caused by an agent. It is often triggered by random fluctuations that are amplified by positive feedbacks. The resulting organization is wholly decentralized or distributed across all of the components of the system. As such, it is typically very robust and able to survive and self-repair substantial damage or perturbations (Zhang, 2013, 2014d; Wikipedia, 2014b).

Unlike other organizations, the self-organizing system arises only from the interactions between the basic components of a system, without external instructions and forces. During the process of self-organization, some structural components can interact and cooperate to display the behaviors that only a group will have. The dynamic interactions between low-level components typically include attraction and repulsion, that is, positive and negative feedbacks (Zhang, 2016r).

Overall, self-organization arises from the increase in complexity or information. According to the thermodynamic laws, this situation will only occur in the open systems far from thermodynamic equilibrium. For most systems, it means energy supply to the system is needed for generating and maintaining a certain mode. In an abstract sense, self-organization is a dynamic process that makes an open system change from the disorder to ordered states and thus reduces the system's entropy by absorbing "negative entropy" from outside the system (Glansdorff and Prigogine, 1971; Nicolis and Prigogine, 1977).

From the perspective of systematic theory, self-organization is an irreversible dynamic process. Each component in the system will spontaneously aggregate to produce an organic entity without outside instructions. From the view of mathematics and physics, self-organization means the dimensional reduction of state space or the reduction of degrees of freedom, i.e., the system converges spontaneously to one or more stable states, i.e., attractors. In such a system, the local interactions between the basic components of the system can change the modes of the system's organization, and the global behaviors of the system cannot be understood intuitively. They cannot be understood by simply observing existing laws and behaviors of between-component interactions (Zhang, 2012e, 2013, 2014d). The global properties of self-organizing systems are not predictable.

Self-organization usually requires to be based on three conditions (Bonabeau *et al.*, 1999): (1) strong nonlinear dynamic interactions, even though they do not necessarily correlate to positive or negative feedbacks; (2) an equilibrium between development and exploration, and (3) complex and diverse interactions.

Prigogine believed that conditions for self-organization include the following: (1) the system must be an open and dissipative system, where the dissipative system is a thermodynamically open system, which is operating out of, and often far from, thermodynamic equilibrium in an environment with which it exchanges matter/energy; (2) the system is far from the thermodynamic equilibrium, which allows for entering the nonlinear zone; (3) nonlinear interactions exist between components, and (4) some parameters fluctuate, and if fluctuations reach some thresholds the system will change to unstable from stable state, catastrophe will occur, and the system may exhibit a highly ordered state. For example, we often try to make a sand dune taller, but after a certain height, the addition of a little amount of sand will cause landslide of a large amount of sand in the sand dune, and the sand dune cannot be further piled higher. In fact, before landslide occurs, the sand dune has reached a critical state (i.e., threshold), and thus small perturbations can lead to instability.

13.2   Theories and Principles of Self-Organization

13.2.1   *Thermodynamic theory of self-organization*

The spontaneous formation of new structures, for example, crystallization process, Bénard phenomenon, Belousov–Zhabotinsky reaction (Sun and Lin, 2004), etc., are all self-organization processes, i.e., the formation of a structure or a mode does not need to be imposed by any external force. It seems that the components of these systems are arranged into a more ordered pattern by themselves. It appears that self-organization violates the second law of thermodynamics. The law holds that the entropy of a closed system can only increase rather than decrease. In other words, the second law of thermodynamics means that an isolated system should evolve in a uniform, simple, difference-eliminating way, which is in fact an evolution to a low-level organization (Zhang, 2016r).

For self-organizing systems, which are not in their equilibrium states, it is hard to determine whether the second law of thermodynamics is true or not. Prigogine has started to study the systems far from thermodynamic equilibrium states since the 1950s and proposed the dissipative structure theory (Glansdorff and Prigogine, 1971; Nicolis and Prigogine, 1977), in which the most used models to explain the dynamic self-organization process are Bénard cells and the Brusselator. The dissipative structure theory tries to address such problems as, under what conditions a system will be able to evolve from the disorder to the ordered and form a new, stable, and internally dynamic structure. The theory concludes that such a structure must be an open system, i.e., there are energy/matter flows in the system; the system will continuously generate entropy, but at the same time the entropy will be actively dissipated from the system or output from the system. Thus, at the cost of environmental disorder, the system will be able to increase order on its own. The system will be able to follow the second law of thermodynamics simply by removing its excess entropy. This dissipation can be mostly found in life systems. Plants and animals obtain energy and matter by absorbing light or food with low entropy, and then output energy and matter by draining the metabolic waste of high entropy. This will reduce its internal entropy to offset the degradation process required by the second law of thermodynamics.

The output of entropy cannot explain why and how self-organization happened. Prigogine held that self-organization would mostly occur in the nonlinear systems far from thermodynamic equilibrium.

13.2.2 *Principles of self-organization*

In the first symposium on self-organization held in 1959 in Chicago, British cybernetic expert, Ashby, proposed "principle of the self-organizing dynamic system" (Ashby, 1947). The principle states that any deterministic dynamic system, ignoring its classification or composition, will automatically evolve towards an equilibrium state, i.e., an attractor. It will leave behind all nonattractor states, i.e., the attractor's basin, and select the attractor out of all others (Wikipedia, 2014a, 2014b). The further evolution of the system is thus constrained to remain in the attractor. This theory reduced our uncertainty on system's state and solved the problem of entropy. It is equivalent to the self-organization, which finally reaches the equilibrium, and the final equilibrium can be considered to be a state of mutual adaptation of all components in the system. Another cybernetic expert, Heinz von Forster, proposed the principle of order from noise (Foerster, 1960, 1996). It states that self-organization is facilitated by random perturbations (i.e., noise) which allow the system to explore a variety of states in the state space. It increases the chance that the system would arrive in the basin of a "strong" or "deep" attractor, from which it would then quickly enter the attractor itself. This idea is very simple: The larger the state space that a system moves through, the faster it will reach the attractor. If the system stays at its initial state, it will not reach the attractors, and self-organization will thus not occur (Foerster, 1996).

Generally, there are multiple attractors in a nonlinear system. The attractor theory maintains that the behavioral trajectories of a complex system in the state space can be represented by the dynamic equations. These dynamic equations are always determined by a set of "attractors". What attractor the system will move towards depends on the attraction domain that the initial state falls into. What attractor the system eventually reaches is uncertain. To find attractors of the system is the key to understand the self-organization of the system. Small fluctuation of some parametrical values will cause the system to change. Thus Prigogine proposed a principle, i.e., order through fluctuations.

Both fluctuation and dissipation are necessary for the self-organization of a system. They correspond to attraction and repulsion in the system respectively. Dissipation is the dominant factor for self-conservation of system, and fluctuation is the dominant factor for self-innovation of system. The system with fluctuation and without dissipation will lose its stability, and the system with dissipation and without fluctuation will quickly terminate its evolution.

The system will evolve steadily once the dissipation is dominant. The system will be governed by random process, and what final state is reached depends on which fluctuation has been initiated at first and what environmental selection was made (Zhang, 2016r).

13.2.3 *Major theories on self-organization*

Self-organization theory is generally believed to mainly contain three parts, dissipative structure theory (Glansdorff and Prigogine, 1971; Nicolis and Prigogine, 1977; Brogliato *et al.*, 2007), synergetics (Haken, 1978, 2004), and catastrophe theory (Zeeman, 1976, 1977; Woodcock and Davis, 1978; Saunders, 1980; Thompson and Michael, 1982). Some basic thoughts and theoretical kernels of self-organization can be derived from dissipative structure theory and synergetics (Zhang, 2016r).

(1) Dissipative structure theory

Prigogine proposed the theory of dissipative structures in a symposium on theoretical physics and biology in 1969 (Nicolis and Prigogine, 1977). The theory was proposed mainly to explain the exchange of matter/energy between the system and the environment and its effect on the self-organizing system. The structure, established on the basis of the exchange of matter/energy between the system and the environment, is a dissipative structure, e.g., living organisms, convection, cyclones, hurricanes, etc. A dissipative structure has a reproducible stable state. This stable state may be reached by systematic evolution (or, by artifice). It is characterized by the spontaneous appearance of symmetry breaking and the formation of complex or chaotic structures where interacting particles exhibit long-range correlations (Wikipedia, 2014a). Far from the thermodynamic equilibrium, the openness of the system, and nonlinear interactions between different components in the system are three conditions for the formation of a dissipative structure. Far from the thermodynamic equilibrium refers to that the distribution of matter/energy in different areas of a system is extremely uneven. The dissipative structure theory is mostly used to treat the evolution of complex systems. It uses two levels of approaches, i.e., deterministic and stochastic approaches in the analysis of systematic evolution. The deterministic approach uses macroscopic physical variables to describe dynamics and features of the system. The stochastic approach treats macroscopic physical variables as the average of corresponding random variables. Analyzing random variables will not only produce the averaged values but also help understand fluctuation characteristics of the system.

(2) Synergetics

Inspired by the laser theory, Haken proposed the concept of "Synergy" in 1976, and another science on self-organization, synergetics, was thus established (Haken, 1978, 1983, 2004; Graham and Wunderlin, 1987). It is a science that explains the formation and self-organization of patterns and structures in open systems far from their thermodynamic equilibriums. Synergetics mainly discusses the coordination (synergy) mechanism between internal components of the system studied. It holds that the coordination between various components in the system is the basis of self-organization process. In synergetics, Haken held that on the one hand, in a system many subsystems interact to produce the structure and function at the macroscopic scale; on the other hand, there are many different scientific disciplines cooperating to explore the general principles for governing self-organizing systems. The order parameters generate and govern subsystems by competition and cooperation between various subsystems. Here the order parameters can be treated as the amplitudes of the unstable modes that determine the macroscopic pattern of the system. Serving of various subsystems to order parameters reinforces order parameters themselves and further promotes the serving of subsystems to order parameters, so that the system can spontaneously organize by itself (Haken, 1978, 2004). Competition and cooperation between order variables will result in different forms of evolution of self-organization. Competition and cooperation between order parameters of the system are direct forces for the formation of new structures. Because of the independent evolution of components in the system, various local and collaborative evolution, as well as random interferences by environmental factors, the actual state of the system always deviates from the average. The magnitude of such a deviation is called fluctuation. When the system is in its transition from one stable state to another, and if the independence evolution and collaborative evolution between system components move into a balance, any small fluctuation will be amplified and quickly spread to the whole system. The resultant giant fluctuation will promote the system to move into an ordered state. In addition, Haken proposed the concept of "functional structure", i.e., the function and structure are dependent on each other. If the energy or matter flow is cut off, the physical and chemical system will lose their structure, while a biological system is mostly able to maintain a fairly long time. Such biological systems seem to combine nondissipative and dissipative structures together (Haken, 1978, 2004).

(3) Catastrophe theory

Catastrophe theory was first proposed by the French mathematician, Thom R, in 1969. Since the 1970s, Zeeman and other scientists had further enriched catastrophe theory by considering a special case where the long-run stable equilibrium can be identified with the minimum of a smooth, well-defined potential function (i.e., Lyapunov function), and applied it to various aspects of physics, biology, ecology, medical science, economics, and sociology, and produced significant impacts (Zeeman, 1976; Zhao and Yu, 1987; Piyaratne *et al.*, 2014). Catastrophe theory is in essence a branch of bifurcation theory in the study of dynamical systems and is a special case of singularity theory in geometry (Zhang, 2016r). Bifurcation theory studies phenomena characterized by sudden shifts in behavior arising from small changes in circumstances and analyses how the qualitative nature of equation solutions depends on the parameters that appear in the equation (Wikipedia, 2014c). The theory was built on the basis of the stability theory.

The theory considers a catastrophe process as a process that transits towards a new stable state through an unstable state from an original stable state. Small changes in certain parameters of a nonlinear system can cause equilibria to appear or disappear, or to change from attracting to repelling and vice versa, leading to large and sudden changes of the behavior of the system (Wikipedia, 2014c). In mathematical view, this means the changes of values of a set of parameters and mathematical functions that denote the states of the system. Thus, it is a theory describing the phenomena that the continuous change of parameters will lead to the discontinuous change of the states of system.

It addresses the systems with almost sure structural stability in the state space but there is some structural instability on some point sets of measure 0. The basic characteristics of catastrophe systems include: multiple stable states; reachability; jumping; lagging; and divergence (which reflects the sensitivity of evolutionary trajectory to the path of control parameters). Catastrophe theory holds that different outcomes may occur even if it is the same process that corresponds to the same controlling factors and critical values; different new stable states may be achieved at different probabilities (Saunders, 1980). In general, catastrophe theory itself does not reveal the mechanism to produce catastrophe phenomenon. It just provides a reasonable mathematical model to describe the phenomenon of catastrophe in the real world and classifies various catastrophe types.

In addition to the theories above, there are other theories on self-organization, for example, Eigen's super circle theory. Super circle theory is a self-organization theory on molecular evolution.

13.2.4 *Properties of self-organization*

A self-organizing system can evolve to produce new structures and functions when it interacts with the environment. Unlike conventional mechanical systems, it owns its unique properties. Those properties can be used as part of the definition of self-organization and self-organizing systems, for example, no centralized control, and continuous adaptation to the changing environment, etc., (Zhang, 2016r).

(1) Local interactions facilitate global order

The most obvious change of the self-organizing system is the formation of global order. Local interactions follow basic physical rules; any impact from a region to another one must first explore all transition regions. When the system explores transition regions, all the processes will be disturbed by the turmoils occurring in the transition regions. First, assume the system is disordered and all components of the system evolve randomly. The impact of any past event will be quickly dispersed and ultimately destructed by these random turbulences. The result is thus that, starting at a chaotic state, the distant parts of the system are actually independent: they do not affect each other. During the self-organization process, all the components are closely connected. Understanding the structure of a local component will be valuable to know the structure of components of its adjacent regions.

(2) Distributed control

A highly organized system is commanded and controlled by external or internal forces. This control is called centralized control. In a self-organizing system, the control distributes across the whole system. All components contribute to the final arrangement of the states of system. Despite some of the advantages of centralized control with respect to distributed control, at some levels the centralized control must be based on distributed control. For example, the function of human brain disperses over the network of interacting neurons. Different brain regions perform specific functions, but no neuron or group of neurons has all the ability to control brain. This is just a result of self-organization (Zhang, 2016r).

(3) Nonlinearity and feedbacks

Nonlinearity means the whole system is not equal to the simple sum of its parts, i.e., the superposition principle is not met. Suppose a system is represented by a function: $y = f(x)$. If the following condition is satisfied: $f(\alpha x_1 + \beta x_2) = \alpha f(x_1) + \beta f(x_2)$, where $\alpha, \beta \in R$, then it is a linear system, otherwise it is a nonlinear system (Zhang, 2010, 2012e). Judging from the mechanical movement, a linear phenomenon is generally manifested as a smooth motion in time and space; it can be described by the functions with good performance, and the functions are continuous and differentiable. The nonlinear phenomenon is a movement from regular motion to irregular motion, with obvious jumping and intermitting features. From the view of perturbation and parameter theory, the response of a linear system is smooth and proportional changes, but a nonlinear system will exhibit substantial changes in some of the key points because of the small changes in parameters, and form and maintain spatially regular and ordered structures. Linear relationship is independent of each other, while the nonlinear relationship is an interactive one, which makes it violate the superposition principle and produce gain or loss. In nonlinear systems, there are feedbacks between system components; each component affects the others, and other components in turn affect it. The positive feedback plays a role similar to the input so that the system's deviation increases, and the system's oscillation is thus amplified. The negative feedback causes reverse outcome as compared with input's role, so that the system's output error can be reduced and the system is thus stabilized. This is just the case of complex biological systems.

In complex self-organizing systems, there are often several chains of positive and negative feedbacks, so a change can be amplified in a certain direction but suppressed in the other directions. It will result in the very complex behaviors difficult to predict.

(4) Far from the thermodynamic equilibrium

Equilibrium is a special state of a system. At this state, the measurable macroscopic physical properties of the system are uniform throughout (so that there is no macroscopic irreversible process inside the system). At the equilibrium state, the system follows the first law of thermodynamics: $d_E = d_Q - pd_V$, i.e., the increment of energy inside system is equal to the absorbed heat subtracting the outward work done by the system. It is also coincident with the second law of thermodynamics: $d_S/d_t \geq 0$, that is, the spontaneous evolution of system is always towards the direction of entropy's increase. For the system in the thermodynamic equilibrium state, it must remove its extra energy; it will remain in the minimum energy state without the input of external energy. A system will likely move across a nonlinear region if the system is far from the thermodynamic equilibrium. The system far from the thermodynamic equilibrium is more sensitive and more vulnerable to environmental changes due to its dependence on external energy input. But it is more powerful to respond to changes.

On the other hand, the surplus of external input energy allows the system for amplifying the self-organization process, for example, off-setting small perturbation or maintaining positive feedbacks longer in the aid of strong interactions. This makes the system more vigorous and more adaptive to external changes (Zhang, 2016r).

(5) Systematic termination and organizational hierarchy

The interactions between individual components of the self-organizing system can be, to some extent, defined as an ordered structure. However, the order does not mean organization. Organization is an ordered structure and can achieve a particular function. In a self-organizing system, this function is to maintain a particular structure under various disturbances. The general characteristics of self-sufficiency can be understood as a closure. A process with causal relationship can be represented as a chain or a sequence: A → B → C → D → ..., where A initiates B, B initiates C, C initiates D, and so on. Overall, this will lead to a continuous change. However, there may be its own termination of a link in the chain, for example, O returns J, so the cycle of the system becomes J, K, L, M, N, O, J, K, L. Thus, the corresponding arrangement of the system will always be maintained or recycled. In addition, if the loop is placed in a negative feedback region, it is relatively unaffected by the impact of external interferences (Foerster, 1960).

A self-organizing system may contain a lot of autonomous and organization-closed subsystems. Those subsystems will interact in a more indirect way. They will also adapt to the structure for termination and determine subsystems at a higher level. Newly generated subsystems will contain the original subsystems as their components. Each self-organizing system constitutes a series of subsystems. A self-organizing system thus forms a layered structure. Each self-organizing system belongs to the high-level self-organizing system and contains low-level self-organizing systems. It interacts with other self-organizing systems at the same level. Therefore, the hierarchization of system's structure and function is also a characteristic of self-organizing systems (Zhang, 2016r). Hierarchical subnetworks of a complex network are examples of such subsystems (Zhang, 2016c).

13.3 Existing Algorithms of Self-Organization

As mentioned earlier, it is hard to predict the complex behaviors of self-organizing systems, we usually use mathematical modeling and computer simulation to describe these systems. They also help people understand how these systems work. A mathematical modeling method for self-organization is to use differential equations, and another method is to use cellular automata (Wolfram, 2002; Ballestores and Qiu, 2012; Zhang *et al.*, 2011; Zhang, 2015a, 2016r).

Many optimization algorithms can be considered as a self-organization system because optimization aims to find the optimal solution to a problem. If the solution is considered as a state of the iterative system, the optimal solution is essentially the selected, converged state or structure of the system, driven by the algorithm based on the system landscape (Yang *et al.*, 2013; Yang, 2014). Actually, we can treat an optimization algorithm as a self-organization system.

In the sense of optimization, the existing algorithms of self-organization can be classified into such hierarchies, Monte Carlo method, heuristic method, meta-heuristic method (Mirjalili *et al.*, 2014, Zhao, 2014), etc. The meta-heuristic method includes such methods as evolution- and population-based method, physics-based method, Swarm Intelligence (SI)-based method, hyper-heuristic method, etc., (Zhang, 2016r).

13.4 Self-Organization in Life Sciences

Self-organization is popular in nature and human society. It is very popular in biological systems (networks), whether at subcellular level or at ecosystem level (Hess and Mikhailov, 1994; Misteli, 2001; Camazine, 2003; Clyde *et al.*, 2003; Motegi *et al.*, 2011).

The self-organization of ecosystems is a fundamental theory in ecology. The essential difference between ecosystems and nonbiological systems is its ability in self-organization. The evaluation of self-organizing capacity of ecosystems has become one of the most important methods to reveal the complexity and uncertainty of ecosystems (Zhang, 2012e, 2015b, 2016r).

In the field of life sciences, there is a rapid growing emphasis on the phenomena of self-organization *in vivo*. In biological systems, self-organization is a process at global level. The system is generated only from the interactions between components at low levels. Implementing the rules of between-component interactions only requires local information rather than global information (Camazine, 2003). Increasing evidences are proving that many biological systems are close to what's called a critical point: they sit on a knife-edge, precariously poised between ordered and disordered. This strategy is believed to increase the flexibility to deal with a complex and unpredictable environment (Ball, 2014).

Almost all biological systems (networks) are self-organizing systems (Hess and Mikhailov, 1994; Misteli, 2001; Clyde *et al.*, 2003; Motegi *et al.*, 2011), for example, (1) the self-assembly of proteins, as well as the formation of other biological macromolecules and lipid bilayers; (2) homeostasis, which is a self-organization from cell to tissue; (3) pattern formation and morphogenesis, i.e., the growth and differentiation of living organisms; the interface between two different types of cell will trigger the formation of a third kind of cell at their boundary; an embryo can construct complex tissues this way, with different cell types in all the right places (Davies, 2014); (4) human motion; (5) creation of structures by gregarious animals, such as social insects, bees, ants, etc.; (6) group behaviors (the most typical examples can be found in birds and fish), and (7) in the super cycle theory and autocatalytic theory, life itself is originated from the self-organizing chemical systems (Zhang, 2016r).

Phase transitions often occur in biological systems. For example, the lipid bilayer formation, the coil–globule transition in protein folding and DNA melting, liquid crystal-like transitions in DNA condensation, and cooperative ligand binding to DNA and proteins with the character of phase transition (Lando and Teif, 2000).

Gel to liquid crystalline phase transitions is critical in physiological role of biomembranes. Due to the low fluidity of membrane lipid fatty-acyl chains, in gel phase, membrane proteins have restricted movement and are restrained. Plants depend on photosynthesis by chloroplast thylakoid membranes exposed to cold environmental temperatures. Thylakoid membranes retain innate fluidity even at relatively low temperatures due to high degree of fatty-acyl disorder allowed by their high content of linolenic acid (YashRoy, 1987).

Molecular self-assembly is fundamental for constructing macromolecules in cells of the living organism, including the self-assembly of lipids to form the membrane, the formation of double helical DNA through hydrogen bonding of the individual strands, and the self-assembly of proteins to form quaternary structures. Molecular self-assembly of nanoscale structures is important in the growth of β-keratin lamellae/setae/spatulae structures which are used to endow geckos the ability to climb walls and adhere to ceilings and rock overhangs (Wikipedia, 2014d; Zhang, 2016r).

In recent years, some scientists have attempted to interpret the origin of life from the view of self-organization. Scientists use a set of biomolecules to show the way in which life might have started. They hold that different chemicals come together due to many forces that act on them and become a molecular machine capable of even more complex tasks. Each living cell is full of these molecular machines. These molecular machines haven't done much on their own. When they add fatty chemicals, which form a primitive cell membrane, it got the chemicals close enough to react in a highly specific manner. Molecules and cells interact according to simple rules, creating a whole that is greater than the sum of its parts (Davies, 2014). This form of self-organization may be popular for the origin of life on both earth and in other planets (Zhang, 2016r).

To interpret the origin of life, Stano and his colleagues chose an assembly that consists of 83 different molecules including DNA, which was programmed to produce a special green fluorescent protein (GFP) that could be observed under a confocal microscope (Lehn, 1988, 1990). The assembly can only produce proteins when its molecules are close enough together to react with each other. When the assembly is diluted with water, they can no longer react. This can explain why the cell is so compact: to allow the chemicals to work.

In order to recreate this molecular crowding, Stano added a fatty molecule, POPC, to the dilute water solution, and these molecules then automatically form liposomes that have a very similar structure to the membranes of living cells. They found that many of these liposomes trapped some molecules of the assembly. Around 5 in every 1,000 liposomes had all 83 of the molecules needed to produce a protein. These liposomes produced large amount of GFP and glowed green under a microscope. Surprisingly, computer calculations reveal that even by chance, 5 liposomes in 1,000 could not have trapped all 83 molecules of the assembly. The calculated probability for even one such liposome to form is essentially zero. This means some quite unique mechanism behind it, and self-organization is one of the reasons (Zhang, 2016r).

Davies (2014) showed how from these interactions we can deduce "rules" of embryo development. For example, cells communicate with each other and tweak their behavior in response to changes in their environments. This is what puts the "adaptive" into adaptive self-organization, ensuring that development can cope with noise or disruption. An example is the way tiny blood vessels called capillaries manage to cater to different kinds of tissue, even while these tissues are moving and growing (Davies, 2014). A feedback loop exists between oxygen and a cell protein called HIF-1-alpha. Oxygen normally causes HIF-1-alpha to be destroyed. If a tissue lacks oxygen, HIF-1-alpha levels rise, triggering a cellular signal encouraging capillaries to grow. This brings in oxygen, which shuts down HIF-1-alpha and halts capillary development. Should the tissue then grow, oxygen levels will fall again, and the loop is set in motion once more (Zhang, 2016r).

13.5　Selforganizology

Self-organization is a universe phenomenon. A lot of theories and methods have been established to describe self-organization. The scientific discipline and terminology, selforganizology, has been proposed by Zhang (2013, 2016r) to find an independent science on self-organization and to facilitate the development of self-organization science. Some of the principles and methods in selforganizology are redescribed below (Zhang, 2016r).

In a self-organization system, the basic structure of behavioral rules includes: (1) IF-THEN-ELSE rule; (2) GO TO rule; (3) DO WHILE rule; (4) SWITCH CASE DO rule; (5) LET rule; (6) AND/OR rules; (7) RANDOMIZE rule; (8) other equation/model/ algorithm/statement based rules, etc. Using these rules for all components at all hierarchical levels will probably produce any complex behaviors of the self-organizing system (Zhang, 2016r). Mathematical equations and models (e.g., differential equations) can be used in the simulation and modeling of self-organization phenomena (Zhang, 2015b). An algorithm, a problem-solving procedure, etc., can be represented by a structure of hierarchical rules in which each rule has both parent and son rules in exception of the top rule and bottom rules. It is expected that many existing self-organization algorithms, such as Swarm Intelligence algorithms, can be resolved into simple rules-based algorithms. We should maximally resolve complex rules into simple rules. Differential equations, for example, can be rewrite to difference equations, and if it's possible, further resolved into many simple rules, until no further simpler rules can be found. Compared to complex rules, simple rules are more useful in exploiting the most fundamental mechanism of self-organization. In general, the simpler rules mean the more effective the simulation method used (Zhang, 2016r).

Some methods, in particular agent-based modeling (Topping *et al.*, 2003; Griebeler, 2011; Zhang, 2012e, 2014a), can be considered as the methodological basis of simulation and modeling of self-organization. These methods will not only help propose hypothesis on behaviors and mechanism of a self-organizing system but also help propose management strategies on the self-organizing system.

It should be noted that the concepts, definitions, principles, and methodology in classic theory of agent-based modeling can be further revised and improved to reasonably describe and model self-organizing systems. For example, when used in self-organization, agents in agent-based modeling are components (Zhang, 2016r).

Chapter 14

Agent-based Modeling

Agent-based modeling (ABM) is considered as the third methodology with the exception of deduction and induction. As pointed out earlier, it can be used in modeling self-organization and network evolution (Zhang, 2016r). More details on ABM can be found in Zhang (2012e, 2016r).

14.1 Complex Systems

14.1.1 *Properties of complex systems*

Complex systems (networks) exhibit such properties as follows (Li *et al.*, 2007; Zhang, 2012e, 2016r):

(1) **Nonlinearity.** Complex systems always behave nonlinearly. The global behavior of a complex system cannot be directly derived from the standalone behaviors of individuals.

(2) **Emergency.** Emergency refers to a dynamic process from low level to high level, from local to global, and from microlevel to macrolevel. Emergency stresses that the interactions between individuals lead to different functions and properties and behaviors from that of individuals. It results in a system with certain functional characteristics and purposeful behavior different from individuals.

(3) **Dynamicness.** A complex system always changes with the time.

(4) **Modularity and hierarchy.** A complex system is the aggregation of interactive and correlative individuals/objects, i.e., modularity. A complex system shows a hierarchical structure (i.e., hierarchy). The objects at the hierarchical level will aggregate to produce the objects at its parent hierarchical level.

(5) **Information flow and associativity.** In a complex system, there will be large amounts of information/matter interactions among individuals/objects. The changes of any individual/ object may affect other individuals/objects or even the entire system.

(6) **Incomputability.** Incomputability of complex systems refers to, (A) the behavior of a complex system cannot be described by using deduction, induction, or other formalization methods, and (B) the process of a complex system can be approximated by the inference system based on some rules.

14.1.2 *Modeling complex systems*

To study and describe complex systems, we can choose to use differential equations, artificial neural networks (ANNs), ABM, network analysis, etc.

14.1.2.1 *Modeling with differential equations*

Differential equations can be used to model complex systems (networks), as described in previous chapters.

Differential equations are developed in a top-down way (Hraber and Milne, 1997). These models are always differential or finite difference equations with one or more dependent and independent variables (Zhang *et al.*, 2011; Zhang, 2016r). In differential equations, the types of between-variable interactions should be predefined. Pairwise interaction coefficients in the equations describe the interactions between variables. All variables are assumed to interact with equal probability, which is a property of the interactive system with fully mixed or mean field (Zhang, 2012e).

The differential equation model is to a certain extent the agent-based model with uniform field. For example, in differential equations, we usually represent spatial heterogeneity of interactions by defining adjacent interactions and including some factors. These models can be used to predict which variable will dominate system as time. The parameters of differential equations can approximate the apparent properties of a system. However, the parameters in differential equations approximate the phenomenological properties of a system without identifying internal mechanisms or considering between-individual variation, and therefore it is hard to describe complex interactions and hard to include various behaviors simulated. In addition, the complexity and incomputability of differential equations will exponentially increase with the system's complexity. They can only be used to describe the systems with lower complexity.

14.1.2.2   *Modeling with artificial neural networks*

ANNs are a kind of network. They are used to simulate complex and nonlinear ecological problems (Zhang, 2010). ANNs are learning models that need enough sample data to train and learn from. Learning from samples, ANNs store the intrinsic mechanism of a system or data sets as the connected weights of network. ANNs are such models that lie between empirical models and mechanistic models.

There are many types of ANNs. Their capability and effectiveness depend on three factors: (1) complexity of neuron models, (2) velocity and efficiency of leaning rules, and (3) topological structure of network.

In a sense, ANNs have the characteristic of fully mixed or mean field systems.

14.1.2.3   *Modeling with network models*

Network analysis treats individuals/objects as nodes in the network, which allows people to analyze the large-scale structure of a complex system (Zhang, 2012e). It makes easier the examination of the effects of interactions, characterization of important elements, and description of global structure of the system. However, the current methodology is in general insufficient to describe the complex dynamics of systems.

14.2 Principle and Methods of Agent-based Modeling

ABM has its roots in the modeling of complex adaptive systems (CAS) (Zhang, 2012e, 2016r). A complex adaptive system may self-organize spontaneously and reconstruct its components dynamically in order to survive in the environment. Holland (1975) held that a complex adaptive system should exhibit such properties as below:

(1) **Nonlinearity.** Simple and linear extrapolation is invalid in the prediction of system's behavior.
(2) **Aggregation.** It allows for the formation of a population.
(3) **Flows.** Flows allow for the transition and transformation of resources and information between components.
(4) **Diversity.** Different agents may have different behaviors, which is usually conducive to the robustness of the system.

The mechanisms of complex adaptive systems include (1) labels: it allows agents to be defined and identified; (2) internal pattern: it allows agents to make inference to their world, and (3) establishment of blocks: it allows components and the entire system to be constructed from simple components at various hierarchies (Zhang, 2012e, 2016r).

The earlier ABM originated from cellular automata (CA) (Gardner, 1970). In answering the question: if a machine can be programmed to reproduce itself, proposed by Von Neumann, the physicist Stanislaw Ulam proved it to be true by using CA. Wolfram (2002) found that the rules for CA will produce astonishing spontaneous patterns. These patterns directly correspond to extensive algorithms and logic systems. He believed that simple rules will result in a complexity similar to the true world.

ABM is a bottom-up modeling method. It is widely used to model the complex systems containing spontaneous and interactive agents (Topping *et al.*, 2003; Qi, 2003; Qi and Zhang, 2002; Qi *et al.*, 2004;

Li and Ma, 2006; Chen *et al.*, 2008). It models the dynamics of adaptive systems according to the adaptation mechanism of individuals. ABM can be used in these cases (Bonabeau, 2002): (1) interactions between agents are complex, nonlinear, and discrete; (2) interactions are complex and heterogeneous; (3) spatial factors are very important and the locations of agents are not fixed; (4) population is heterogeneous, and each individual is different from others; and (5) agents show complex and diverse behaviors, including learning and adaptation. As the systems become more and more complex, conventional simulation tools show obvious shortages. So far, it has been successfully used in organizational simulation (risks, organizational design, molecular self-organizing), diffusion simulation (diffusion dynamics), flow simulation (traffic, flow management), market simulation (stock market), etc., (Bonabeau, 2002).

14.2.1 *Agent and behaviors*

Agents can be defined in various ways. Any independent component, like a node, a model, an individual, etc., can be considered as an agent (Bonabeau, 2002). The behavior of an independent component might be a simple and primitive response and decision, or even a complex adaptive intelligence. Overall, the behavior of an agent should be self-adaptive, and the agent can learn from its changing environment (Mellouli *et al.*, 2003). Casti (1997) pointed out that the behavioral rules of an agent must include two parts, basic rules, and the high-level rules that govern basic rules (rule-changing rules). Basic rules define necessary responses to the environment, and rule-changing rules define adaptation. Jennings (2000) maintained that the basic characteristic of an agent is that it can make decisions independently.

Different from differential equations, combining new agents and new agent types into ABM is easy. At lower levels, differential equations can be included to approximate system's behaviors. At higher levels, the interactions between agents are allowed to define aggregative behaviors. Aggregative behaviors are spontaneous behaviors, generated by between-agent or agent–environment interactions (Zhang, 2012e, 2016r).

ABM allows an agent and the entire system to contain more simple agents at various levels. Unlike artificial intelligence agent, the environmental responses and influences of an agent in ABM are always not so complex.

Agents should satisfy these criteria (Macal and North, 2005):

(1) An agent is an independent and identifiable individual which possesses a set of properties and rules that govern its behavior and decision ability. An agent is self-contained and independent. It has a boundary, through which we can easily discern between outside and inside the agent or shared characteristic.

(2) Each agent locates in a certain spatial position and interacts with its adjacent agents. An agent has a set of protocols that govern its interactions with other agents, such as communication protocol, the capability to affect its environment, etc. An agent is able to identify and discern the characteristics of other agents.

(3) Agents are goal-directed. An agent behaves to achieve some goals.

(4) Agents are independent, autonomous, and self-guided. At least within a finite range, an agent can independently operate in its environment.

(5) Agents are flexible. An agent is capable of accumulating experiences and learning from the environment and adjusting its behavior. This requires some form of memory. An agent possesses some high-level rules to adjust its low-level behavioral rules. It has diverse properties and behavioral rules.

To define agents, exactly specifying their behaviors and reasonably representing interactions between agents are fundamental. Once agents are defined, we should exactly define their behaviors. First, it is necessary to determine a theory on behaviors, and existing behavioral theories can be used. Agents may use various behavioral models, including if-then rules and threshold rules.

14.2.2 *Procedures of Agent-based Modeling*

The main procedures of ABM include (Zhang, 2012e, 2016r):

(1) Determine various types of agents and define the behaviors of those agents.

(2) Identify relationships between agents, and construct interaction types between agents.

(3) Choose the platforms and environments for ABM, and set the strategies for ABM.

(4) Acquire necessary data for ABM.

(5) Validate the patterns of agents' behaviors and the system's behavior.

(6) Run ABM model and analyze the output from the standpoint of linking the microscale behaviors of agents to the macroscale behavior of the system.

To design the ABM, the key is software design and model development. At the design phase, the structure and function of the model should be defined. At implementation phase, we should develop the model based on the designed plan. The model will be used during the practical operation. These phases are usually repeated several times to generate a more detailed model.

Grimma *et al.* (2006) has proposed a standard protocol for describing ABM. The core of the protocol is to build the information about ABM in the same sequence. This sequence consists of seven elements which can be grouped in three blocks, overview, design concepts, and details (Grimma *et al.*, 2006). (1) The overview consists of three elements, including purpose, state variables and scales, and process overview and scheduling. It provides an overview of the overall purpose and structure of the model. It includes the declaration of all objects (classes) describing the models entities (i.e., different types of individuals or environments, etc.) and the scheduling of the model's processes. (2) The design concepts describe the general concepts that underlie the design of the model. The purpose of this element is to link model design to general concepts that have been identified in the field of complex systems. These concepts include questions about emergence, the type of interactions among individuals (agents), whether individuals consider predictions about future conditions, or why and how stochasticity is considered. (3) The details include three elements, i.e., initialization, input, and submodels, which present the details that were omitted in the overview. The submodels implementing the model's processes are particularly described in detail. All information required to completely reimplement the model and run the baseline simulations should be provided (Zhang, 2012e, 2016r).

The logic behind the protocol sequence is, context and general information is provided first (overview), followed by more strategic considerations (design concepts), and finally more technical details.

Part 4
Flow Analysis

Chapter 15

Flow/Flux Analysis

In this chapter, flow analyses for both molecular networks and ecological networks are discussed. The terminologies of flow and flux are used without substantial differentiation.

15.1 Flux Balance Analysis

15.1.1 *Principle of FBA*

The product of between-node correlation matrix by between-node flux matrix is 0, which is called flux balance state. The model analysis at flux balance state of the biological system is called Flux Balance Analysis (FBA). FBA assumes that through natural selection, flux is balanced without disturbance. If there is perturbation, the system (network) will be regulated so that the flux after the disturbance maximally approximates the flux at the resting state.

15.1.2 *Basic procedures of FBA*

Generally, Flux Balance Analysis includes the following procedures:

(1) **Define the system** (network), including nodes and between-node interactions in the system.
(2) **Quantity conservation.** Develop the ordinary differential equations of all interactions according to flows and their mathematical relationships that coincide with quantity conservation principle (Chen *et al.*, 2010).

(3) **Define measurable metabolic flows.** Because of the robustness and redundancy of biological networks, the number of interactions is always greater than the number of nodes. Therefore, biological networks are generally represented by a group of indefinite equations, with fewer cases of balanced equations and over-determined equations. It is theoretically impossible to solve all flows. It is therefore necessary to add some constraints that generally give the values or ranges of certain flows as constraints.

(4) **Goal optimization.** Given the objective function, use the method of linear or nonlinear programming to find the optimal flow distribution in the solution space.

15.1.3 *FBA of biological networks*

15.1.3.1 *FBA of cellular metabolic networks*

The growth and differentiation of cells involves a large number of metabolic interactions. In a cell, a considerable number of metabolic pathways interact with each other and constitute a complex network. Metabolic system is the terminal system that plays a role in the regulation of cells, and it is one of the most complex biochemical systems. Metabolic network is so complex that it has to make more effective use of the model to guide the research.

(1) FDP metabolic network model *Saccharomyces cerevisiae* (Ying and Ouyang, 2000)

Ying and Ouyang (2000) established the FDP synthetic metabolic model of *Saccharomyces cerevisiae* using a flux balance model (see Fig. 1). They used the method of Fredrickson (1976) to calculate the rate of accumulation of metabolite i based on material balance principle (Luo, 2007)

$$dc_i/dt = r_{f,i} - \mu c_i$$

Here $\mu = 0$, we have

$$dc_i/dt = r_{f,i}$$

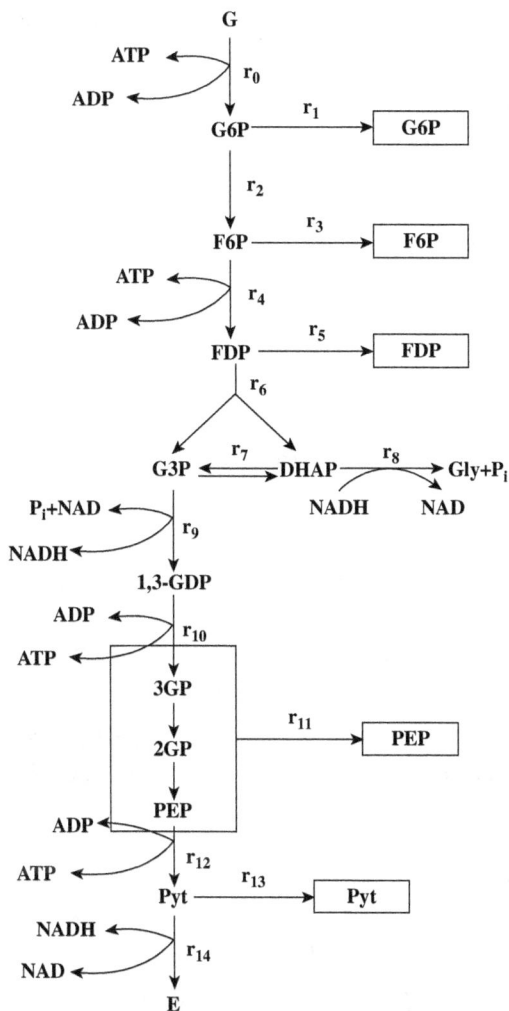

Fig. 1 *S. cerevisiae* FDP metabolic network model (Ying and Ouyang, 2000).

Because

$$r_{f,i} = \sum_j v_{ij} v_j$$

thus

$$dc_i/dt = \sum_j v_{ij} v_j$$

For the material balance of the extracellular metabolites in the batch reactor, we have

$$d(V_R C_i)/dt = V_R R_{f,i}$$

i.e.,

$$dC_i/dt = R_{f,i} - (C_i/v_R)(dv_R/dt)$$

The synthesis rate of metabolite i can be written as

$$R_{f,i} = x \sum_j v_{ij} v_j$$

The cultivate volume is a constant, i.e.,

$$dV_R/dt = 0$$

therefore

$$dC_i/dt = x \sum_j v_{ij} v_j$$

Balance equations of all substances in glycolysis process can be listed based on the last equation. In addition, assume (1) the cell is in non-growing phase, and (2) the concentrations of substances inside and outside the cell are the same. Finally, all chemical reaction equations are achieved and the metabolic flux of every node can be obtained

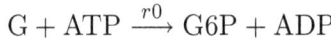

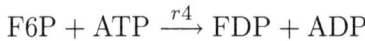

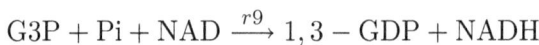

$$G + ATP \xrightarrow{r0} G6P + ADP$$

$$G6P \xrightarrow{r2} F6P$$

$$F6P + ATP \xrightarrow{r4} FDP + ADP$$

$$FDP \xrightarrow{r6} DHAP + G$$

$$DHAP \xrightarrow{r7} G3P$$

$$DHAP + NADH + H2O \xrightarrow{r8} Gly + Pi + NAD$$

$$G3P + Pi + NAD \xrightarrow{r9} 1,3 - GDP + NADH$$

$$1,32GDP + ADP \xrightarrow{r10} PEP + ATP$$

$$\text{PEP} + \text{ADP} \xrightarrow{r12} \text{Pyr} + \text{ATP}$$

$$\text{Pyr} + \text{NADH} \xrightarrow{r14} \text{Eth} + \text{CO} + \text{NAD}$$

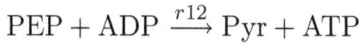

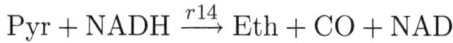

Calculate the following metabolic rate based on the above reaction equations:

$$dx/dt = r_p - r_c$$

where x: concentration of the metabolite, r_p: production rate of the metabolite, and r_c: consumption rate of the metabolite. Finally, a group of 11 equations was established, in which there are 16 unknown variables and its degree of freedom is 5.

(2) Available methods for linear optimization of metabolic networks

The flux distribution of metabolic networks can be optimized using linear optimization methods (Varma and Palsson, 1994). The major hypothesis of these methods is that the cell possesses the capacity to expand its flux distribution under the control of external constraints, so that metabolic flux distribution can be maximized to achieve the metabolic goal. Actually, the organism possesses some control mechanisms to ensure optimal growth conditions (such as maximum biomass yield) in response to specific environmental conditions (Edwards and Palsson, 2000; Edwards et al., 2001).

So far, numerous methods for optimal functions under external constraints have been proposed, e.g., maximization of ATP production (Ramakrishna et al., 2001), minimization of nutrient consumption (Savinell and Palsson, 1992), minimization of metabolic adjustment (MOMA) (Segre et al., 2002), and maximization of cellular biomass production (Varma and Palsson, 1993, 1994).

(A) Basic model

The basic model is based on the constrained maximization of cellular biomass production. For a metabolic network with n metabolites and m chemical reactions (interactions), its equation for metabolic balancing is (Luo, 2007)

$$\text{max}\quad C = v_{\text{biomass}}$$

$$\sum_{j=1}^{m} s_{ij}v_j = b_i$$

$$b_i \in R, v_j \in R_+$$

where v_{biomass}: metabolic flux of cellular biomass, s_{ij}: the ith metabolite of the jth reaction, v_j: reaction rate of the jth reaction, C: cellular biomass production. Cellular biomass consists of some essential biomass components (such as amino acids, nucleosides, etc.) at the appropriate biological ratio (Neidhardt *et al.*, 1996). The degree of freedom of the model $f = m - n$.

(B) Simplified basic model

The basic model above is an indefinite equation. Some constraints should be used to achieve the solution. Assume $b_i = 0$, i.e., the cell is at its quasi-steady state, therefore $v_{\text{min}} \leq v_j \leq v_{\text{max}}$. The model above is simplified as follows:

$$\text{max}\quad C = v_{\text{biomass}}$$

$$\sum_{j=1}^{m} s_{ij}v_j = 0$$

$$v_{\text{min}} \leq v_j \leq v_{\text{max}}$$

If some gene is knocked out and the corresponding enzyme reaction is thus interrupted, let $v_k = 0$. We have

$$\text{max}\quad C = v_{\text{biomass}}$$

$$\sum_{j=1}^{m} s_{ij}v_j = 0$$

$$v_{\text{min}} \leq v_j \leq v_{\text{max}}$$

$$v_k = 0, k \in Q$$

where Q is the set of interrupted enzyme reactions.

(C) MOMA

The MOMA model is

$$\min \; C = \sum_{i=1}^{m} (v_i - w_i)^2$$

$$\sum_{j=1}^{m} s_{ij} v_j = 0$$

$$v_{\min} \leq v_j \leq v_{\max}$$

$$v_k = 0, \, i, k \in Q$$

where w_i and v_i are metabolic flux before and after the external control, respectively.

The goal function can be revised as

$$\min C = \sum_{i=1}^{m} ((v_i - w_i)/w_i)^2$$

The goal function represents the cell's capacity of self-regulation and self-repairing.

(D) MOMA with maximization of growth capacity (Luo, 2007)

In the absence of metabolic and genetic engineering modifications, the natural production of many microbes is often far below their theoretical maximum yield. This is because the cell's metabolic network mainly responds to past natural selection pressure, rather than producing the largest amount of certain compounds. Luo (2007) used the dual optimization architecture to analyze how the cell controls the distribution of metabolic flux through metabolic regulation, and how the distribution is controlled by internal cell target. The internal cell target is the maximization of cellular biomass production and the MOMA.

Luo (2007) argued that MOMA does not consider the cell's growth capacity, and the basic model considers the theoretical maximum production only. Suppose the cell possesses the capacity of both the maximization of cellular biomass production and the minimization of metabolic adjustment. We can combine the two methods to enhance the constrained optimization of metabolic flux, i.e., we have

$$\min \ C = \sum_{i=1}^{m} ((v_i - w_i)/w_i)^2 / v_{\text{biomass}}$$

$$\sum_{j=1}^{m} s_{ij} v_j = 0$$

$$v_{\min} \leq v_j \leq v_{\max}$$

$$v_k = 0, \quad i, k \in Q$$

It is actually a problem of two-layer optimization.

(3) *S. cerevisiae* metabolic network model (Luo, 2007)

Based on the overall understanding of cell metabolic network (Nissen and Schulze, 1997; Ying and Ouyang, 2000; Förster *et al.*, 2002; Cakir *et al.*, 2004), Luo (2007) established a central carbon metabolism network model of the yeast (*S. cerevisiae*) cell and Metabolic Flux Analysis (MFA) was conducted based on the model. Metabolic Balance Analysis was used to establish mathematical model of cellular metabolic network (see Fig. 2).

The metabolic network model is as follows (Luo, 2007):

1. Substrate Uptake

(1) 1 GLUC + 1 ATP→1 GLUC6P + ADP
GLK1, HXKG
(2) 1 GAL + 1 ATP→ 1 GLUC6P + ADP
GAL1, 5, 7
(3) 1 ETOH + 1 NADcyt → 1 ACAL +1 NADHcyt
ADH2E

2. Glycolysis and Gluconeogenesisa

(4) 1 GLUC6P ↔ 1 FRUC6P
PGI1

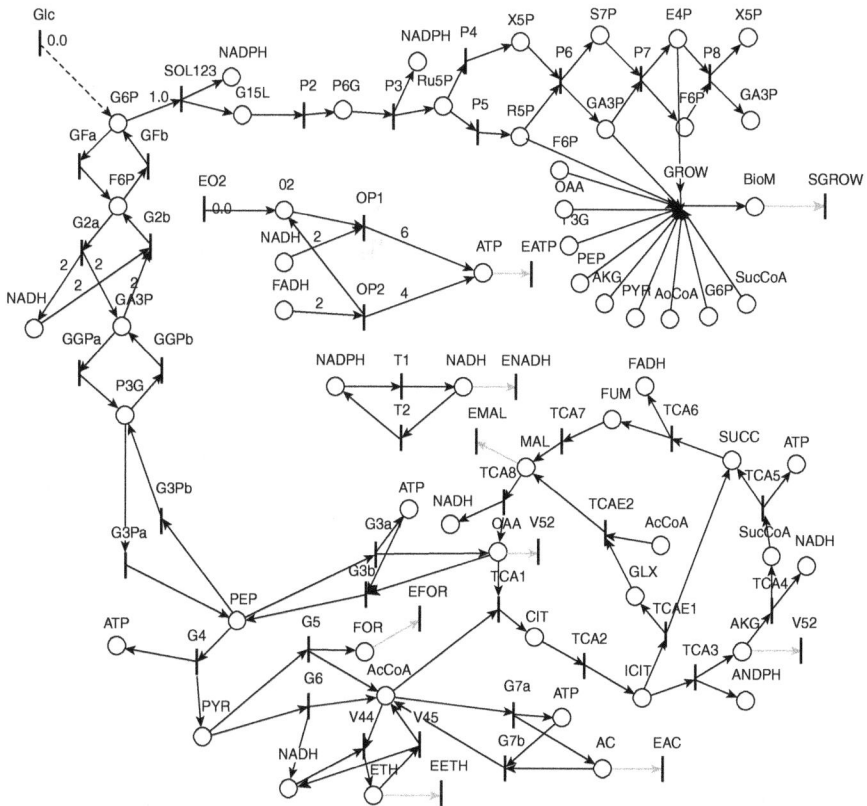

Fig. 2 Model graph of *S. cerevisiae* cell's metabolic network (Luo, 2007).

(5) 1 FRUC6P + 1 ATP → 1 FRUCDP + ADP
    PFK1, 2G

(6) 1 FRUCDP → 1 FRUC6P
    FBP1E

(7) 1 FRUCDP ↔ 1 GA3P + 1 DHAP
    FBA1

(8) 1 DHAP ↔ 1 GA3P
    TPI1

(9) 1 GA3P + 1 NADcyt ↔ 1 P13G + 1 NADHcyt
    TDH1, 2, 3

(10) 1 P13G + 1 ADP ↔ 1 P3G + 1 ATP
     PGK1

(11) 1 P3G $\leftrightarrow$ 1 P2G
GPM1, 2, 3

(12) 1 P2G $\leftrightarrow$ 1 PEP
ENO1, 2

(13) 1 PEP + 1 ADP $\rightarrow$ 1 PYR + 1 ATP
PYK1, 2G

(14) 1 DHAP + 1 NADHcyt $\rightarrow$ 1 GOH3P +1 NADcyt
GPD1, 2

(15) 1 GOH3P $\rightarrow$ 1 GOH
GPP

(16) 1 PYR $\rightarrow$ 1 ACAL + 1 CO2
PDC1, 2, 5

(17) 1 ACAL + 1 NADHcyt$\rightarrow$ 1 ETOH +1 NADcyt
ADH1, 4G

(18) 1 ACAL + 1 NADPcyt$\rightarrow$ 1 AC +1 NADPHcyt
ALD6

(19) 1 ACAL + 1 NADmit $\rightarrow$ 1 AC +1 NADHmit
ALD4

(20) 1 AC + 2 ATP $\rightarrow$ 1 ACCOAcyt +2 ADP
ACS1, 2

(21) 1 PYR + 1 NADmit$\rightarrow$ 1 ACCOAmit +1 NADHmit + 1 CO2
PDA1, 2, PDB

(22) 1 PYR + 1 ATP + 1 CO2 $\rightarrow$ 1 OAC +1 ADP
PYC1, 2

(23) 1 OAC + 1 ATP $\rightarrow$ 1 PEP + 1 ADP +1 CO2
PCK1E

3. Pentose Phosphate Pathway

(24) 1 GLUC6P + 1 NADPcyt $\rightarrow$ 1 G15L +1 NADPHcyt
ZWF1

(25) 1 G15L $\rightarrow$ 1 P6G
SOL1, 2, 3, 4

(26) 1 P6G + 1 NADPcyt$\rightarrow$ 1 RIBL5P +1 NADPHcyt + 1 CO2
GND1, 2

(27) 1 RIBL5P $\leftrightarrow$ 1 RIB5P
RKI1

(28) 1 RIBL5P ↔ 1 XYL5P
RPE1

(29) 1 RIB5P + 1 XYL5P ↔1 SED7P +1 GA3P
TKL, TKI

(30) 1 SED7P + 1 GA3P ↔ 1 FRUC6P +1 E4P
TAL1

(31) 1 XYL5P + 1 E4P↔ 1 FRUC6P +1 GA3P
TKI, TKL

4. Citric Acid Cycle

(32) 1 OAC + 1 ACCOAmit → 1 CIT
CIT1, 3

(33) 1 CIT ↔ 1 ISOCIT
ACO1

(34) 1 ISOCIT + 1 NADmit→ 1 AKG +1 NADHmit + 1 CO2
IDH1, 2

(35) 1 ISOCIT + 1 NADPmit → 1 AKG +1 NADPHmit + 1 CO2
IDP1

(36) 1 ISOCIT + 1 NADPcyt → 1 AKG +1 NADPHcyt + 1 CO2
IDP2

(37) 1 AKG + 1 NADmit → 1 SUCCOA +1 NADHmit + 1 CO2
KGD1, 2

(38) 1 SUCCOA + 1 ADP↔ X 1 SUC +1 ATP
LSC1, 2

(39) 1 SUC + 1 FAD → 1 FUM +1 FADH2
SDH1, 2, 3

(40) 1 FUM + 1 FADH2 → 1 SUC +1 FAD
OSM1

(41) 1 FUM ↔ 1 MAL
FUM1

(42) 1 MAL + 1 NADmit ↔ 1 OAC +1 NADHmit
MDH1

(43) 1 MAL + 1 NADPmit → 1 PYR +1 CO2 + 1 NADPHmit
MAE1

(44) 1 ACCOAcyt → 1 ACCOAmit
CAT2

(45) 1 NADHcyt + 1 NADmit →1 NADcyt +1 NADHmit
ShuttleX

5. Glyoxylate Shunt

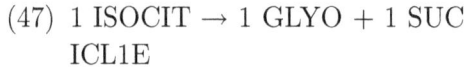

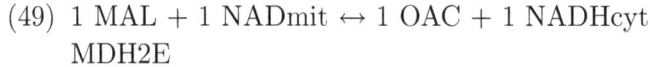

(46) 1 OAC + 1 ACCOAcyt → 1 CIT
CIT2E
(47) 1 ISOCIT → 1 GLYO + 1 SUC
ICL1E
(48) 1 GLYO + 1 ACCOAcyt → 1 MAL
MLS1E
(49) 1 MAL + 1 NADmit ↔ 1 OAC + 1 NADHcyt
MDH2E

6. Oxidative Phosphorylation

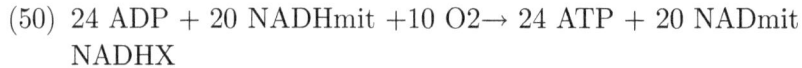

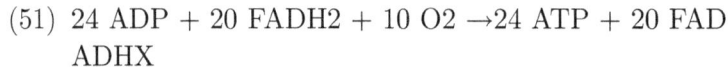

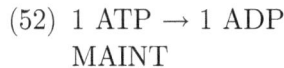

(50) 24 ADP + 20 NADHmit +10 O2→ 24 ATP + 20 NADmit
NADHX
(51) 24 ADP + 20 FADH2 + 10 O2 →24 ATP + 20 FAD
ADHX
(52) 1 ATP → 1 ADP
MAINT

7. Biomass Formation

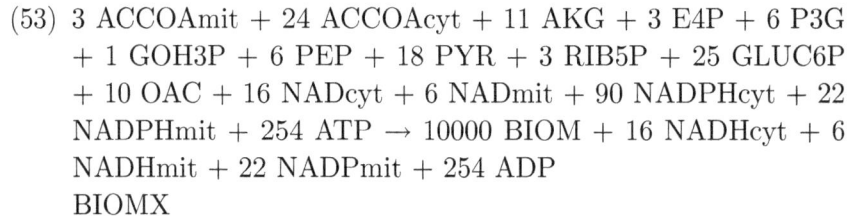

(53) 3 ACCOAmit + 24 ACCOAcyt + 11 AKG + 3 E4P + 6 P3G
+ 1 GOH3P + 6 PEP + 18 PYR + 3 RIB5P + 25 GLUC6P
+ 10 OAC + 16 NADcyt + 6 NADmit + 90 NADPHcyt + 22
NADPHmit + 254 ATP → 10000 BIOM + 16 NADHcyt + 6
NADHmit + 22 NADPmit + 254 ADP
BIOMX

Let GLC=9.5 and oxygen=12.9. Perform FBA on the basis of maximization of cellular biomass production. The resulting distribution of metabolic flux is indicted in Fig. 3.

Suppose the PDB gene in the yeast cell mutates and inactivates, or it is knocked out. The corresponding metabolic flux distribution,

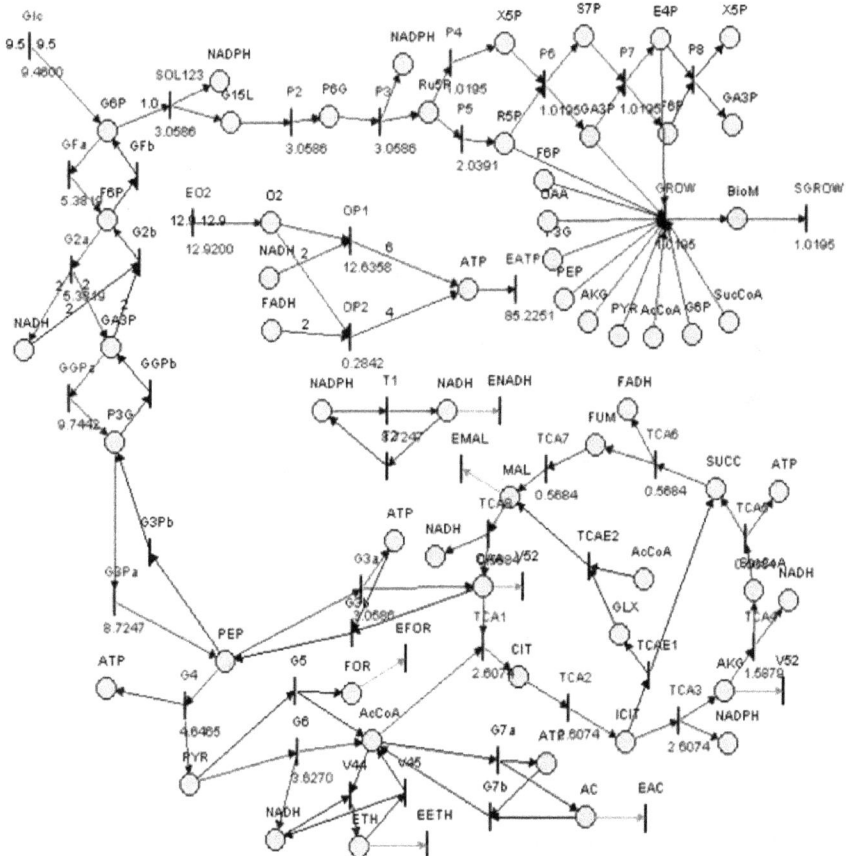

Fig. 3 Flux distribution map of *S. cerevisiae* cell's FBA model under the over-grow constraint (Luo, 2007).

obtained using several methods described above, is shown in Figs. 4–6.

15.1.3.2 *FBA of ecological networks*

Here are some of the advanced models or methods on flow analysis of ecological networks (Libralato *et al.*, 2006; Zhang, 2012e).

(1) Ecopath Model

Ecopath is the free software to construct and analyze mass-balanced flow networks (Zhang, 2012e). It is the software based on a method

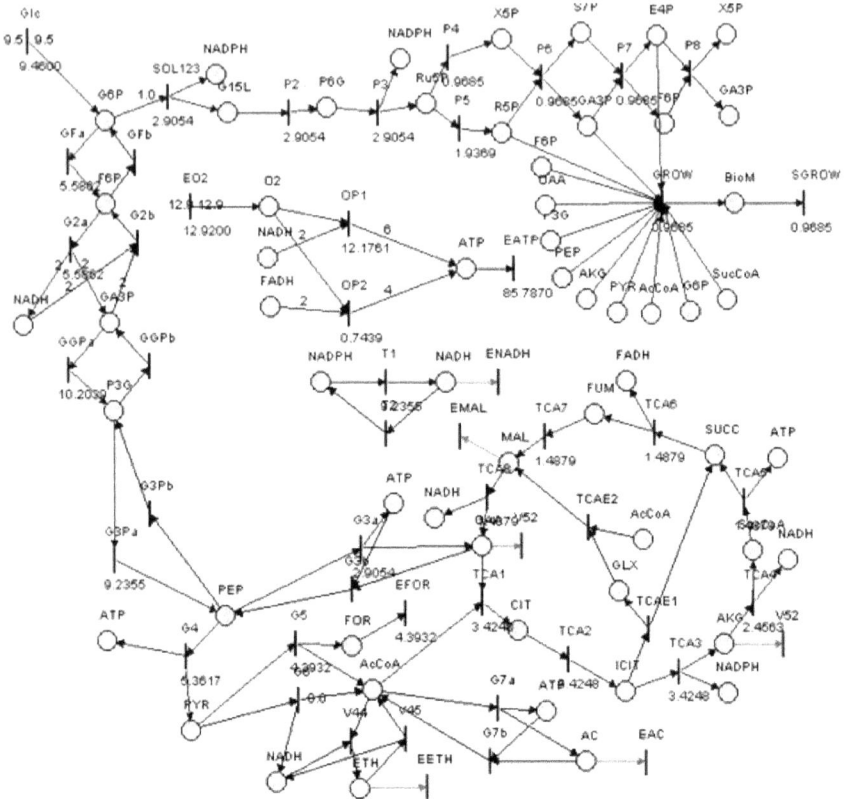

Fig. 4 Flux distribution network of *S. cerevisiae* cell's FBA model under the overgrow constraint if the PDB gene is knocked out (Luo, 2007).

proposed by Polovina (1984), and was upgraded later with some more methods (Christensen and Pauly, 1992; Walters *et al.*, 1997, 2000; Pauly *et al.*, 2000; Christensen and Walters, 2004). Actually, Ecopath is the core program of the Ecopath with Ecosim (EwE) (Libralato *et al.*, 2006). The interactions in the network represent trophic links at the species or functional level. It allows users to input ecosystem data such as total mortality estimates, consumption estimates, diet compositions, and fishery catchers (Christensen and Walters, 2004). The model uses two main equations, one regarding production (production = catch + predation + net migration + biomass accumulation + other mortality), and the other regarding

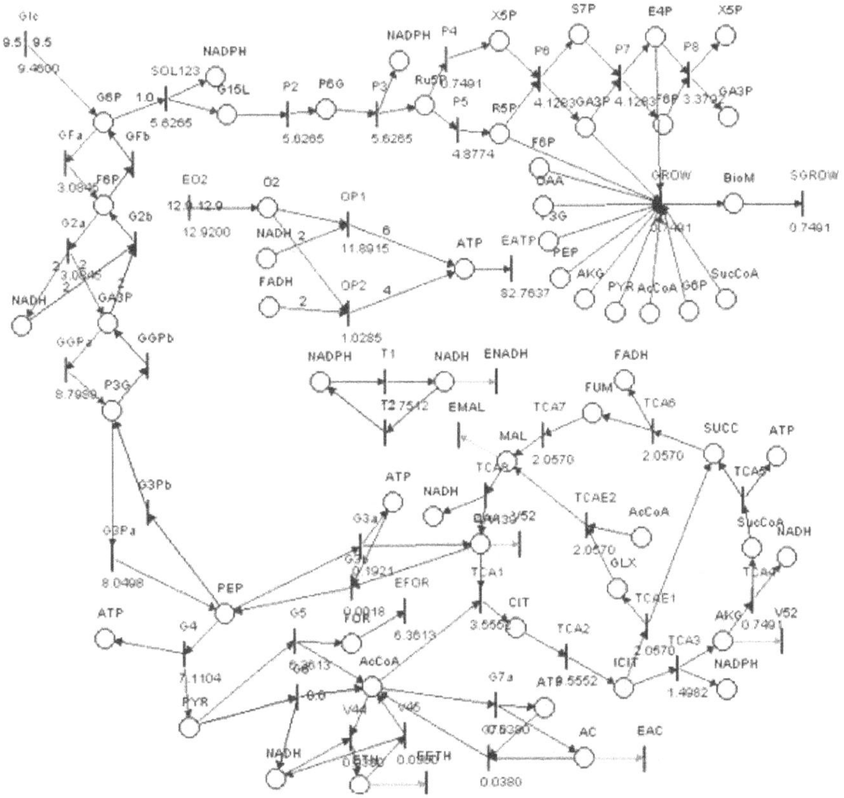

Fig. 5 Flux distribution network of *S. cerevisiae* cell's MOMA model if the PDB gene is knocked out (Luo, 2007).

consumption(consumption = production + respiration + unassimilated food).

According to Libralato *et al.* (2006), the balance of mass (or energy or nutrients) of any functional group i in the network can be achieved by setting its production equal to the sum of the consumption terms as follows:

$$(P/B)_i B_i = \sum_{j=1}^{n} (Q/B)_j B_j DC_{ij} + E_i + Y_i + BA_i$$

$$+ (P/B)_i B_i (1 - EE_i)$$

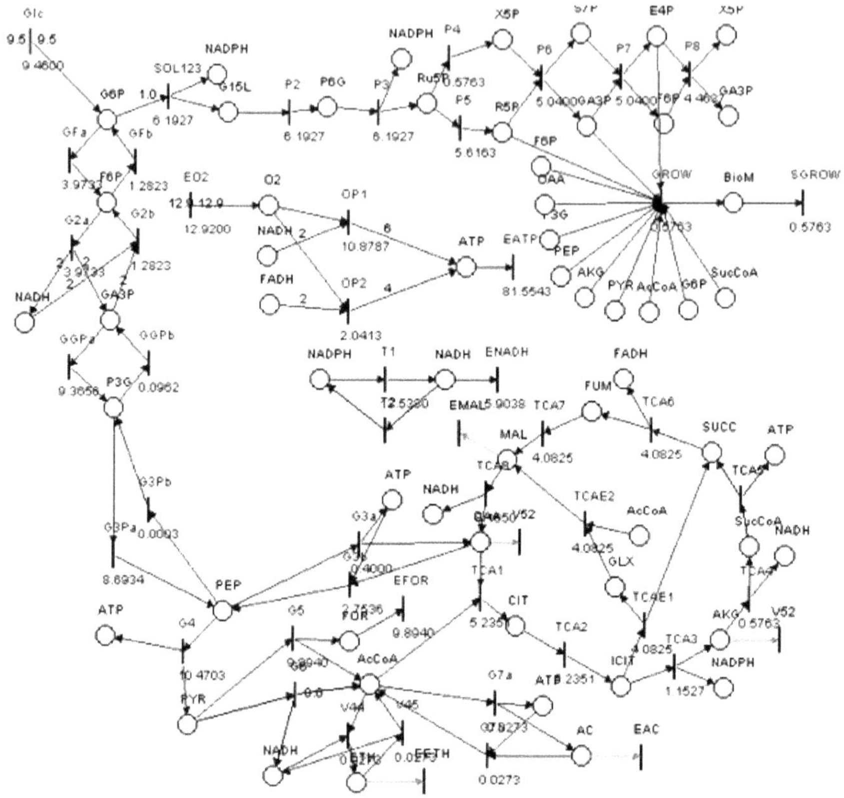

Fig. 6 Flux distribution map of *S. cerevisiae* cell's MOMA model under the overgrow constraint if the PDB gene is knocked out (Luo, 2007).

where the production (i.e., the left side of the equation) is the product of the production *vs.* biomass ratio, $(P/B)_i$, and the biomass, B_i ; the right-hand side terms are the sum of the predation terms: the product of the consumption *vs.* biomass ratio, $(Q/B)_j$, the biomass of the predators B_j, and the proportion of the prey i in the diet of the predator j, DC_{ij}; the net flow through the boundaries of the system, E_i; the fishing exploitation, Y_i; the accumulation or depletion of biomass, BA_i; and the last term, nonpredation natural mortality, where EE_i is the ecotrophic efficiency.

The flows within a trophic web can be achieved by solving the equations above.

Given the mass balance model of a trophic network, the mixed trophic impact of each pair of functional groups (i, j) in the network, m_{ij} (i.e., direct and indirect impacts that each group i has on any group j of the network. It is the first-order partial derivative in terms of biomass of the master equation above), can be estimated based on the net impact matrix (Libralato *et al.*, 2006). Positive (negative) values of m_{ij} denote the increase (decrease) of biomass of the group j due to a slight increase of biomass of the group i.

First, the net impact of i on j, q_{ij}, is given by the difference between positive effects, quantified by the fraction of the prey i in the diet of the predator j, d_{ji}, and negative effects, evaluated through the fraction of total consumption of i used by predator j, f_{ij} (Ulanowicz and Puccia, 1990). By doing so, the resulting matrix of the net impacts, $Q = (q_{ij})$ is

$$q_{ij} = d_{ji} - f_{ij}$$

The mixed trophic impact m_{ij} is the product of all the net impacts q_{ij} for all the possible pathways in the network that link the functional groups i and j. As an easier way, the matrix of the mixed trophic impacts M, can be obtained by the inverse of the matrix Q (Ulanowicz and Puccia, 1990).

According to Libralato *et al.* (2006), the negative elements in matrix M indicate a prevailing of negative effects, or the effects of the predator on the prey, and positive elements in M indicate prevailing effects of the prey on the predator. Negative elements in M can thus be associated with prevailing top-down effects and positive ones to bottom-up effects.

(2) Ecosim Method

Ecosim is the dynamic program of the EwE (Walters *et al.*, 1997, 2000). It is based on a group of differential equations derived from the Ecopath equation above. This allows a dynamic representation of the system variables like biomasses, predation, production, etc., (Libralato *et al.*, 2006). Ecosim uses Lotka–Volterra models that have been modified to include foraging arena theory (Walters *et al.*, 1997, 2000), which avoids Lotka–Volterra's unrealistic assumption of uniform or random distribution of species interactions. In the foraging arena theory, the biomass of the prey available to predators is only a vulnerable fraction of total biomass, with exchanges rates between the vulnerable and the invulnerable states calculated using vulnerability coefficients (Christensen and Walters, 2004; Libralato *et al.*, 2006; Zhang, 2012e).

The Ecosim master equation takes the following form (Pinnegar *et al.*, 2005)

$$dB_i/dt = g_i \sum_{j=1}^{n} Q_{ji} - \sum_{j=1}^{n} Q_{ij} + I_i - (M_i + F_i + e_i)B_i$$

where dB_i/dt is the growth rate of group i during the time interval dt, g_i is the net growth efficiency (production/consumption ratio), M_i is the nonpredation natural mortality rate, F_i the fishing mortality rate, e_i the emigration rate, I_i the immigration rate. In the equation, consumption rates Q_{ji}, are calculated based on a "foraging foraging arena" idea, in which the B_i's are divided into vulnerable and invulnerable components (Walters *et al.*, 1997). The size of the transfer rate v_{ij}, i.e., the rate at which prey move between being "vulnerable" and "not vulnerable", determines if control is top-down, bottom-up or an intermediate type.

(3) Ecoexergy

Jorgensen and Fath (2006) concluded that some network changes are important for the power and ecoexergy of the network; however, changes that only move energy flows from one place to another are not important. Changes that increase flows always enhance the power and the ecoexergy. Natural selection of ecological networks should be further studied using various methods (Park *et al.*, 2006).

Some advances achieved in the past years include (Editorial, 2007): (1) increased network input leads to a proportional increase in ecoexergy and power with unchanged network structure; (2) additional network links will only influence the power and ecoexergy, if it gives additional exergy or energy transfer; (3) length increase in the food chain has positive effect on the power and ecoexergy of the network; (4) reduction of loss of exergy to the environment by respiration or as detritus produces a higher power and ecoexergy of the network; (5) faster cycling, through either faster detritus decomposed or increased transfer rates between two tropic levels, yields higher power and exergy, and (6) input of additional exergy or energy cycling flows has greater effect if the addition takes place earlier in the food chain.

15.2 Flow Indices for Ecological Network Analysis

Some standardized indices and matrices (Latham, 2006; Zhang, 2012e) for ecological network analysis are described as below.

15.2.1 *Ascendency*

Ascendency index (A) was proposed by Ulanowicz and Norden (1990) and Ulanowicz (1997)

$$A = \text{AMI}^* T_{..}$$

AMI was believed to increase during a system's development toward a climax community, due to the pruning of less efficient pathways in the network and the increasing autocatalytic between-compartment interconnection (Ulanowicz, 1980, 1997).

A suggested measure for comparing ascendency across networks is the development ascendency, which is represented as (Latham, 2006)

$$T_{..} = -A \bigg/ \left[\sum_{i=1}^{n+2} \sum_{j=0}^{n} T_{ij} \log_2(T_{ij}/T_{..}) \right]$$

Ulanowicz and Norden (1990) presented four measures to analyze specific aspects of networks (Latham, 2006) as follows:

Network import

$$A_0 = \sum_{i=1}^{n} T_{i0} \log_2(T_{i0} T_{..}/(T_{i.} T_{.0}))$$

$$\Phi_0 = -\sum_{i=1}^{n} T_{i0} \log_2(T_{i0}^2/(T_{i.} T_{.0}))$$

$$C_0 = -\sum_{i=1}^{n} T_{i0} \log_2(T_{i0}/T_{..})$$

Network internal

$$A_{\mathrm{I}} = \sum_{i,j=1}^{n} T_{ij} \log_2(T_{ij}T_{..}/(T_{i.}T_{.j}))$$

$$\Phi_{\mathrm{I}} = - \sum_{i,j=1}^{n} T_{ij} \log_2(T_{ij}^2/(T_{i.}T_{.j}))$$

$$C_{\mathrm{I}} = - \sum_{i,j=1}^{n} T_{ij} \log_2(T_{ij}/T_{..})$$

Network export

$$A_{\mathrm{E}} = \sum_{j=1}^{n} T_{n+1j} \log_2(T_{n+1j}T_{..}/(T_{n+1.}T_{.j}))$$

$$\Phi_{\mathrm{E}} = - \sum_{j=1}^{n} T_{n+1j} \log_2(T_{n+1j}^2/(T_{n+1.}T_{.j}))$$

$$C_{\mathrm{E}} = - \sum_{j=1}^{n} T_{n+1j} \log_2(T_{n+1j}/T_{..})$$

Network dissipation

$$A_{\mathrm{S}} = \sum_{j=1}^{n} T_{n+2j} \log_2(T_{n+2j}T_{..}/(T_{n+2.}T_{.j}))$$

$$\Phi_{\mathrm{S}} = - \sum_{j=1}^{n} T_{n+2j} \log_2(T_{n+2j}^2/(T_{n+2.}T_{.j}))$$

$$C_{\mathrm{S}} = - \sum_{j=1}^{n} T_{n+2j} \log_2(T_{n+2j}/T_{..})$$

15.2.2 *Average mutual information*

Average Mutual Information (AMI) (Rutledge *et al.*, 1976) is a measure for the average amount of constraint placed on a unit of flow anywhere in the network (Latham and Scully, 2002; Latham, 2006).

An assumption for AMI is that rigid diets and highly constrained flow in an ecosystem may result in a fragile system unable to persist in the changing environmental conditions (Rutledge *et al.*, 1976). As a system grows and matures to form a network of energy and material flow, the AMI should be (Latham, 2006)

$$\text{AMI} = k \sum_{i=1}^{n+2} \sum_{j=0}^{n} (T_{ij}/T_{..}) \log_2 (T_{ij} T_{..}/(T_{i.} T_{.j}))$$

where n: the number of nodes or compartments in the network, but does not include the compartment 0, $n+1$, or $n+2$; k: constant scalar; T_{ij}: the flow from compartments j to i; $T_{i.}$: total inflow for compartment i; $T_{.j}$: total outflow for compartment j; $T_{..}$: total system throughput, i.e., the sum of network links as below

$$T_{..} = \sum_{i=1}^{n+2} \sum_{j=0}^{n} T_{ij}$$

The upper bound of AMI is

$$H_R = - \sum_{j=0}^{n} (T_{.j}/T_{..}) \log_2 (T_{.j}/T_{..})$$

AMI/H_R is most useful for comparing the degree of constraint across systems (Latham, 2006).

15.2.3 *Compartmentalization index*

Compartmentalization index (C) is a measure for the degree of well connected subsystems (which should have similar number of species and connectance) in a network (Pimm and Lawton, 1980)

$$C = \left[\sum_{i=1}^{n} \sum_{\substack{j \neq i \\ j = 1}}^{n} c_{ij}/(n(n-1)) \right]$$

where c_{ij}: the number of species with which both i and j interact divided by the number of species with which either i or j interact.

A higher value of C means the higher compartmentalization of the network.

15.2.4 *Constraint efficiency*

Constraint efficiency (CE) is a measure of a total of constraints that govern flow out of individual compartments (Latham and Scully, 2002)

$$CE = \left[\sum_{i=1}^{n} \log_2(n+2) + \sum_{i=1}^{n+2} \sum_{j=1}^{n} (T_{ij}/T_{\cdot j}) \log_2(T_{ij}/T_{\cdot j}) \right] \Big/ \sum_{i=1}^{n} \log_2(n+2)$$

CE is the sum of all constraints. Thus, local changes of network topography will be directly represented by the system-wide value of CE (Latham and Scully, 2002; Latham, 2006).

15.2.5 *Effective measures*

Zorach and Ulanowicz (2003) presented some measures for weighted networks.

Effective connectivity

$$C_z = \prod_{i,j=1}^{n} (T_{ij}^2/(T_{i\cdot}T_{\cdot j}))^{-(0.5T_{ij}/T_{\cdot\cdot})}$$

Effective connectivity is similar to link density in unweighted networks. However, it has weight for the size of links.

Effective flows

$$F_z = \prod_{i,j=1}^{n} (T_{ij}/T_{\cdot\cdot})^{-T_{ij}/T_{\cdot\cdot}}$$

Effective flows is similar to the number of links in unweighted networks.

Effective nodes

$$N_z = \prod_{i,j=1}^{n} (T^2/(T_{i\cdot}T_{\cdot j}))^{(0.5T{ij}/T_{\cdot\cdot})}$$

Effective nodes is the weighted mean of the normalized throughflow of each node.

Effective roles

$$R_z = \prod_{i,j=1}^{n} (T_{ij}T_{\cdot\cdot}/(T_{i\cdot}T_{\cdot j}))^{T{ij}/T_{\cdot\cdot}}$$

Effective roles is a measure of weighted, differentiated, distinct functions in a network, i.e., sectors that uniquely take input and pass matter/energy to one destination (Latham, 2006).

15.2.6 *Cycling index*

Cycling index was proposed by Finn (1976, 1978, 1980) and Patten *et al.* (1976). It is a matrix procedure for calculating the proportion of cycled throughflow in a system. There are three versions of cycling indices. One of the versions is

$$\text{FCI} = \text{TST}_c/\text{TST}$$

where TST: total network compartmental throughflows; TST_c: total system cycled throughflow (Latham, 2006).

15.2.7 *Homogenization index*

Homogenization index was developed by Fath and Patten (1999a) to measure network homogenization. It measures the evenness of flow in a network. Homogenization degree is determined by comparing the evenness of flow in a matrix accounting cycling and direct flow against the normalized direct flow matrix (Latham, 2006). Cycling generally increases the evenness of flow in a network. Cycling and homogenization are thus related (Fath and Patten, 1999a)

$$\mathrm{CV_N} = \left[\sum_{i,j=1}^{n} (m^{\mathrm{N}} - m_{ij}^{\mathrm{N}})^2 / (n^2 - 1) \right]^{0.5} \Big/ m^{\mathrm{N}}$$

$$\mathrm{CV_G} = \left[\sum_{i,j=1}^{n} (m^{\mathrm{G}} - m_{ij}^{\mathrm{G}})^2 / (n^2 - 1) \right]^{0.5} \Big/ m^{\mathrm{G}}$$

where $\mathrm{CV_N}$: homogenization for the integral matrix (including cycled flow); $\mathrm{CV_G}$: homogenization for the direct flow matrix; $m^{\mathrm{N}}, m^{\mathrm{G}}$: average element from the integral flow matrix N and direct flow matrix G, respectively; $m_{ij}^{\mathrm{N}}, m_{ij}^{\mathrm{G}}$: elements from the integral flow matrix N and direct flow matrix G, respectively.

A lower CV means an even flow in the matrix. In addition, homogenization occurs if $\mathrm{CV_G} > \mathrm{CV_N}$ (Latham, 2006).

15.2.8   *Dominance index of indirect effects*

Higashi and Patten (1986, 1989) and Fath and Patten (1999b) proposed an index to describe the dominance of indirect effects

$$i/d = \sum_{i,j=1}^{n} (n_{ij} - \delta_{ij} - g_{ij}) \Big/ \sum_{i,j=1}^{n} g_{ij}$$

The denominator denotes the direct processes and the numerator of the right term denotes the indirect processes. The dominance degree of indirect processes increases with the index value.

15.2.9   *Amplification matrix and utility matrix*

Amplification matrix is a binary matrix to describe which links are providing more energy/matter than they are directly capable of producing (Fath and Patten, 1999b; Latham, 2006). In the matrix, elements have a value of 1, and any element off of the diagonal in the $(I - G)^{-1}$ matrix has a value greater than 1 — these are the links in which amplification occur.

Both utility matrix and synergism (b/c) are used to describe the degree of positive interactions in the network (Fath and Patten, 1999b). Utility matrix is represented by

$$(I - D_P)^{-1}$$

where $D_P = (d_{ij})$ is nondimensional direct flow-based utility matrix, $d_{ij} = (T_{ij} - T_{ji})/T_i$. The positive elements of the matrix denote positive interactions (mutualism) and negative elements denote negative interactions (competition). The b/c index is the ratio of positive to negative interactions.

Part 5
Link and Node Prediction

Chapter 16

Link Prediction:
Sampling-based Methods

Theoretically, there is no (either direct or indirect) interaction between two nodes if they do not correlate with each other. Conversely, an interaction may likely exist if there are some correlation(s) between two nodes. A correlation can be a linearly, pseudo-linearly, or nonlinearly dependent mathematical relationship. Whether an interaction is direct or indirect should be further determined according to the property of the correlation. In most cases, the quantitative correlation between two nodes can be estimated by statistically sampling two nodes. In the present chapter, various sampling-based methods for link prediction (Zhang, 2015c, 2015d; Zhang and Li, 2015a, 2015b) are described. Here, the links and interactions have the same meaning. It should be noted that most of these methods are also used to construct biological networks. In a sense, the sampling-based network construction is highly similar to the sampling-based link prediction.

16.1 Linear Correlation Analysis for Finding Interactions

Researchers usually use Pearson linear correlation to find interactions (Goh, *et al.*, 2000; Pazos and Valencia, 2001; Tu, 2006; Zhang and Li, 2015b). In a series of earlier studies (Zhang, 2007, 2011c, 2012b, 2012e), the methodology for constructing ecological networks by correlation analysis of community sampling data have been proposed. I have mentioned that a statistically significant Pearson linear correlation means an indirect or direct interaction (link) between two taxa (i.e., nodes), and a statistically significant partial (net, or pure) correlation based on Pearson linear correlation means a candidate direct interaction between two taxa (e.g., species, families). For the studies of ecological communities and ecosystems, the interactions refer to predation, parasitism, competition, amensalism, mutualism, protocooperation, commensalism, etc. Two taxa may interact by acting on the same resource, or by changing the environment of opposite sides, etc. An interaction represents an interdependency relationship in state changes of two taxa (direct interaction). Conversely, a seemingly dependent relationship in state changes of two taxa does not necessarily mean an interaction (i.e., it is an indirect interaction).

Some nondeterministic interactions may be found and some candidate direct interactions may be missed when using this method, as pointed out by Zhang (2011c) and Zhang and Li (2015b).

16.1.1 *Method*

As pointed out by Zhang and Li (2015b), a network can be a linear network, quasi-linear network, or nonlinear network. Moreover, a network that changes in a local domain (a short time, or a small extent, etc.) can be approximated with a linear network, i.e., in the local domain, all between-node changes can be treated as linear ones. That is, suppose x_i is the state of node i, $i = 1, 2, \ldots, m$. We have (Zhang and Li, 2015b)

$$dx_i(t)/dt = a_{ij}dx_j(t)/dt, \quad i, j = 1, 2, \ldots, m$$

or

$$dx_i(l)/dl = a_{ij}dx_j(l)/dl, \quad i, j = 1, 2, \ldots, m$$

where t: time; l: space length; a_{ij}: constants, $i, j = 1, 2, \ldots, m$, and $a_{ii} = 1$, $i = 1, 2, \ldots, m$. Under these situations, linear correlation measures can be used.

Pearson linear correlation is the most often used measure to describe the linear dependence between two taxa. A statistically significant Pearson linear correlation manifests a direct or indirect linear interaction between two taxa. Partial (net, or pure) linear correlation is based on Pearson linear correlation. It has eliminated the indirect effects produced by the remaining taxa. A statistically significant partial linear correlation represents a candidate direct linear interaction between two taxa. Here, we treat the linear interactions, predicted by partial linear correlation, as candidate direct interactions (Zhang and Li, 2015b).

The following are Matlab codes for calculation and statistical test of Pearson linear correlation and partial linear correlation, and for finding interactions (Zhang and Li, 2015b):

```
% X is m*n raw data matrix. m: number of taxa; n: number of samples.
str=input('Input the file name of raw data matrix (e.g., raw.txt, raw.xls,
etc. The file has m rows (taxa) and n columns (samples)):','s');
X=load(str);
sig=input('Input significance level(e.g., 0.01):');
dim=size(X);
m=dim(1); n=dim(2);
r=corr(X');
disp('Correlation matrix')
r
tvalues=abs(r)./sqrt((1-r.^2)/(n-2));
alpha=(1-tcdf(tvalues,n-2))*2;
sigmat=alpha<sig;
sigmat=sigmat.*r-eye(m);
sigmatr=sigmat;
disp('Pairs with statistically significant correlation')
if (sigmat~=ones(m))
[pairx,pairy,rvalues]=find(sigmat);
temp1=pairx; temp2=pairy;
pairxs=pairx(temp1<temp2);
pairys=pairy(temp1<temp2);
rvaluess=rvalues(temp1<temp2);
PairsAndCorrelations=[pairxs pairys rvaluess]
else
disp('No significant pairs')
end
inversr=inv(r);
for i=1:m-1; for j=i+1:m;
parr(i,j)=-inversr(i,j)/sqrt(inversr(i,i)*inversr(j,j));end;end;
for i=1:m-1; for j=i+1:m; parr(j,i)=parr(i,j);end;end;
for i=1:m; parr(i,i)=1;end;
disp('Partial correlation matrix')
```

```
parr
if (n>m)
tvalues=abs(parr)./sqrt((1-parr.^2)/(n-m));
alpha=(1-tcdf(tvalues,n-m))*2;
else
disp('The number of samples is not enough to support the required statistic
test (DF=n-m) of partial correlations. Here use the statistic test with
DF=n-2 (not recommended). Please input the proportion of statistically
significant pairs based on DF=n-m vs. statistically significant pairs based
on DF=n - 2 (y, %) as the following. The estimation formula, y=88.748exp
(-0.045m), is suggested for use (Zhang WJ. 2015. Selforganizology, 2(4):
55-67). If it is hard to be estimated, the full percent, 100, can be
input.')
y=input('Input the proportion (a value between 0 and 100):')
tvalues=abs(parr)./sqrt((1-parr.^2)/(n-2));
alpha=(1-tcdf(tvalues,n-2))*2;
end
sigmat=alpha<sig;
sigmat=sigmat.*parr-eye(m);
if (n<=m) threshr=rrank(sigmat,y); sigmat=sigmat>=threshr; sigmat=sigmat.
*parr;
end;
sigmatparr=sigmat;
disp('Pairs with statistically significant partial correlation')
if (sigmat =ones(m))
[pairx,pairy,rvalues]=find(sigmat);
temp1=pairx; temp2=pairy;
pairxs=pairx(temp1<temp2);
pairys=pairy(temp1<temp2);
rvaluess=rvalues(temp1<temp2);
PairsAndPartialCorrelations=[pairxs pairys rvaluess]
else
disp('No significant pairs')
end
x=sigmatparr & (~sigmatr);
y=(~sigmatparr) & sigmatr;
z=sigmatparr & sigmatr;
for i=1:3;
switch i
    case 1
        mat=x; s='Significant partial correlation but insignificant linear
correlation';
    case 2
        mat=y; s='Significant linear correlation but insignificant partial
correlation';
    case 3
        mat=z; s='Significant both partial correlation and linear
correlation';
end;
[pairx,pairy]=find(mat);
```

```
temp1=pairx; temp2=pairy;
pairxs=pairx(temp1<temp2);
pairys=pairy(temp1<temp2);
disp([s])
SignificantPairs=[pairxs pairys]
end;
```

The M function file, rrank.m, is as follows:

```
function threshr = rrank(mat,percent)
dim=size(mat); m=dim(1);
len=(m*m-m)/2;
vec=zeros(1,len);
n=0;
for i=1:m-1; for j=i+1:m;
if (mat(i,j)~=0) n=n+1; vec(n)=mat(i,j); end;
end; end;
num=round(percent/100*n);
vecc=sort(vec,'descend');
if (num~=0) threshr=vecc(num); else threshr=1;
end;
```

16.1.2 *Application example*

Data of various biological networks were obtained from that of Zhang (2011c). These biological networks are different in countries, years, seasons, types of taxa, and number of taxa. Therefore, we expect the wide representativeness of conclusions drawn from them.

The results show that partial linear correlation (y) is approximately half the Pearson linear correlation (x)

$$y = -0.0064 + 0.4785x \ (r^2 = 0.173, p < 0.00001, n = 1447)$$

As indicated in Table 1, in all predicted candidate interactions based on partial linear correlation, about 34.35% (x, 0–100%) of them are not successfully detected by linear correlation. In all predicted interactions by Pearson linear correlation, 50.58% (y, 0–100%) of them are nondeterministic interactions, i.e., not successfully detected by partial linear correlation (Zhang and Li, 2015b). In all predicted interactions by Pearson linear correlation, 49.42% (z, 0–100%) of them are candidate direct interactions, i.e., successfully detected by partial linear correlation also (Fig. 1).

Results show that the precisely predicted (z) candidate direct interactions by Pearson linear correlation analysis are not necessarily

Table 1 Comparison of results of Pearson linear correlation (PLC) and partial linear correlation (Partial PLC) (Zhang and Li, 2015b).

Network ID (Data set)	No. Taxa (m)	No. Partial PLC > PLC	No. Absolute Partial PLC > Absolute PLC (NAP)	NAP/N (%)	No. SS Yes Partial PLC but SS Not PLC (SSN)	$x = SSN/(SSN+SYY)$ (%)	No. SS Yes PLC but SS Not Partial PLC (SPN)	$y = SPN/(SPN+SYY)$ (%)	No. SS Yes PLC & SS Yes Partial PLC (SYY)	$z = SYY/(SPN+SYY)$ (%)
CN-06Sep	4	0	0	0.00	1	100	1	50.00	1	50.00
CN-06Sep	4	2	3	50.00	2	100	0	0.00	2	100.00
CN-06Oct	4	2	3	50.00	2	100	1	50.00	1	50.00
PH-Mar	21	99	126	60.00	0	0	6	66.67	3	33.33
PH-Apr	20	57	106	55.79	4	33.33	14	63.64	8	36.36
PH-Sep	21	70	122	58.10	1	25	10	76.92	3	23.08
PH-Oct	21	98	100	47.62	0	0	8	72.73	3	27.27
PH-Mar	7	10	12	57.14	1	50	0	0.00	1	100.00
PH-Apr	7	4	6	28.57	0	0	5	71.43	2	28.57
PH-Sep	7	5	7	33.33	0	0	5	83.33	1	16.67
PH-Oct	7	11	13	61.90	0	0	0	0.00	2	100.00
CN-06Sep	23	120	174	68.77	3	25	10	52.63	9	47.37
CN-06Oct	23	116	127	50.20	1	14.29	10	62.50	6	37.50
CN-06Oct	27	168	248	70.66	5	33.33	14	58.33	10	41.67
Mean						34.35		50.58		49.42
± ($p \leq 0.05$)						77.76		58.19		58.19

Notes: No. SS Yes Partial PLC but SS Not PLC: Total No. of statistically significant Partial PLC ($p \leq 0.01$) but statistically not significant PLC ($p \leq 0.01$); No. SS Yes PLC but SS Not Partial PLC: Total No. of statistically significant PLC ($p \leq 0.01$) but statistically not significant Partial PLC ($p \leq 0.01$); No. SS Yes PLC & SS Yes Partial PLC: Total No. of both statistically significant PLC ($p \leq 0.01$) and statistically significant Partial PLC ($p \leq 0.01$).

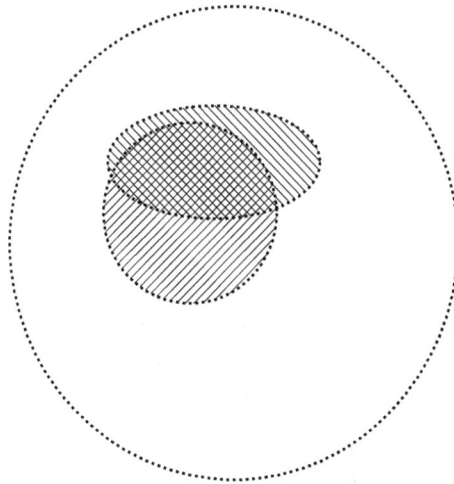

Fig. 1 Proportion of maximal possible direct interactions (inside outer circle), direct interactions detected by both linear correlation measures (⊠, 49.42%), direct interactions detected by partial linear correlation but not Pearson linear correlation (◺, 34.35%), and direct interactions detected by Pearson linear correlation but not partial linear correlation (◹, 50.58%) (Zhang and Li, 2015b).

those with the highest Pearson linear correlations. For example, the predicted interactions by Pearson linear correlation analysis (and Pearson linear correlations in parentheses) of network CN-06Sep (taxon: family) are as follows, of which italic interactions are candidate direct interactions (Zhang and Li, 2015b):

$(1,2)(0.3962)$

$(2,3)(0.4021)$

$(3,4)(0.3877)$

$(4,10)(0.3531)$ $(4,16)(0.4379)$ $(4,18)(0.4954)$

$(5,12)(0.4095)$ $(5,13)(0.6502)$

$(6,17)(0.43)$

$(8,11)(0.4017)$ $(8,12)(0.5605)$

$(10,12)(0.3611)$ $(10,18)(0.4883)$

$(11,12)(0.3784)$ $(11,18)(0.451)$

(12,13)(0.4703)

(14,22)(0.4201)

(16,21)(0.3541)

(17,20)(0.43)

Therefore, we should not try to choose a portion (e.g., 49.42% (z), as calculated above) of predicted interactions with the greatest Pearson linear correlations as candidate direct interactions.

The proportions of missed (x), mis-predicted (y), and precisely predicted (z) candidate direct interactions by Pearson linear correlation analysis have significant biological meaning, for example, if we want to detect candidate direct interactions between species by Pearson linear correlation analysis, a false conclusion may sometimes be drawn (y) (Zhang and Li, 2015b). For example, Tu (2006) found that some true interactions (i.e., the interactions already confirmed by experiments. A candidate direct interaction is a true interaction if it has been confirmed by experiments) between proteins are not statistically significant Pearson linear correlation. In medical science (biological control, biodiversity conservation), assume there is a true direct interaction between two proteins (a predator and a prey, two species), A and B. We want to design a medicine (release A, release A) to affect A or B (to control B, to balance B) directly, and hope that the measure will make A affect (control, balance) B. However, we know that A may likely not affect (control, balance) B due to the proportion x (medicine failure, failure of biological control, failure of balancing plan).

According to Tu (2006), the proportions of missed (x) and precisely predicted (z) true direct interactions by Pearson linear correlation analysis in small-cell lung cancer (SCLC; $m = 19$ protein sequences) were 25% and 75%, respectively. For non-small-cell lung cancer (NSCLC; $m = 92$ protein sequences), the proportions of missed (x) and precisely predicted (z) true direct interactions by Pearson linear correlation analysis were 42.5% and 57.5%, respectively. The results (means of SCLC and NSCLC: $x = 33.8\%, z = 66.3\%$) coincide better with our conclusions ($x \approx 34.35\%; y \approx 49.42\%$). The results of Tu indicated that true direct interactions by Pearson linear correlation analysis are not necessarily those with the highest Pearson linear correlations, which confirmed our conclusion also.

In ecological networks (systems), the phenomena above are popular. For example, it is well known that ladybird *Coccinella septempunctata* is the natural enemy of cotton aphid *Aphis gossypii* Glover, and thus there is a true (direct) interaction between the two species. However, in some cotton fields, the ladybird lacks prey species and mainly feeds on the cotton aphid, and the Pearson linear correlation between them is thus likely significant (the interaction belongs to z). In other cotton fields, the ladybird may have many available prey species, and so the Pearson linear correlation between them is thus likely insignificant (the interaction belongs to x). In this situation, the ladybird, as a natural enemy of cotton aphid, is ineffective.

16.2   Partial Correlation of General Correlation Measures

To silence the effect of indirect between-attribute (e.g., between-node) interactions of other attributes (e.g., nodes) not being tested, the partial linear correlation of Pearson linear correlation measure can be used. Unfortunately, for most of the correlation measures, partial correlation measures are not available. This section introduces partial correlation measures proposed for some correlation measures (Zhang, 2015d).

16.2.1   *Methods*

16.2.1.1   *Correlation measures*

There are a variety of correlation measures (Zhang, 2007, 2011c, 2012b, 2012e; Zhang *et al.*, 2014b). Furthermore, Zhang (2015d) proposed three correlation measures, revised Dice coefficient, overlap coefficient, and proportion correlation, and used more correlation measures to derive their partial correlations, with the exception of Pearson linear correlation which has been affiliated with partial linear correlation, of which Jaccard correlation, revised Dice coefficient, overlap coefficient, and point correlation are for binary attributes (e.g., nodes, taxa, variables, etc.), and Spearman rank correlation and proportion correlation are for interval value attributes (e.g., nodes, taxa, variables, etc.).

Suppose there are m attributes (e.g., nodes, taxa, variables, etc.) and n samples (e.g., components, members, etc.), and the raw data is a matrix (x_{ij}), $i = 1, 2, \ldots, m; j = 1, 2, \ldots, n$.

(1) Jaccard correlation

$$r_{ij} = (e - (c + b))/(e + c + b), \quad i, j = 1, 2, \ldots, m$$

where $-1 \leq r_{ij} \leq 1$, c is the number of sample pairs of 1 for attribute i but not for j; b is the number of sample pairs of 1 for attribute j but not for i; e is the number of sample pairs of 1 for both attribute i and attribute j.

(2) Revised Dice coefficient

$$r_{ij} = 4^* e/(b + c) - 1, \quad i, j = 1, 2, \ldots, m$$

where $-1 \leq r_{ij} \leq 1$, b is the number of sample of 1 for attribute i; c is the number of sample of 1 for attribute j; e is the number of sample pairs of 1 for both attribute i and attribute j.

(3) Point correlation

$$r_{ij} = (ad - bc)/((a + b)(c + d)(a + c)(b + d))^{1/2},$$
$$i, j = 1, 2, \ldots, m$$

where $-1 \leq r_{ij} \leq 1$, both attribute i and attribute j take values 0 or 1. a is the number of both attribute i and attribute j take value 0, b is number of attribute i taking 0 and attribute j taking 1, c is number of attribute i taking 1 and attribute j taking 0, and d is number of both attribute i and attribute j taking value 1.

(4) Overlap coefficient

$$r_{ij} = 2e/(e + b) - 1, \quad i, j = 1, 2, \ldots, m$$

where $-1 \leq r_{ij} \leq 1$, b is the number of sample pairs with different values; e is the number of sample pairs of 1 for both attribute i and attribute j.

(5) Proportion correlation

$$r_{ij} = 2 \sum_{k=1}^{n} (p_{ik}p_{jk}) \Bigg/ \left(\sum_{k=1}^{n} p_{ik}^2 \sum_{k=1}^{n} p_{jk}^2 \right)^{1/2} - 1, \quad i,j = 1,2,\ldots,m$$

where $-1 \le r_{ij} \le 1$, and $p_{il} = x_{il}/\sum_{k=1}^{n} x_{ik}$, $p_{jl} = x_{jl}/\sum_{k=1}^{n} x_{jk}$, $l = 1,2,\ldots,n$. It is used specifically for the situations of $r_{ij} \ge 0, i = 1,2,\ldots,m; \ j = 1,2,\ldots,n$.

(6) Spearman rank correlation (Spearman, 1904; Schoenly and Zhang, 1999)

$$r_{ij} = 1 - 6^* \sum d^2/[n(n^2-1)]i, \quad j = 1,2,\ldots,m$$

where $-1 \le r_{ij} \le 1, d = r(i) - r(j)$, and $r(i)$ and $r(j)$ are rank of an element in the attribute i and attribute j, from the smaller to the larger values in n samples.

For the correlation measures, calculate $t = |r_{ij}|/[(1 - r_{ij}^2)/(n - 2)]^{1/2}$, and if $t > t_\alpha(n-2)$, the correlation (either positive or negative correlation) between attributes i and j is statistically significant.

16.2.1.2 *Partial correlations*

The partial correlation between two attributes eliminates the indirect effects produced by the remaining attributes. Let $R = (r_{ij})$, where r_{ij} are correlations calculated above. Calculate

$$\sum_{k=1}^{m} r_{ik}r_{kj}^- = 0, \quad i,j = 1,2,\ldots,m; \ i \ne j$$

$$\sum_{k=1}^{m} r_{ik}r_{ki}^- = 1, \quad i = 1,2,\ldots,m$$

That is, calculate $R^- = (r_{ij}^-)$ such that $RR^- = I$ and I is the unit matrix. Partial correlation between attributes i and j is

$$\mathrm{parr}_{ij} = -r_{ij}^-/(r_{ii}^{-*}r_{jj}^-)^{1/2}, \quad i,j = 1,2,\ldots,m; \ i \ne j$$

where $-1 \le \mathrm{parr}_{ij} \le 1$. A singular or ill-condition matrix, R, will result in wrong results.

For the partial correlation measures, calculate $t = |\text{parr}_{ij}|/$ $[(1 - \text{parr}_{ij}^2)/(n - m)]^{1/2}$, and if $t > t_\alpha(n - m)$, the partial correlation (either positive or negative correlation) between attributes i and j is statistically significant, and there is a direct interaction/interdependence between attributes i and j in terms of the correlation measure being used.

The Matlab codes of the full algorithm above and of finding interactions are listed as follows, of which the Pearson linear correlation measure (for interval value attributes) is included also (Zhang, 2015d):

```
str=input('Input the file name of raw data matrix (e.g., raw.txt, raw.xls,
etc. The file has m rows (taxa) and n columns
(samples)):','s');
X=load(str);
measure=input('Input a number for correlation measure (1: Pearson linear
correlation; 2: Spearman rank correlation; 3: Proportion correlation; 4:
Point correlation; 5: Jaccard correlation; 6: Revised Dice coefficient; 7:
Overlap coefficient)');
sig=input('Input significance level(e.g., 0.01)');
%X is m*n raw data matrix. m: number of attributes (taxa, etc.); n: number
of samples.
dim=size(X);
m=dim(1); n=dim(2);
r=zeros(m);
switch measure
    case 1
        for i=1:m-1; for j=i+1:m;  r(i,j)=linearcorre(X(i,:)',X(j,:)');
            end;end;
    case 2
        for i=1:m-1; for j=i+1:m;  r(i,j)=spearman(X(i,:)',X(j,:)');end;end;
    case 3
        for i=1:m-1; for j=i+1:m;  r(i,j)=propcorre(X(i,:)',X(j,:)');end;end;
    case 4
        for i=1:m-1; for j=i+1:m;  r(i,j)=pointcorre(X(i,:)',X(j,:)');end;end;
    case 5
        for i=1:m-1; for j=i+1:m;  r(i,j)=jaccard(X(i,:)',X(j,:)');end;end;
    case 6
        for i=1:m-1; for j=i+1:m;  r(i,j)=reviseddice(X(i,:)',X(j,:)');
            end;end;
    case 7
        for i=1:m-1; for j=i+1:m;  r(i,j)=overlap(X(i,:)',X(j,:)');end;end;
end
for i=1:m-1; for j=i+1:m;  r(j,i)=r(i,j);end;end;
for i=1:m; r(i,i)=1;end;
disp('Correlation matrix')
r
```

```
tvalues=abs(r)./sqrt((1-r.^2)/(n-2));
alpha=(1-tcdf(tvalues,n-2))*2;
sigmat=alpha<sig;
sigmat=sigmat.*r-eye(m);
sigmatr= sigmat;
disp('Pairs with statistically significant correlation')
if (sigmat =ones(m))
[pairx,pairy,rvaluess]=find(sigmat);
temp1=pairx; temp2=pairy;
pairxs=pairx(temp1<temp2);
pairys=pairy(temp1<temp2);
rvaluess=rvalues(temp1<temp2);
PairsAndCorrelations=[pairxs pairys rvaluess]
else
disp('No significant pairs')
end
inversr=inv(r);
for i=1:m-1; for j=i+1:m;
parr(i,j)=-inversr(i,j)/sqrt(inversr(i,i)*inversr(j,j));end;end;
for i=1:m-1; for j=i+1:m; parr(j,i)=parr(i,j);end;end;
for i=1:m; parr(i,i)=1;end;
disp('Partial correlation matrix')
parr
if (n>m)
tvalues=abs(parr)./sqrt((1-parr.^2)/(n-m));
alpha=(1-tcdf(tvalues,n-m))*2;
else
disp('The number of samples is not enough to support the required statistic
test (DF=n-m) of partial correlations. Here use the statistic test with
DF=n-2 (not recommended). Please input the proportion of statistically
significant pairs based on DF=n-m vs. statistically significant pairs based
on DF=n - 2 (y, %) as the following. For Pearson linear correlation measure,
the estimation formula, y=88.748 exp(-0.045m), is suggested for use, and for
Spearman rank correlation measure, y=120.687exp(-0.045m), is suggested,
where m is the number of attributes (taxa, etc.). If it is hard to be
estimated, the full percent, 100, can be input.')
y=input('Input the proportion (a value between 0 and 100)')
tvalues=abs(parr)./sqrt((1-parr.^2)/(n-2));
alpha=(1-tcdf(tvalues,n-2))*2;
end
sigmat=alpha<sig;
sigmat=sigmat.*parr-eye(m);
if (n<=m) threshr=rrank(sigmat,y); sigmat=sigmat>=threshr; sigmat=sigmat.
   *parr;
end;
sigmatparr= sigmat;
disp('Pairs with statistically significant partial correlation')
if (sigmat~=ones(m))
[pairx,pairy,rvalues]=find(sigmat);
temp1=pairx; temp2=pairy;
```

```
pairxs=pairx(temp1<temp2);
pairys=pairy(temp1<temp2);
rvaluess=rvalues(temp1<temp2);
PairsAndPartialCorrelations=[pairxs pairys rvaluess]
else
disp('No significant pairs')
end
x=sigmatparr & (∼sigmatr);
y=(∼sigmatparr) & sigmatr;
z=sigmatparr & sigmatr;
for i=1:3;
switch i
    case 1
        mat=x; s='Significant partial correlation but insignificant linear
correlation';
    case 2
        mat=y; s='Significant linear correlation but insignificant partial
correlation';
    case 3
        mat=z; s='Significant both partial correlation and correlation';
end;
[pairx,pairy]=find(mat);
temp1=pairx; temp2=pairy;
pairxs=pairx(temp1<temp2);
pairys=pairy(temp1<temp2);
disp([s])
SignificantPairs=[pairxs pairys]
end;
```

Eight Matlab function files, linearcorre.m, spearman.m, prop-corre.m, pointcorre.m, jaccard.m, reviseddice.m, and overlap.m are as follows:

```
function linearcor = linearcorre(x,y)     %x and y: two column vectors to be
    tested.
m=max(size(x));
if (m∼=max(size(y)))
  error('Array sizes do not match.');
end
if ((min(size(x))∼=1) | (min(size(y))∼=1))
  error('Both x and y are vectors');
end
xbar=mean(x);
ybar=mean(y);
aa=sum((x-xbar).*(y-ybar));
bb=sum((x-xbar).^2);
cc=sum((y-ybar).^2);
linearcor=aa/sqrt(bb*cc);

function spearm =spearman(x,y)     %x and y: two column vectors to be tested.
if (max(size(x))∼=max(size(y)))
```

```
    error('Array sizes do not match.');
end
if ((min(size(x))~=1) | (min(size(y))~=1))
    error('Both x and y are vectors');
end
n=max(size(x));
for i=1:n
rx(i)=0;ry(i)=0;xx(i)=0;yy(i)=0;
end
for j=1:n
nx=1;ny=1;
for i=1:n
if (x(i)<x(j)) nx=nx+1; end
if (y(i)<y(j)) ny=ny+1; end
end
rx(j)=nx;
ry(j)=ny;
end
for j=1:n
if (rx(j)==(n+1)) continue; end
nx=rx(j);
ntie=-1;
for i=1:n
if (rx(i)~=nx) continue; end
ntie=ntie+1;
xx(i)=rx(i);
rx(i)=0;
end
for i=1:n
if (rx(i)~=0) continue; end
xx(i)=xx(i)+(ntie*0.5);
rx(i)=n+1;
end
end
for j=1:n
if (ry(j)==(n+1)) continue; end
ny=ry(j);
ntie=-1;
for i=1:n
if (ry(i)~=ny) continue; end
ntie=ntie+1;
yy(i)=ry(i);
ry(i)=0;
end
for i=1:n
if (ry(i)~=0) continue; end
yy(i)=yy(i)+ntie*0.5;
ry(i)=n+1;
end
end
```

```
rs=0;
rs=sum((xx-yy).^2);
spearm=1-((6*rs)/(n*(n^2-1)));

function prop = propcorre(x,y)   %x and y: two column vectors to be tested.
m=max(size(x));
if (m~=max(size(y)))
   error('Array sizes do not match.');
end
if ((min(size(x))~=1) | (min(size(y))~=1))
   error('Both x and y are vectors');
end
xp=x/sum(x);
yp=y/sum(y);
aa=sum(xp.*yp);
bb=sum(xp.^2);
cc=sum(yp.^2);
prop=2*aa/sqrt(bb*cc)-1;

function pointcor = pointcorre(x,y)   %x and y: two column vectors to be
   tested.
if (max(size(x))~=max(size(y)))
   error('Array sizes do not match.');
end
if ((min(size(x))~=1) | (min(size(y))~=1))
   error('Both x and y are vectors');
end
aa=sum((x==0) & (y==0));
bb=sum((x==0) & (y~=0));
cc=sum((x~=0) & (y==0));
dd=sum((x~=0) & (y~=0));
pointcor=(aa*dd-bb*cc)/sqrt((aa+bb)*(cc+dd)*(aa+cc)*(bb+dd));

function jac = jaccard(x,y)       %x and y: two column vectors to be tested.
if (max(size(x))~=max(size(y)))
   error('Array sizes do not match.');
end
if ((min(size(x))~=1) | (min(size(y))~=1))
   error('Both x and y are vectors');
end
bb=sum((x==0) & (y~=0));
cc=sum((x~=0) & (y==0));
dd=sum((x~=0) & (y~=0));
jac=(dd-(cc+bb))/(dd+cc+bb);

function dic = reviseddice(x,y)   %x and y: two column vectors to be tested.
if (max(size(x))~=max(size(y)))
   error('Array sizes do not match.');
end
if ((min(size(x))~=1) | (min(size(y))~=1))
```

```
    error('Both x and y are vectors');
end
dd=sum((x~=0) & (y~=0));
bb=sum(x~=0);
cc=sum(y~=0);
dic=4*dd/(bb+cc)-1;

function overl = overlap(x,y)    %x and y: two column vectors to be tested.
if (max(size(x))~=max(size(y)))
    error('Array sizes do not match.');
end
if ((min(size(x))~=1) | (min(size(y))~=1))
    error('Both x and y are vectors');
end
bb=sum(x~=y);
dd=sum((x~=0) & (y~=0));
overl=2*dd/(dd+bb)-1;

function threshr = rrank(mat,percent)
dim=size(mat); m=dim(1);
len=(m*m-m)/2;
vec=zeros(1,len);
n=0;
for i=1:m-1;
for j=i+1:m;
if (mat(i,j)~=0) n=n+1; vec(n)=mat(i,j); end;
end;
end;
num=round(percent/100*n);
vecc=sort(vec,'descend');
if (num~=0) threshr=vecc(num);
else threshr=1;
end;
```

16.2.2 *Application examples*

16.2.2.1 *Interval value measures*

An example data matrix of interval values with eight attributes and 15 samples is (Zhang, 2015d)

8.086	9.237	4.679	7.099	3.494	4.031	7.806	8.979	9.522	8.973	4.173	4.657	6.372	2.269	2.666
9.263	3.459	9.396	8.262	0.779	4.744	8.577	3.713	10.097	2.649	8.391	6.628	5.198	6.423	9.539
4.801	2.963	8.367	0.335	1.715	8.500	3.183	6.501	5.818	9.800	6.831	1.292	0.000	3.548	2.698
9.029	7.547	10.988	4.680	8.904	8.480	7.240	2.471	1.818	3.454	5.803	0.432	3.340	7.943	8.667
6.929	10.521	7.038	7.137	7.891	2.905	7.887	8.060	1.722	2.432	9.061	5.016	2.952	0.418	5.178
1.730	5.232	0.000	2.721	3.334	5.610	2.029	3.682	3.150	3.825	3.399	8.984	8.920	7.983	10.128
3.489	8.489	8.913	7.107	10.699	4.618	1.490	5.042	8.660	6.084	8.527	9.167	3.053	5.302	1.999
5.579	1.624	9.796	0.000	0.981	6.859	5.168	7.071	3.519	7.133	8.417	1.106	0.738	2.701	3.171

(1) Spearman rank correlation

The calculated Spearman rank correlations between attributes are

1	0.0036	0.1679	−0.4357	0.175	−0.3964	−0.0036	0.1393
0.0036	1	0.0143	0.1143	−0.1714	−0.3	−0.1571	0.1929
0.1679	0.0143	1	0.1714	−0.1357	−0.2821	0.0321	0.9214
−0.4357	0.1143	0.1714	1	0.1643	−0.3071	−0.1	0.2107
0.175	−0.1714	−0.1357	0.1643	1	−0.325	0.1429	0.0536
−0.3964	−0.3	−0.2821	−0.3071	−0.325	1	−0.175	−0.3393
−0.0036	−0.1571	0.0321	−0.1	0.1429	−0.175	1	−0.0464
0.1393	0.1929	0.9214	0.2107	0.0536	−0.3393	−0.0464	1

The statistically significant attribute pair and corresponding Spearman rank correlation are as follows ($p < 0.01$):

3   8   0.9214

Partial correlation of Spearman rank correlation is

1	0.0634	0.314	−0.7077	0.3017	−0.5402	−0.357	−0.2754
0.0634	1	−0.6399	0.0795	−0.6029	−0.3583	0.0274	0.6346
0.314	−0.6399	1	0.268	−0.7346	−0.0889	0.3137	0.9652
−0.7077	0.0795	0.268	1	0.2899	−0.4961	−0.3631	−0.2199
0.3017	−0.6029	−0.7346	0.2899	1	−0.2061	0.2744	0.6994
−0.5402	−0.3583	−0.0889	−0.4961	−0.2061	1	−0.317	0.0476
−0.357	0.0274	0.3137	−0.3631	0.2744	−0.317	1	−0.3126
−0.2754	0.6346	0.9652	−0.2199	0.6994	0.0476	−0.3126	1

and the statistically significant attribute pair and corresponding partial correlation of Spearman rank correlation is ($p < 0.01$)

3   8   0.9652

(2) Proportion correlation

Using proportion correlation measure, the calculated proportion correlations between attributes are

1	0.6866	0.602	0.5116	0.7181	0.408	0.6786	0.5388
0.6866	1	0.5153	0.6725	0.612	0.5036	0.5921	0.5836
0.602	0.5153	1	0.5541	0.4076	0.1481	0.5426	0.9324
0.5116	0.6725	0.5541	1	0.6947	0.3927	0.5809	0.599
0.7181	0.612	0.4076	0.6947	1	0.312	0.705	0.5056
0.408	0.5036	0.1481	0.3927	0.312	1	0.4377	0.078
0.6786	0.5921	0.5426	0.5809	0.705	0.4377	1	0.4533
0.5388	0.5836	0.9324	0.599	0.5056	0.078	0.4533	1

The statistically significant attribute pairs and corresponding proportion correlations are as follows ($p < 0.01$):

1	2	0.6866
2	4	0.6725
1	5	0.7181
4	5	0.6947
1	7	0.6786
5	7	0.7050
3	8	0.9324

Partial correlations of proportion correlations are

1	0.6115	0.7121	−0.4884	0.751	0.0237	−0.3648	−0.6351
0.6115	1	−0.612	0.4499	−0.4902	0.3324	0.4215	0.6454
0.7121	−0.612	1	0.4058	−0.7929	0.1334	0.6794	0.9611
−0.4884	0.4499	0.4058	1	0.5797	0.1557	−0.228	−0.2963
0.751	−0.4902	−0.7929	0.5797	1	−0.0116	0.6816	0.7432
0.0237	0.3324	0.1334	0.1557	−0.0116	1	0.0612	−0.2382
−0.3648	0.4215	0.6794	−0.228	0.6816	0.0612	1	−0.6406
−0.6351	0.6454	0.9611	−0.2963	0.7432	−0.2382	−0.6406	1

and the statistically significant attribute pair and corresponding partial correlation of proportion correlation is ($p < 0.01$)

3   8   0.9611

16.2.2.2   *Binary (Boolean) value measures*

The example data matrix of binary values with eight attributes and 15 samples is

1	1	0	1	0	0	1	1	1	1	0	0	1	0	0
1	0	1	0	0	0	1	0	1	0	1	1	0	1	1
0	0	0	1	0	1	0	1	0	1	1	0	0	0	0
1	1	1	0	1	0	1	0	0	0	1	0	0	1	1
0	1	0	1	1	0	1	1	0	0	1	1	0	0	0
1	0	1	0	0	0	0	0	0	1	0	0	1	1	1
0	1	0	0	0	0	0	1	0	0	1	1	0	0	0
1	0	1	0	0	1	0	1	0	0	1	0	1	0	0

The calculated point correlations between attributes are

1.0000	−0.3393	0.0945	−0.3393	0.0714	−0.0546	−0.0403	−0.0546
−0.3393	1.0000	−0.4725	0.4643	−0.1964	0.2182	−0.0403	−0.0546
0.0945	−0.4725	1.0000	−0.4725	0.1890	−0.2887	0.2132	0.2887
−0.3393	0.4643	−0.4725	1.0000	0.0714	0.2182	−0.0403	−0.0546
0.0714	−0.1964	0.1890	0.0714	1.0000	−0.7638	0.6447	−0.2182
−0.0546	0.2182	−0.2887	0.2182	−0.7638	1.0000	−0.4924	0.1667
−0.0403	−0.0403	0.2132	−0.0403	0.6447	−0.4924	1.0000	0.1231
−0.0546	−0.0546	0.2887	−0.0546	−0.2182	0.1667	0.1231	1.0000

The statistically significant attribute pairs and corresponding point correlations are as follows ($p < 0.01$):

5	6	−0.7638
5	7	0.6447

Partial correlations of point correlations are

1.0000	−0.1882	−0.1497	−0.3168	0.1984	0.1549	−0.1232	0.0246
−0.1882	1.0000	−0.2813	0.3081	−0.2678	−0.1202	0.2186	−0.0692
−0.1497	−0.2813	1.0000	−0.3430	0.1231	−0.0425	−0.0123	0.3384
−0.3168	0.3081	−0.3430	1.0000	0.5152	0.4144	−0.2422	0.1883
0.1984	−0.2678	0.1231	0.5152	1.0000	−0.6919	0.5796	−0.3316
0.1549	−0.1202	−0.0425	0.4144	−0.6919	1.0000	0.1016	−0.0049
−0.1232	0.2186	−0.0123	−0.2422	0.5796	0.1016	1.0000	0.3555
0.0246	−0.0692	0.3384	0.1883	−0.3316	−0.0049	0.3555	1.0000

No significant attribute pairs were found with partial correlations of point correlations.

16.2.3   *Comparison of statistic test of partial correlation with DF = n − m and DF = n − 2*

Comparisons between statistic tests of partial correlations with DF = $n − m$ and DF = $n − 2$, based on the data sets described in Zhang (2011c) were made ($p < 0.01$) (Zhang, 2015d). Results of statistic tests with different degrees of freedom (DFs) are somewhat different, but the statistically significant pairs with DF= $n − m$ fall in the scope of that with DF = $n − 2$. Thus, for nonstrict uses, the number of samples can be less than the number of attributes (i.e., $n < m$) and the statistic test with DF = $n − 2$ can be carefully used in this situation.

The proportion of statistically significant pairs based on DF = $n − m$ *vs.* statistically significant pairs based on DF = $n − 2$ (y, %) is related to significance level p (in present study, $p < 0.01$) and the number of attributes (m). For this case study, the regression relationships are as follows (Zhang, 2015d)

$$y = 88.748e^{-0.045\,m}, \quad r^2 = 0.579, \quad p = 0.047$$
$$\text{(for Pearson partial linear correlation)}$$

Table 2 Relationship between the proportion of statistically significant pairs based on DF = nm *vs.* statistically significant pairs based on DF = $n - 2$ $(y, \%)$ and number of taxa (m).

No. Taxa (m)	10	20	30	40	50	60	70	80	90	100	150	200
For Pearson partial linear correlation $(y, \%)$	56.59	36.08	23.01	14.67	9.35	5.96	3.80	2.42	1.55	0.99	0.10	0.01
For Spearman partial rank correlation $(y, \%)$	76.52	48.79	31.11	19.84	12.65	8.06	5.14	3.28	2.09	1.33	0.14	0.01

$$y = 120.687e^{-0.045\,m}, \quad r^2 = 0.956, \quad p = < 0.00001$$

(for partial Spearman rank correlation)

As an example, if Pearson linear correlation is used and $m = 50$, we have $y = 9.4\%$. That is, the first approximately 9.4% of statistically significant pairs based on DF = $n - 2$, which have the greatest partial correlations, should be the statistically significant pairs based on DF = $n - m$. Table 2 lists the $y - m$ relationship based on the two regression equations above.

16.3 Combined Use of Linear Correlation and Rank Correlation: A Hierarchical Method for Finding Interactions

As stated earlier, in ecological communities and ecosystems, there are generally two types of interactions, direct interactions and indirect interactions. Direct interactions refer to predation, parasitism, competition, amensalism, mutualism, protocooperation, commensalism, etc., (Zhang, 2015c). Two taxa may interact by acting on the same resource, or by changing the environment of opposite sides, etc. Definition of interactions in other biological networks (metabolic regulatory networks, cancer networks, etc.) can be found in related references. In this section, I describe an earlier work on joint use of Pearson linear correlation measure and Spearman rank correlation measure and their partial correlations for finding interactions (Zhang, 2015c).

We have known that a network changed in a local domain can be approximated with a linear network, i.e., in the local domain, all between-node (or -taxon, -component, etc.,) changes are treated as linear ones, and Pearson linear correlation measure can be used (Zhang, 2015d). For a wider domain, the quasi-linear measure, Spearman rank correlation (Spearman, 1904; Schoenly and Zhang, 1999) can be used also. We can jointly use Pearson linear correlation measure and Spearman rank correlation measure, and their partial correlations to find interactions. First, define the sets of interactions as follows (Zhang, 2015c).

A: Interactions detected by statistically significant partial correlations for Pearson linear correlation measure;
B: Interactions detected by statistically significant partial correlations for Spearman rank correlation measure;
C: Interactions detected by statistically significant Pearson linear correlations;
D: Interactions detected by statistically significant Spearman rank correlations.

Some principles are followed in developing the rules for finding interactions: (1) Pearson linear correlation measure is more reliable than Spearman rank correlation measure; (2) partial correlation measure is more reliable than correlation measure; (3) joint use of Pearson linear correlation measure and Spearman rank correlation measure is more reliable than the single use of two measures, and (4) joint use partial correlation measure and correlation measure is more reliable than the single use.

According to the above principles and the rules of set operation, the reliability levels (basically, from the set of most reliable interactions, L1, to the sets of most unreliable interactions, L61 and L62) are defined as follows (Zhang, 2015c)

L1: A∩B∩C∩D;	(candidate direct interactions)
L21: A∩C; L22: B∩D;	(candidate direct interactions)
L3: A∩B;	(candidate direct interactions)
L41: A; L42: B;	(candidate direct interactions)
L5: C∩D;	(indirect/direct interactions)
L61: C; L62: D.	(indirect/direct interactions)

Known the raw data is a matrix X, with m rows, i.e., m attributes (e.g., taxa, proteins, genes, etc.), and n columns, i.e., n samples, the Matlab codes of the full algorithm above are as follows (Zhang, 2015c):

```
% X is the m*n raw data matrix. m: number of attributes (taxa, proteins,
genes, % etc.); n: number of samples.
str=input('Input the file name of raw data matrix (e.g., raw.txt, raw.xls,
etc. The file has m rows (taxa) and n columns (samples)):','s');
X=load(str);
sig=input('Input significance level (e.g., 0.01)');
dim=size(X);
m=dim(1); n=dim(2);
if (n<=m)
disp('The number of samples is not enough to support the required statistic
test (DF=n-m) of partial correlations. Here use the statistic test with
DF=n-2 (not recommended). Please input the proportion of statistically
significant pairs based on DF=n-m vs. statistically significant pairs based
on DF=n-2 (y, %) as the following. For Pearson linear correlation measure,
the estimation formula, y1=88.748 exp(-0.045m), is suggested for use, and
for Spearman rank correlation measure, y2=120.687exp(-0.045m), is suggested,
where m is the number of attributes (taxa, proteins, genes, etc.). If it is
hard to be estimated, the full percent, 100, can be input.')
y1=input('Input the proportion for Pearson linear correlation measure
    (a value between 0 and 100):')
y2=input('Input the proportion for Spearman rank correlation measure
    (a value between 0 and 100):')
end;
r=corr(X');
disp('Pearson linear correlation matrix')
r
for i=1:m-1; for j=i+1:m;
spr(i,j)=spearman(X(i,:)',X(j,:)');end;end;
for i=1:m-1; for j=i+1:m; spr(j,i)=spr(i,j);end;end;
for i=1:m; spr(i,i)=1;end;
disp('Spearman rank correlation matrix')
spr
for k=1:2;
if (k==1) rr=r; else rr=spr; end;
inverse=inv(rr);
for i=1:m-1; for j=i+1:m;
parr(i,j)=-inverse(i,j)/sqrt(inverse(i,i)*inverse(j,j));end;end;
for i=1:m-1; for j=i+1:m; parr(j,i)=parr(i,j);end;end;
for i=1:m; parr(i,i)=1;end;
if (k==1)
disp('Partial correlation matrix for Pearson linear correlation')
peparr=parr;
peparr
else
disp('Partial correlation matrix for Spearman rank correlation')
spparr=parr;
spparr
```

```
end;
end;
for k=1:2;
if (k==1)
tvalues=abs(r)./sqrt((1-r.^2)/(n-2));
alpha=(1-tcdf(tvalues,n-2))*2;
sigmat=alpha<sig;
sigmatr=sigmat.*r;
if (n>m)
partvalues=abs(peparr)./sqrt((1-peparr.^2)/(n-m));
paralpha=(1-tcdf(partvalues,n-m))*2;
else
partvalues=abs(peparr)./sqrt((1-peparr.^2)/(n-2));
paralpha=(1-tcdf(partvalues,n-2))*2;
end
parsigmat=paralpha<sig;
parsigmatr=parsigmat.*peparr;
if (n<=m) threshr=rrank(parsigmatr,y1); parsigmatr=parsigmatr>=threshr;
parsigmatr=parsigmatr.*peparr; end;
else
tvalues=abs(spr)./sqrt((1-spr.^2)/(n-2));
alpha=(1-tcdf(tvalues,n-2))*2;
spsigmat=alpha<sig;
spsigmatr=spsigmat.*spr;
if (n>m)
partvalues=abs(spparr)./sqrt((1-spparr.^2)/(n-m));
paralpha=(1-tcdf(partvalues,n-m))*2;
else
partvalues=abs(spparr)./sqrt((1-spparr.^2)/(n-2));
paralpha=(1-tcdf(partvalues,n-2))*2;
end;
spparsigmat=paralpha<sig;
spparsigmatr=spparsigmat.*spparr;
if (n<=m) threshr=rrank(spparsigmatr,y2); spparsigmatr=spparsigmatr>=threshr;
spparsigmatr=spparsigmatr.*spparr; end;
end;
end;
L1=(parsigmatr & spparsigmatr & sigmatr & spsigmatr);    % candidate direct
    interactions
L21=(parsigmatr & sigmatr); L22=(spparsigmatr & spsigmatr);    % candidate
    direct interactions
L3=(parsigmatr & spparsigmatr);    % candidate direct interactions
L41=parsigmatr; L42=spparsigmatr;    % candidate direct interactions
L5=(sigmatr & spsigmatr);    % indirect/direct interactions
L61=sigmatr; L62=spsigmatr;    % indirect/direct interactions
for i=1:9;
switch (i)
  case 1
      mat=L1; s='L1';
```

```
   case 2
      mat=L21; s='L21';
   case 3
      mat=L22; s='L22';
   case 4
      mat=L3; s='L3';
   case 5
      mat=L41; s='L41';
   case 6
      mat=L42; s='L42';
   case 7
      mat=L5; s='L5';
   case 8
      mat=L61; s='L61';
   case 9
      mat=L62; s='L62';
end;
disp(['Interactions (the interactions found in higher levels are also
listed) with statistically significance for reliability level:' s])
[pairx,pairy]=find(mat);
temp1=pairx; temp2=pairy;
pairxs=pairx(temp1<temp2);
pairys=pairy(temp1<temp2);
InteractionPairs=[pairxs pairys]
end;
```

Two M function files, spearman.m, and rrank.m, are as follows:

```
function spearm =spearman(x,y)     %x and y: two column vectors to be tested.
if (max(size(x))~=max(size(y)))
   error('Array sizes do not match.');
end
if ((min(size(x))~=1) | (min(size(y))~=1))
   error('Both x and y are vectors');
end
n=max(size(x));
for i=1:n
rx(i)=0;ry(i)=0;xx(i)=0;yy(i)=0;
end
for j=1:n
nx=1;ny=1;
for i=1:n
if (x(i)<x(j)) nx=nx+1; end
if (y(i)<y(j)) ny=ny+1; end
end
rx(j)=nx;
ry(j)=ny;
end
for j=1:n
if (rx(j)==(n+1)) continue; end
nx=rx(j);
```

```
ntie=-1;
for i=1:n
if (rx(i)~=nx) continue; end
ntie=ntie+1;
xx(i)=rx(i);
rx(i)=0;
end
for i=1:n
if (rx(i)~=0) continue; end
xx(i)=xx(i)+(ntie*0.5);
rx(i)=n+1;
end
end
for j=1:n
if (ry(j)==(n+1)) continue; end
ny=ry(j);
ntie=-1;
for i=1:n
if (ry(i)~=ny) continue; end
ntie=ntie+1;
yy(i)=ry(i);
ry(i)=0;
end
for i=1:n
if (ry(i)~=0) continue; end
yy(i)=yy(i)+ntie*0.5;
ry(i)=n+1;
end
end
rs=0;
rs=sum((xx-yy).^2);
spearm=1-((6*rs)/(n*(n^2-1)));

function threshr = rrank(mat,percent)
dim=size(mat); m=dim(1);
len=(m*m-m)/2;
vec=zeros(1,len);
n=0;
for i=1:m-1;
for j=i+1:m;
if (mat(i,j)~=0) n=n+1; vec(n)=mat(i,j); end;
end;
end;
num=round(percent/100*n);
vecc=sort(vec,'descend');
if (num~=0) threshr=vecc(num);
else threshr=1;
end;
```

The final results for candidate direct interactions achieved above are true direct interactions, and are confirmed by experiments and observations. The method provides a prompt and relatively reliable tool for primeval and batch screening of possible interactions.

In a sense, the significance level is a reliability measure. To avoid missing candidate direct interactions as much as possible, i.e., coarse screening of interactions, the significance level can be adjusted to a reasonable value, for example, $p = 0.05$, or even $p = 0.1$ (Zhang, 2015c).

The validity of taxa by sample-based interaction finding, as presented in this study, is dependent upon the representativeness of samples, reasonable distribution of samples over space or time, number of samples (sample size), etc. How to take a set of statistically representative samples for correlation analysis is one of the keys for interaction finding (Zhang, 2015c).

Chapter 17

Link Prediction: Structure- and Perturbation-based Methods

Link prediction aims to estimate the likelihood of the existence of a link (connection) between two nodes, based on observed links and the attributes of nodes (Zhou, 2015; Zhou *et al.*, 2009; Zhang, 2016e). Many biological networks, such as food webs, protein–protein interaction networks, and metabolic networks, are incomplete networks due to missing links (Zhang, 2015e). For example, 80% of the molecular interactions in cells of yeast (Yu *et al.*, 2008) and 99.7% of human (Amaral, 2008) are still unknown. An incomplete network occurs due to our limited knowledge on a complete network, or the network is in evolution and thus more links or even nodes are expected with time. Link prediction can considerably reduce the experimental costs for link finding (Zhang, 2015e) Moreover, link prediction algorithms can be used to predict the links that may appear in the future of evolving networks (Lü and Zhou, 2011; Lü *et al.*, 2012; Zhou, 2015). So far, link prediction has attracted wide attention. Numerous papers on this topic have been published (Clauset *et al.*, 2008; Guimera and Sales-Pardo, 2009; Barzel and Barabási 2013; Bastiaens *et al.*, 2015; Lü *et al.*, 2015; Zhang, 2015c, 2015e; Zhang and Li, 2015b; Zhao *et al.*, 2015; Zhou, 2015). A lots of structure- and perturbation-based methods for link prediction were proposed, such as the common neighbor (CN) index (Lorrain and White, 1971),

Adamic–Adar (AA) index (Adamic and Adar, 2003), stochastic block model (SBM) based on maximum likelihood method (Airoldi *et al.*, 2008), structural perturbation method (SPM) (Lü *et al.*, 2015), power-law model generation (Zhang, 2016i), random perturbation (Zhang, 2016e), etc. All of these methods and assessments will be discussed in this chapter.

17.1 Methods

17.1.1 *Similarity methods*

17.1.1.1 *Common neighbors*

CN is the simplest similarity index for link prediction (Lorrain and White, 1971). In this method, there is likely an interaction (link) between two nodes if they share more interactions (CN). For the node v_x in a network, the set of all its neighbors is $\Gamma(x)$. The similarity score for the two nodes v_x and v_y is

$$s_{xy} = |\Gamma(x) \cap \Gamma(y)|$$

where s_{xy} is the similarity score between nodes v_x and v_y, and the absolute value means the number of elements (i.e., neighbors, or links) in the set. The following are Matlab codes of CN:

```
%CN
a=input('Input the excel file name of adjacency matrix (e.g., raw.xls, etc.
Adjacency matrix is d=(dij)m*m, where m is the number of nodes in the
network. dij=1, if vi and vj are adjacent, and dij=0, if vi and vj are not
adjacent; i, j=1,2,..., m:','s');
a=xlsread(a);
v=size(a,2);
b=a*a;
c=((a==0) & (b~=0)).*b;
for i=1:v
for j=i:v
c(i,j)=0;
end; end
[pairx,pairy,scores]=find(c');
temp1=pairx; temp2=pairy;
pairxs=pairx(temp1<temp2);
pairys=pairy(temp1<temp2);
scoress=scores(temp1<temp2);
scorelist=[pairxs pairys scoress];
```

```
ranking=sortrows(scorelist,-3);
fprintf('Node Node Score\n')
disp([ranking])
```

17.1.1.2   Adamic — Adar index

AA index is proposed on the basis of CN index. AA gives a weight for each node according to the degree of a CN of two nodes. The CN with a small degree contributes much than the CN with a greater degree. If there are many paths that pass through a node, the node is considered to be important, and its adjacent nodes may distribute in different paths, and these nodes may have lower probability for interacting. AA is defined as (Adamic and Adar, 2003)

$$s_{xy} = \sum_{z \in \Gamma(x) \cap \Gamma(y)} (1/\log k_z)$$

where k_z is the degree of node z. The following are Matlab codes of AA

```
%AA
a=input('Input the excel file name of adjacency matrix (e.g., raw.xls, etc.
Adjacency matrix is d=(dij)m*m, where m is the number of nodes in the
network. dij=1, if vi and vj are adjacent, and dij=0, if vi and vj are not
adjacent; i, j=1,2,..., m:','s');
a=xlsread(a);
v=size(a,2);
b=a./repmat(log(sum(a,2)),[1,size(a,1)]);
b(isnan(b))=0;
b(isinf(b))=0;
c=a*b;
d=((a==0) & (c~=0)).*c;
for i=1:v
for j=i:v
d(i,j)=0;
end; end
[pairx,pairy,scores]=find(d');
temp1=pairx; temp2=pairy;
pairxs=pairx(temp1<temp2);
pairys=pairy(temp1<temp2);
scoress=scores(temp1<temp2);
scorelist=[pairxs pairys scoress];
ranking=sortrows(scorelist,-3);
fprintf('Node Node Score\n')
disp([ranking])
```

17.1.1.3 *Resource allocation*

Resource allocation (RA) index treats the CNs between two nodes as the media of transferring information. It is similar to AA and is defined as (Zhou *et al.*, 2009)

$$s_{xy} = \sum_{z \in \Gamma(x) \cap \Gamma(y)} (1/k_z)$$

The following are Matlab codes of RA:

```
%RA
a=input('Input the excel file name of adjacency matrix (e.g., raw.xls, etc.
Adjacency matrix is d=(dij)m*m, where m is the number of nodes in the
network. dij=1, if vi and vj are adjacent, and dij=0, if vi and vj are not
adjacent; i, j=1,2,..., m:','s');
a=xlsread(a);
v=size(a,2);
b=a./repmat(sum(a,2),[1,size(a,1)]);
b(isnan(b))=0;
b(isinf(b))=0;
c=a*b;
d=((a==0) & (c~=0)).*c;
for i=1:v
for j=i:vv
d(i,j)=0;
end; end
[pairx,pairy,scores]=find(d');
temp1=pairx; temp2=pairy;
pairxs=pairx(temp1<temp2);
pairys=pairy(temp1<temp2);
scoress=scores(temp1<temp2);
scorelist=[pairxs pairys scoress];
ranking=sortrows(scorelist,-3);
fprintf('Node Node Score\n')
disp([ranking])
```

17.1.1.4 *Katz index*

CN, AA, and RA consider the information of local links around the two nodes. Katz index was proposed from the global perspective, which considers all the paths of the network. It is defined as (Katz, 1953)

$$s_{xy} = \alpha A_{xy} + \alpha^2 (A^2)_{xy} + \alpha^3 (A^3)_{xy} +$$

where $0 < \alpha < 1$, A is the adjacency matrix, $(A^2)_{xy}$ is the number of paths of length 2 between nodes v_x and v_y, etc. α determines the weights of paths, and a longer path has a lower weight. Thus, the shorter paths contribute more in between-node similarity than the longer paths. The following are Matlab codes of Katz

```
%Katz
a=input('Input the excel file name of adjacency matrix (e.g., raw.xls, etc.
Adjacency matrix is d=(dij)m*m, where m is the number of nodes in the
network. dij=1, if vi and vj are adjacent, and dij=0, if vi and vj are not
adjacent; i, j=1,2,..., m:','s');
alpha=input('Input the parameter alpha (0<alpha<1):');
a=xlsread(a);
v=size(a,2);
c=inv(sparse(eye(size(a,1)))-alpha*a);
c=c-sparse(eye(size(a,1)));
d=((a==0) & (c~=0)).*c;
for i=1:v
for j=i:v
d(i,j)=0;
end; end
[pairx,pairy,scores]=find(d');
temp1=pairx; temp2=pairy;
pairxs=pairx(temp1<temp2);
pairys=pairy(temp1<temp2);
scoress=scores(temp1<temp2);
scorelist=[pairxs pairys scoress];
ranking=sortrows(scorelist,-3);
fprintf('Node Node Score\n')
disp([ranking])
```

17.1.1.5 *Average commute time*

Average Commute Time (ACT) is defined on the basis of random walk process. Suppose $m(x, y)$ is the average walking steps required by a random particle walking from nodes v_x and v_y. The ACT between v_x and v_y is thus $n(x, y) = m(x, y) + m(y, x)$. ACT is defined as (Klein and Randi, 1993)

$$s_{xy} = M/n(x, y)$$

where M is the total number of links of the network. The shorter ACT between two nodes means the closer distance, and thus an interaction will more easily occur between them. The following is the Matlab function of ACT (Zhou, 2016b):

```
function thisauc=ACT(train, test)
D=sparse(eye(size(train,1)));
D(logical(D))=sum(train,2);
pinvL=sparse(pinv( full(D - train) ));
```

```
clear D;
Lxx=diag(pinvL);
Lxx=repmat(Lxx, [1,size(train,1)]);
sim=1./(Lxx+Lxx'-2*pinvL);
sim(isnan(sim))=0; sim(isinf(sim))=0;
thisauc=CalcAUC(train,test,sim,10000);
end
```

17.1.1.6   *Structural perturbation method*

SPM is proposed inspiring from 1st-order matrix perturbation in quantum mechanics (Lü *et al.*, 2015). Some of links are used for perturbation, and their adjacency matrix is ΔA. The adjacency matrix of the rest of the links is A^R. Known eigenvalues λ_k and eigenvector x_k of A^R, such that

$$A^R = \sum_{k=1}^{N} \lambda_k x_k x_k^{\mathrm{T}}$$

use ΔA to perturb A^R, and after a series of calculations, the following relationship holds:

$$\tilde{A} = \sum_{k=1}^{N} (\lambda_k + x_k^{\mathrm{T}} \Delta A x_k / x_k^{\mathrm{T}} x_k) x_k x_k^{\mathrm{T}}$$

and SPM similarity score is

$$s_{xy} = \tilde{A}_{xy}$$

The following are Matlab codes of SPM:

```
%SPM
a=input('Input the excel file name of adjacency matrix (e.g., raw.xls, etc.
Adjacency matrix is d=(dij)m*m, where m is the number of nodes in the
network. dij=1, if vi and vj are adjacent, and dij=0, if vi and vj are not
adjacent; i, j=1,2,..., m:','s');
ar=input('Input the excel file name of adjacency matrix from which missing
links are ready to be predicted (e.g., raw.xls, etc. Adjacency matrix is
d=(dij)m*m, where m is the number of nodes in the network. dij=1, if vi and
vj are adjacent, and dij=0, if vi and vj are not adjacent; i, j=1,2,...,
m:','s');
a=xlsread(a);
ar=xlsread(ar); v=size(ar,2);
deltaa=a & ~ ar;
[x,eigenval]=eig(ar);
lamda=sum(eigenval);
```

```
abar=zeros(v);
for k=1:v
deltalamda(k)=((x(:,k)'*deltaa)*x(:,k))/(x(:,k)'*x(:,k));
abar=abar+(lamda(k)+deltalamda(k))*x(:,k)*x(:,k)';
end
u=ones(v);
for i=1:v
u(i,i)=0;
end
newmat=(u & ~ar).*abar;
[pairx,pairy]=find(newmat);
pairval=zeros(size(pairx,1),1);
for i=1:size(pairx,1)
pairval(i)=abar(pairx(i),pairy(i));
end
temp1=pairx; temp2=pairy;
pairxs=pairx(temp1<temp2);
pairys=pairy(temp1<temp2);
pairvals=pairval(temp1<temp2)
res=[pairxs pairys pairvals];
ires=sortrows(res,-3);
disp([ires])
```

However, Zhang (2016e) pointed out that SPM is not a reliable method.

17.1.1.7   *CAR index*

CAR index was originally based on the idea of Ahn *et al.* (2010). Ravasi *et al.* argued that a network tends to form a local community structure. They defined a new index CAR

$$\text{CAR}_{xy} = \text{CN}_{xy} * \text{LCL}_{xy}$$

where CN_{xy} is just the s_{xy} of Common Neighbors, as described above. LCL_{xy} is the number of local community links of x and y, that is, the number of links between all CNs of x and y.

17.1.2   *Link prediction with generated power-law degree distribution model*

The power-law distribution function is as below (Goemann, 2011; Zhang, 2011c, 2012e, 2016i)

$$p(x) = x^{-\lambda}$$

where λ is a known constant. Suppose there are v nodes in the network being built.

Given the adjacency matrix of an original network, i.e., the network with which missing links are prepared to be predicted. Suppose the mean of node degrees of the original network is m'. The procedures of the algorithm are as follows (Zhang, 2016i):

(1) Let the mean of node degrees of generated network D, $m = m'$ $(1 + per)$, where per is the perturbation rate, and $per = 0.2$, 0.3, etc., which represents a percentage increment of mean in the network perturbation or evolution.
(2) Use m to empirically calculate parameter λ (Zhang, 2016i)

$$\lambda = 1.6347 - 0.1401m + 0.0019v + 0.0038m^2$$

where m is the mean of node degrees, v is the number of nodes in the network.
(3) Run the algorithm of Zhang (2016i) (with power-law distribution), the resultant adjacency matrix G, is achieved.
(4) Swap rows and columns of G according to the ranking of node degrees of D such that the same row (column) of G and D has the same ranking of node degree in the respective matrices. By doing so, the final adjacency matrix of the predicted network, H, transformed from G, is achieved.
(5) Compare H and D, the link pairs in original network only, link pairs in predicted (i.e., generated) network only, and link pairs in both original and predicted networks, etc., are thus achieved.
(6) If simulation times are achieved, go to (7), otherwise return to (3).
(7) Calculate means of all indices and mean number (likelihood) of links in original network only, mean number (likelihood) of links in generated network only, and mean number (likelihood) of links in both networks. The percentage of correctly predicted links *vs.* true links is defined as

$$100x/(x + y)$$

where x is the number of links in both networks, and y is the number of links in original network only. The percentage of correctly predicted links *vs.* predicted links is defined as

$$100x/(x+z)$$

where x is the number of links in both networks, and z is the number of links in predicted (generated) network only.

The following are Matlab codes of the algorithm, connectionFinding.m. The functions, netConstrDistr.m and netGen.m can be found in Zhang (2016i):

```
clear
choice=input('Input the type (1 or 2) of data file of the network from which
missing links are ready to be predicted (1: adjacency matrix; 2: two
array):');
disp('Adjacency matrix: d=(dij)m*m, where m is the number of nodes in the
network. dij=1, if vi and vj are adjacent, and dij=0, if vi and vj are not
adjacent; i, j=1,2,..., m');
disp('Two array: there are two columns, A1 and A2, in the data file; an
element of A1 stores a node of a link and the corresponding element of A2
stores another node of the link.');
if (choice==1)
adjstr=input('Input the file name of adjacency matrix from which missing
links are ready to be predicted (e.g., raw.txt, raw.xls, etc. Adjacency
matrix is d=(dij)m*m, where m is the number of nodes in the network. dij=1,
if vi and vj are adjacent, and dij=0, if vi and vj are not adjacent; i,
j=1,2,..., m:','s');
end
if (choice==2)
adjstr=input('Input the file name of two array of the network from which
missing links are ready to be predicted (e.g., raw.txt, raw.xls, etc. There
are two columns, A1 and A2, in the data file; an element of A1 stores a node
of a link and the corresponding element of A2 stores another node of the
link:','s');
end
pro=input('Input perturbation rate to increase the mean of node degrees of
    the network (e.g., 0.2, 0.3, etc.):');
simu=input('Input the simulation times (e.g., 20, 30, etc.):');
if (choice==1) adjmat=load(adjstr); v=size(adjmat,2); end
if (choice==2)
twoarray=load(adjstr);
nn=size(twoarray,1);
v=max(max(twoarray));
for i=1:nn
adjmat(twoarray(i,1),twoarray(i,2))=1;
adjmat(twoarray(i,2),twoarray(i,1))=1;
end; end
```

```
degr=sum(adjmat);
meanmat=sum(degr)/v;
m=meanmat*(1+pro);
c=1.6347-0.1401*m+0.0019*v+0.0038*m^2;
er=0.05;
sim=v*5;
fa=1.2;    %It can be replaced with, e.g.,
fa=22.6712-34.4086*c+0.0764*v-0.0423*c*v+13.0632*c^2-0.6
percentCorrect=zeros(1,simu);
percentPrediCorrect=zeros(1,simu);
degd=zeros(simu,v);
summ=v*(v-1)/2;
su1=zeros(summ,2*simu);
su2=zeros(summ,2*simu);
su3=zeros(summ,2*simu);
adj=zeros(v);
adjj=zeros(v);
pdegr=zeros(1,v);
pdeg=zeros(1,v);
fdegr=zeros(1,v);
fdeg=zeros(1,v);
fd=zeros(1,v);
pd=zeros(1,v);
persum=zeros(1,3);
for siml=1:simu
for fac=fa:0.3:v
in=zeros(1,v);
i=1;
in(1)=1;
while (v>0)
i=i+1;
in(i)=in(i-1)*fac;
if (in(i)>v) in(i)=v; break; end
end;
deg=in(1:i);
class=i-1;
ff=zeros(1,class);
p=zeros(1,class);
su=0;
for i=1:class-1
p(i)=netConstrDistr(1,c,in(i),in(i+1));
ff(i)=p(i)*v;
su=su+ff(i);
end
ff(class)=v-su;
id=0;
minerr=1e+10;
tot=1;
while (v>0)
[adj0,fff0,error0]=netGen(class,deg,ff);
```

```
if (error0<minerr)
adj=adj0; fff=fff0; minerr=error0;
if (minerr<=er) id=1; break; end
end
if (tot>=sim) break; end;
tot=tot+1;
end
if (id==1) break; end
end
degg=sum(adj);
for h=1:2
if (h==1) fd=degr; end
if (h==2) fd=degg; end
for i=1:v
pd(i)=i;
end
for i=1:v-1
k=i;
for j=i:v-1
if (fd(j+1)>fd(k)) k=j+1; end
end
l=pd(i); pd(i)=pd(k); pd(k)=l;
u=fd(i); fd(i)=fd(k); fd(k)=u;
end
if (h==1) pdegr=pd; fdegr=fd; end
if (h==2) pdeg=pd; fdeg=fd; end
end
for i=1:v
adjj(pdegr(i),:)=adj(pdeg(i),:);
end
for i=1:v
adj(:,pdegr(i))=adjj(:,pdeg(i));
end
degg=sum(adj);
degd(siml,:)=degg;
fprintf('\nAdjacency matrix of the original network\n')
disp([adjmat])
fprintf('\nNode degrees of adjacency matrix of the original network\n')
disp([degr])
fprintf(['\nMean of node degrees of the original network:'
  num2str(mean(degr)) '\n'])
fprintf('\n\nAdjacency matrix of the generated network with power-law
  distributed node degrees\n')
disp([adj])
fprintf('\nNode degrees of adjacency matrix of the generated network\n')
disp([degg])
fprintf(['\nMean of node degrees of the generated network:'
  num2str(mean(degg)) '\n'])
fprintf(['Fitting error of expected and practical distribution of node
  degrees of the generated network:' num2str(minerr) '\n\n'])
```

```
xx=adjmat & (~adj);
yy=(~adjmat) & adj;
zz=adjmat & adj;
for i=1:3
switch i
  case 1
    mat=xx; s='Link pairs in original network only (x)';
  case 2
    mat=yy; s='Link pairs in predicted network only (y)';
  case 3
    mat=zz; s='Link pairs in both original and predicted networks (z)';
end;
[pairx,pairy]=find(mat);
temp1=pairx; temp2=pairy;
pairxs=pairx(temp1<temp2); pairys=pairy(temp1<temp2);
ConnectionPairs=[pairxs pairys];
persum(i)=size(ConnectionPairs,1);
dm=size(ConnectionPairs,1);
if (i==1) su1(:,siml*2-1)=[pairxs;zeros(summ-dm,1)];
su1(:,siml*2)=[pairys;zeros(summ-dm,1)]; end
if (i==2) su2(:,siml*2-1)=[pairxs;zeros(summ-dm,1)];
su2(:,siml*2)=[pairys;zeros(summ-dm,1)]; end
if (i==3) su3(:,siml*2-1)=[pairxs;zeros(summ-dm,1)];
su3(:,siml*2)=[pairys;zeros(summ-dm,1)]; end
end
percentageCorrect(siml)=persum(3)/(persum(1)+persum(3))*100;
percentagePrediCorrect(siml)=persum(3)/(persum(2)+persum(3))*100;
disp(s)
disp([ConnectionPairs])
disp('Percentages of correctly predicted links (%):')
percentageCorrect(siml)
 end
disp('---------------------------Summary----------------------------')
disp('Node degrees of original networks:')
degr=degr
disp('Averaged node degrees of generated networks under different
  simulations:')
dega=round(sum(degd)/simu+0.5)
disp('Chi square of node degrees between node degrees of original networks
  and averaged node degrees of generated networks:')
chi2=sum(((degr-dega).^2)./degr)
disp('Mean percentage of correctly predicted links vs. true links (%):')
meanPercent=mean(percentageCorrect)
disp('Standard deviation of mean percentage of correctly predicted links vs.
  true links (%):')
standardDevi=std(percentageCorrect)
disp('Mean percentage of correctly predicted links vs. predicted links (%):')
meanPercentPredi=mean(percentagePrediCorrect)
disp('Standard deviation of mean percentage of correctly predicted links vs.
  predicted links (%):')
```

```
standardDeviPredi=std(percentagePrediCorrect)
prop1=zeros(v);
prop2=zeros(v);
prop3=zeros(v);
%su2(:,k*2-1),su2(:,k*2)
for i=1:v-1
for j=i+1:v
for k=1:simu
for l=1:v*(v-1)/2
if ((su1(1,k*2-1)==i) & (su1(1,k*2)==j)) prop1(i,j)=prop1(i,j)+1; break; end
if ((su2(1,k*2-1)==i) & (su2(1,k*2)==j)) prop2(i,j)=prop2(i,j)+1; break; end
if ((su3(1,k*2-1)==i) & (su3(1,k*2)==j)) prop3(i,j)=prop3(i,j)+1; break; end
end; end; end; end
disp('---------------------------------------------------------------------')
disp('Mean number (Likelihood) of links in both networks:')
disp('Node Node Likelihood')
[pairx,pairy]=find(prop3);
s=0;
for i=1:v-1
for j=i+1:v
if (prop3(i,j)~=0) s=s+1;pairvalue(s)=prop3(i,j)/simu; end;
end; end
result=[pairx pairy pairvalue'];
ires=sortrows(result,-3);
disp([ires])
clear pairvalue
disp('Mean number (Likelihood) of links in original network only:')
disp('Node Node Likelihood')
[pairx,pairy]=find(prop1);
s=0;
for i=1:v-1
for j=i+1:v
if (prop1(i,j)~=0) s=s+1;pairvalue(s)=prop1(i,j)/simu; end;
end; end
result=[pairx pairy pairvalue'];
ires=sortrows(result,-3);
disp([ires])
clear pairvalue
disp('Mean number (Likelihood) of predicted links in generated
  network only:')
disp('Node Node Likelihood')
[pairx,pairy]=find(prop2);
s=0;
for i=1:v-1
for j=i+1:v
if (prop2(i,j)~=0) s=s+1;pairvalue(s)=prop2(i,j)/simu; end;
end; end
result=[pairx pairy pairvalue'];
ires=sortrows(result,-3);
disp([ires])
```

Table 1 Simulation of some tumor-related networks (Zhang, 2016i).

	x *vs.* y	x *vs.* z
Ras	3.4 ± 2.3	20.8 ± 7.2
$p53$	0.1 ± 0.3	16.8 ± 4.5
Akt	3.3 ± 1.7	17.7 ± 3.7
HGF	2.6 ± 2.4	17.4 ± 5.4
JAK-STAT	2.4 ± 2.0	14.6 ± 4.0
JNK	2.6 ± 1.6	21.2 ± 3.7
PPAR	3.1 ± 2.7	15.9 ± 5.4
TGF-β	2.7 ± 2.5	15.8 ± 5.6
TNF	0.4 ± 0.9	17.8 ± 4.9

Notes: x: Mean percentage of correctly predicted links. y: true links with random networks (%) ($\pm$standard deviation). z: true links with generated networks (%) ($\pm$standard deviation).

Some of the summarized results for link prediction of tumor-related networks (signaling pathways) are listed in Table 1, in which the mean percentages of correctly predicted links with random networks (%) are given also (Zhang, 2016i).

Compared to the mean percentages of correctly predicted links *vs.* true links with random networks, it is obvious that the results of generated networks are effective. Therefore the algorithm is effective in predicting missing links of biological networks.

17.1.3 *Link prediction with random perturbation method*

Link prediction is closely correlated with network evolution. Following the principle of network evolution of Zhang's model (Zhang, 2016f), Zhang (2016e) proposed the random perturbation method to predict missing links in the network. It was assumed that a node with more existing links harbors more missing links, which is a reasonable assumption because new nodes tend to connect the nodes with more links (Barabasi and Albert 1999; Zhang, 2012e; Zhang, 2016f).

Assume there are totally v nodes in the network being predicted, and adjacency matrix of the network is $d = (d_{ij})$, $i, j = 1, 2, \ldots, v$, where $d_{ij} = d_{ji}$, $d_{ii} = 0$, and if $d_{ij} = 1$ or $d_{ji} = 1$, there is a link (connection) between nodes i and j. The adjacency matrix of the network for missing links only is $D = (D_{ij})$, $i, j = 1, 2, \ldots, v$. The procedures are as follows (Zhang, 2016e):

(1) Calculate the expected missing links to be predicted, $m = m' \times per$, where m' is the total links of the network, per is the perturbation rate, and $per = 0.2, 0.3$, etc., which represents a percentage increment of links in the network perturbation.

(2) Calculate node degree, $a_i(t)$, $i = 1, 2, \ldots, v$. The cumulative attraction strength of node 1 to node i is

$$p_i(t) = \sum_{j=1}^{i} a_j(t)^{\lambda(t, a_j)} \bigg/ \sum_{j=1}^{v} a_j(t)^{\lambda(t, a_j)}$$

where λ is attraction factor, $\lambda > 0$, e.g., $\lambda = 1.2$, 1.5, etc.

(3) Generate missing links. Let $p_0 = 0$, and generate two random values w and u. For $p_0, p_1, p_2, \ldots, p_v$ one of the following two rules can be used

Rule 1: if $(j - 1)/v \le w \le j/v$, $p_{k-1} \le u \le p_k, k \neq j$ and $d_{kj} = d_{jk} = 0$, let $D_{kj} = 1$ and $D_{jk} = 1$, i.e., there is a missing link between nodes k and j.

Rule 2: if $p_{j-1}(t) \le w \le p_j(t), p_{k-1}(t) \le u \le p_k(t)$, $k \neq j$, and $d_{kj} = d_{jk} = 0$, let $D_{kj} = 1$ and $D_{jk} = 1$, i.e., there is a missing link between nodes k and j.

Rule 1 means that a randomly chosen node tends to connect to the node with greater degree. Rule 2 means that a link tends to be created between two nodes with greater degrees. By doing so, a new link is created. Repeat the procedure m times to produce m (missing) links. By doing so, an adjacency matrix of the network for missing links only, $D = (D_{ij})$, $i, j = 1, 2, \ldots, v$, is generated.

(4) Return to (3) to perform the next prediction, until the desired simulation times are achieved.

(5) Calculate mean number (i.e., likelihood) of predicted missing links, and rank the likelihood from greater to smaller. The first m links are the predicted missing links with maximal likelihood.

The following are Matlab codes of the algorithm, links Prediction.m:

```
clear
choice=input('Input the type (1 or 2) of data file of the network from which
missing links are ready to be predicted (1: adjacency matrix; 2: two
array):');
disp('Adjacency matrix: d=(dij)m*m, where m is the number of nodes in the
network. dij=1, if vi and vj are adjacent, and dij=0, if vi and vj are not
adjacent; i, j=1,2,..., m');
disp('Two array: there are two columns, A1 and A2, in the data file; an
element of A1 stores a node of a link and the corresponding element of A2
stores another node of the link.');
if (choice==1)
adjstr=input('Input the file name of adjacency matrix from which missing
links are ready to be predicted (e.g., raw.txt, raw.xls, etc. Adjacency
matrix is d=(dij)m*m, where m is the number of nodes in the network. dij=1,
if vi and vj are adjacent, and dij=0, if vi and vj are not adjacent; i,
j=1,2,..., m:','s');
end
if (choice==2)
adjstr=input('Input the file name of two array of the network from which
missing links are ready to be predicted (e.g., raw.txt, raw.xls, etc. There
are two columns, A1 and A2, in the data file; an element of A1 stores a node
of a link and the corresponding element of A2 stores another node of the
link:','s');
end
rule=input('Input the rule type (1 or 2) used in the algorithm:');
pro=input('Input perturbation rate to increase missing links of the network
   (e.g., 0.2, 0.3, etc.):');
lamda=input('Attraction factor of nodes (lamda>0; e.g., 1.3, 1.5, etc.)=');
simu=input('Input the simulation times (e.g., 100, 200, etc.):');
if (choice==1) adjmat=load(adjstr); v=size(adjmat,2); end
if (choice==2)
twoarray=load(adjstr);
nn=size(twoarray,1);
v=max(max(twoarray));
for i=1:nn
adjmat(twoarray(i,1),twoarray(i,2))=1;
adjmat(twoarray(i,2),twoarray(i,1))=1;
end; end
degr=sum(adjmat);
m=round(sum(degr)/2*pro);
fprintf('\nAdjacency matrix of the original network\n')
```

```
disp([adjmat])
fprintf('\nNode degrees of adjacency matrix of the original network\n')
disp([degr])
fprintf(['\nMean of node degrees of the original network:'
  num2str(mean(degr)) '\n\n'])
cnow=(sum(degr)/2)/((v^2-v)/2);
fprintf(['\nConnectance=' num2str(cnow) '\n'])
summ=sum(degr);
summa=sum(degr.*(degr-1));
h=v*summa/(summ*(summ-1));
fprintf(['\nAggregation index (AI) of node degrees=' num2str(h) '\n'])
cv=(std(degr))^2/mean(degr);
fprintf(['\nCoefficient of variation (CV) of node degrees='
  num2str(cv) '\n'])
summ=v*(v-1)/2;
su=zeros(summ,2*simu);
prop=zeros(1,v);
proptot=zeros(v);
degrr=degr.^lamda;
prop(1)=degrr(1)/sum(degrr);
for i=2:v;
prop(i)=prop(i-1)+degrr(i)/sum(degrr);
end
for siml=1:simu
adj=zeros(v);
temp=zeros(m,2);
mm=1;
while (v>0)
rep=0;
while (v>0)
propp=prop;
if ((rep==0) & (rule==1))
for i=1:v;
propp(i)=i/v;
end; end
ran=rand();
for j=1:v
if (j==1) st=0; end
if (j>=2) st=propp(j-1); end
if ((ran>=st) & (ran<propp(j))) rep=rep+1; id(rep)=j; break; end
end
if ((rep>=2) & (id(rep)~=id(1)))
tab=0;
for i=1:mm
if (((id(1)==temp(i,1)) & (id(rep)==temp(i,2))) | ((id(rep)==temp(i,1)) &
(id(1)==temp(i,2)))) tab=1; break; end
end
if (tab==1) continue; end;
temp(mm,1)=id(1); temp(mm,2)=id(rep);
break;
```

```
end; end
if (adjmat(id(1),id(rep))==0) adj(id(1),id(rep))=1; adj(id(rep),id(1))=1;
   mm=mm+1; end;
if (mm==m+1) break; end;
end
fprintf(['Simulation' num2str(siml)])
fprintf('\n\nAdjacency matrix for predicted links only\n')
disp([adj])
[pairx,pairy]=find(adj);
temp1=pairx; temp2=pairy;
pairxs=pairx(temp1<temp2);
pairys=pairy(temp1<temp2);
ConnectionPairs=[pairxs pairys];
dm=size(ConnectionPairs,1);
su(:,siml*2-1)=[pairxs;zeros(summ-dm,1)];
su(:,siml*2)=[pairys;zeros(summ-dm,1)];
disp('Predicted links')
disp([ConnectionPairs])
end
disp('--------------------------Summary---------------------------')
disp(['There are totally' num2str(sum(degr)/2) 'links in the original
   network'])
disp(['You wish to predict ' num2str(m) 'missing links in the original
   network'])
fprintf('\n');
proptot=zeros(v);
for i=1:v-1
for j=i+1:v
for k=1:simu
for l=1:v*(v-1)/2
if ((su(l,k*2-1)==i) & (su(l,k*2)==j)) proptot(i,j)=proptot(i,j)+1;
proptot(j,i)=proptot(i,j); break; end
end; end; end; end
disp('Likelihood (mean number) of predicted links:')
disp(' Node Node Likelihood')
s=0;
for j=1:v
for i=1:v
if (proptot(i,j)~=0) s=s+1;pairvalue(s)=proptot(i,j)/simu; end;
end; end
[pairx,pairy]=find(proptot);
result=[pairx pairy pairvalue'];
results(1,1)=result(1,1); results(1,2)=result(1,2); results(1,3)
  =result(1,3);
su=1;
for i=2:s
lab=0;
for j=1:i-1
if ((result(j,2)==result(i,1)) & (result(j,1)==result(i,2))) lab=1; break;
```

```
     end;
  end
  if (lab==0) su=su+1;results(su,1)=result(i,1); results(su,2)=result(i,2);
  results(su,3)=result(i,3); end
  end
  ires=sortrows(results,-3);
  disp([ires])
```

Zhang (2016e) used the data of tumor-related networks (pathways) (ABCAM, 2012; Huang and Zhang, 2012; Li and Zhang, 2013; Pathway Central 2012). These networks are complete. For each network, some links are removed following reverse process of the algorithm above and are then predicted. The simulation times are set to be 100. The perturbation rate is $per \approx 0.25$. Attraction factor $\lambda = 1.5$. In step (3) of the algorithm, use the Rule 2 for prediction. The results for some pathways are listed in Table 2. The prediction efficiency of Rule 2 is generally higher. Moreover, the prediction efficiency of the algorithm increases dramatically with the network complexity (Table 2).

17.2 Performance Assessment of Link Prediction Methods

17.2.1 *AUC assessment*

Area Under the Receiver Operating Characteristic Curve (AUC) is originally derived from signal detection theory. It can be understood as probability of the score of randomly choosing a link in the testing set being higher than the score of randomly choosing a nonexisting link (i.e., $x > y$). AUC is defined as (Fawcett, 2006)

$$AUC = (n' + 0.5n'')/n$$

where n is the total number for comparisons (e.g., $n = 1000$). n' is the number of $x > y$; n'' is the number of $x = y$. $AUC = 0.5$ is the situation for equivalent probability of $x > y$ and $x < y$. The higher AUC means the better performance of the method used.

The data on signaling pathways are used to assess the methods (ABCAM, 2012; Huang and Zhang, 2012; Li and Zhang,

Table 2   Link prediction of some tumor-related networks (pathways) of missing links with Rule 2 (*per* =∼ 0.25, λ = 1.5).

	Ras	p53	Akt	HGF	JNK	PPAR	TGF-β	TNF
Percentage (%) of correctly predicted links with the algorithm (x)	62.5	92.3	50.0	57.1	75.0	33.3	37.5	62.5
Percentage (%) of correctly predicted links against predicted missing links with the algorithm	0.7	0.9	0.2	1.5	1.2	1.8	1.5	1.7
Total number of predicted links with 100 simulations	314	388	301	300	404	221	304	291
The averaged rank before which all missing links fall in the list of predicted links (y)	74	92	6	78	81	41	63	95
Prediction efficiency (x/y)	0.8446	1.0033	8.3333	0.7321	0.9259	0.8122	0.5952	0.6579
Percentage (%) of correctly predicted links against number of missing links with random network (x)	37.5	53.9	16.7	85.7	62.5	83.3	87.5	75.0
Percentage (%) of correctly predicted links against predicted missing links with random network	1.6	3.1	2.0	1.3	2.9	0.9	0.9	1.7
Total number of predicted links with 100 simulations	411	823	851	412	838	277	478	359
The averaged rank before which all missing links fall in the list of predicted links (y)	77	325	23	213	246	55	184	111
Prediction efficiency (x/y)	0.487	0.1658	0.7261	0.4023	0.2541	1.5145	0.4755	0.6757

Table 3 *AUC* summary of 100 link predictions for 30 signaling pathways.

Pathways	CN	AA	RA	Katz_0.01	Katz_0.001	ACT	SPM
p53	0.5074	0.4956	0.5074	0.5053	0.5056	0.6699	0.7106
JAK-STAT	0.5340	0.4910	0.5351	0.4984	0.4983	0.5925	0.8308
Akt	0.4873	0.4832	0.4890	0.4756	0.4755	0.7943	0.5714
TNF	0.4781	0.4903	0.4782	0.5043	0.5044	0.4753	0.8405
Ras	0.4710	0.4835	0.4707	0.4051	0.4052	0.5643	0.6909
VEGF	0.4798	0.4989	0.4790	0.4236	0.4236	0.4495	0.8679
PPAR	0.4670	0.4886	0.4668	0.4236	0.4235	0.6140	0.7612
STAT3	0.4576	0.4789	0.4578	0.3586	0.3579	0.6314	0.7558
PI3K	0.5358	0.5344	0.5364	0.5239	0.5241	0.5531	0.8595
MA	0.5086	0.5008	0.5085	0.4732	0.4725	0.5947	0.8094
ErbF	0.4742	0.4883	0.4743	0.4155	0.4153	0.5454	0.8725
TGFB	0.4861	0.5006	0.4859	0.4498	0.4497	0.5474	0.7802
EGF	0.5023	0.5034	0.5024	0.4899	0.4900	0.5502	0.7753
If	0.4774	0.4934	0.4774	0.4476	0.4481	0.4164	0.7695
Estro	0.4755	0.4818	0.4753	0.4387	0.4383	0.6122	0.7169
HGF	0.4975	0.4986	0.4974	0.4920	0.4918	0.7738	0.6470
BRCA1	0.4521	0.4637	0.4520	0.3979	0.3984	0.6370	0.6523
CC	0.4500	0.4557	0.4501	0.3760	0.3761	0.7158	0.7363
Andro	0.4690	0.4907	0.4690	0.4028	0.4033	0.6173	0.8769
PTEN	0.4841	0.4791	0.4848	0.4508	0.4508	0.6275	0.7826
JNK	0.5310	0.5351	0.5308	0.4568	0.4566	0.5794	0.8862
MAPK	0.5294	0.5132	0.5304	0.4945	0.4941	0.6485	0.7434
mTOR	0.4812	0.4839	0.4815	0.4194	0.4196	0.5767	0.7243
HIF1A	0.4517	0.4590	0.4516	0.4025	0.4029	0.7415	0.5751
IGF1R	0.4968	0.4909	0.4971	0.4540	0.4542	0.5626	0.8019
FAS	0.5009	0.4919	0.5012	0.4865	0.4864	0.5490	0.7242
ERK	0.4847	0.4853	0.4849	0.3745	0.3747	0.5531	0.8084
cAMP-PKA	0.4730	0.4773	0.4736	0.4382	0.4381	0.7327	0.7312
CA	0.5026	0.5061	0.5029	0.4420	0.4419	0.6734	0.8377
CCCR	0.4712	0.4783	0.4708	0.4726	0.4730	0.5770	0.7786
Average	0.4872	0.4907	0.4874	0.4465	0.4465	0.6059	0.7639
Variance	0.0006	0.0003	0.0006	0.0019	0.0019	0.0078	0.0068

Note: Only average AUC of 100 predictions are listed.

2013; Pathway Central 2012; Zhang, 2016a). Results are shown in Table 3.

The results reveal that in most cases, the values of CN, AA, RA, and Katz are close to or less than 0.5. Adjacency matrices of signaling pathways are always sparse. Most of between-node numbers of CNs are zero. Therefore the situation of $x = y$ in AUC assessment occurs frequently, leading to a calculated value of 0.5 for these indices. CN, AA, and RA are all based on CNs. They differ in weighting node degree. RA restrains the nodes of greater degree, seconded by AA. CN is free of node degree. For the network with more links and different node degrees, e.g., social networks, RA is superior to other methods (Pan *et al.*, 2010).

17.2.2 *Model-based assessment*

Zhang (2016a) proposed a dynamic model to describe dynamics of occurrence probability of missing links in predicted missing link list

$$y = K(1 - e^{-rx/K})$$

where K: the expected total number of true missing links, r: maximum occurrence probability (intrinsic occurrence probability of true missing links), and $K > 0, 0 < r \leq 1$. r denotes the prediction capability or effectiveness of a prediction method, and can thus be used to assess the performance of a prediction method.

Use the previous methods to predict missing links of tumor-related pathways FAS, JAK-STAT, JNK, MARK, and p53 (ABCAM, 2012; Huang and Zhang, 2012; Li and Zhang, 2013; Pathway Central 2012; Zhang, 2016a). Training set and prediction set are set separately. Prediction results are indicated in Table 4. Finally, use the dynamic model above to fit missing link series in predicted missing link list. The results are shown in Table 5.

Jointly considering Tables 4 and 5, in terms of r and total number of true missing links S, we can find that random perturbation method performs significantly better than other methods, seconded by CN. The curves of random perturbation are more similar to the theoretical curve of the dynamic model.

Table 4 Summary of prediction results of six methods for missing links in five tumor-related pathways.

Pathways	Type of links	Random perturbation	Power-law generation	CN	AA	RA	Katz
FAS	Predicted missing links (n)	**640**	866	105	105	105	386
	Predicted true missing links (S)	**9**	11	1	1	1	3
JAK-STAT	Predicted missing links (n)	545	613	**76**	**76**	**76**	736
	Predicted true missing links (S)	5	9	**4**	**4**	**4**	6
JNK	Predicted missing links (n)	**612**	709	201	201	201	759
	Predicted true missing links (S)	**16**	16	2	2	2	12
MARK	Predicted missing links (n)	819	1006	133	133	133	1166
	Predicted true missing links (S)	8	11	2	2	2	9
p53	Predicted missing links (n)	**539**	1067	142	142	142	406
	Predicted true missing link (S)	**12**	13	3	3	3	6

Table 5   Model fitting and statistic test of missing link predictions of six methods for five tumor pathways.

Pathways	Model para	Random perturbation	Power-law generation	CN	AA	RA	Katz
FAS ($L=12$)	K	9.2480	28.2614	-2.6×10^9	1.2×10^9	1.2×10^9	4.0375
true missing	r	**0.0458**	0.0177	0.0141	0.0142	0.0142	0.0190
links	χ^2	330.5661**	92.2104**	57.1607**	169.0797	169.0797	61.7429**
JAK-STAT	K	7.2210	13.9050	3.8933	4.1197	4.1197	5.2×10^6
($L=11$ true	r	0.0158	0.0250	**0.2368**	0.2042	0.2042	0.0070
missing links	χ^2	36.1173**	54.3138**	5.3417**	20.6666**	20.6666**	139.9072**
JNK ($L=16$)	K	16.0194	21.8450	2.2015	4.4009	5.5020	6.8×10^5
true missing	r	**0.1672**	0.0347	0.0359	0.0147	0.0138	0.0172
links	χ^2	43.4027**	127.4522**	12.1910**	21.7042**	21.3651*	695.3245*
MARK	K	3.2×10^5	4.9×10^6	2.2247	2.1745	2.1745	15.7222
($L=15$ true	r	0.0104	0.0119	**0.0460**	0.0296	0.0296	0.0122
missing links	χ^2	117.8132**	71.5835**	36.0525**	69.4835**	69.4835**	496.0534**
p53 ($L=13$	K	11.2094	15.0853	1.7×10^{10}	3.0554	3.0530	6.8874
true missing	r	0.0788	0.0163	0.0039	0.2080	**0.2250**	0.0327
links	χ^2	59.8453**	653.3348**	56.2755**	6.8479**	7.4530**	26.0670**

Notes: **: $p < 0.01$; *: $p < 0.05$.

Chapter 18

Link Prediction: Node-Similarity-based Methods

Zhang (2015e) proposed a link prediction method based on between-node similarity. Later, a method for screening node attributes that significantly influences node centrality in the network was presented (Zhang, 2016p). These methods and application examples are described below.

18.1 Link Prediction Based on Node Similarity

Suppose there is an incomplete network X with m nodes, its adjacency matrix is $d = (d_{ij})_{m \times m}$. $d_{ij} = 1$, if two nodes v_i and v_j are adjacent, and $d_{ij} = 0$, if v_i and v_j are not adjacent; $i, j = 1, 2, \ldots, m$. Adjacency matrix d is a symmetric matrix, i.e., $d = d'$. Known n attributes for m nodes. The raw data matrix is $a = (a_{ij})_{m \times n}$ Pearson correlation measure, cosine measure, and (negative) Euclidean distance measure (the three measures are for interval attributes); contingency correlation measure (for nominal (1, 2, 3...) attributes); and Jaccard coefficient measure (for binary (0, 1) attributes) can be used to measure between-node similarity.

Pearson correlation measure is

$$r_{ij} = \frac{\sum_{k=1}^{n}((a_{ik} - a_{ib})(a_{jk} - a_{jb}))}{(\sum_{k=1}^{n}(a_{ik} - a_{ib})^2 \sum_{k=1}^{n}(a_{jk} - a_{jb})^2)^{1/2}},$$

$$i, j = 1, 2, \ldots, m$$

where $-1 \leq r_{ij} \leq 1$, $a_{ib} = \sum_{k=1}^{n} a_{ik}/n$, $a_{jb} = \sum_{k=1}^{n} a_{jk}/n$, $i, j = 1, 2, \ldots, m$.

Cosine measure is (Zhang, 2007; Zhang, 2012e)

$$r_{ij} = \sum_{k=1}^{n} a_{ik} a_{jk} \bigg/ \left(\sum_{k=1}^{n} a_{ik}^2 \sum_{k=1}^{n} a_{jk}^2 \right)^{1/2},$$

$$i, j = 1, 2, \ldots, m$$

Euclidean distance measure is (Zhang, 2007, 2012e)

$$d_{ij} = \left(\sum_{k=1}^{n} (a_{ik} - a_{jk})^2 \right)^{1/2}$$

Thus, its negative value is used as the similarity measure

$$r_{ij} = -d_{ij}$$

Contingency correlation measure is (Zhang, 2007, 2012e; Zhang *et al.*, 2014b)

$$r_{ij} = 2(h/(s(p-1)))^{1/2} - 1, \quad i, j = 1, 2, \ldots, m$$

where $-1 \leq r_{ij} \leq 1$, and

$$h = s_{..} \left(\sum_{i=1}^{p} \sum_{j=1}^{p} s_{ij}^2/(s_{i.}s_{.j}) - 1 \right)$$

$$s_{..} = \sum_{i=1}^{p} s_{i.}, \quad s_{i.} = \sum_{j=1}^{p} s_{ij}, \quad n_{.j} = \sum_{i=1}^{p} s_{ij}$$

where there are p available nominal values, i.e., $t_1 t_2, \ldots, t_p$, for attributes i and j, s_{kl} is the number of attributes of node i taking value t_k and node j taking value t_l, $k, l = 1, 2, \ldots, p$.

Jaccard coefficient measure is

$$r_{ij} = (e - (c + b))/(e + c + b), \quad i, j = 1, 2, \ldots, m$$

where $-1 \leq r_{ij} \leq 1$, c is the number of node pairs of 1 for attribute i but not for j; b is the number of node pairs of 1 for attribute j but not for i; e is the number of node pairs of 1 for both attribute i and attribute j.

Between-node similarity matrix, $r = (r_{ij})_{m \times m}$, is a symmetric matrix, i.e., $r = r'$. In this algorithm, whether a node v_k can connect to v_i or not depends on the similarity between v_k and v_i, the similarities between v_i and its adjacent nodes, the similarities between v_k and the adjacent nodes of v_i, and the degree of node v_i, and vice versa. The procedures of the algorithm for prediction of missing links are as follows (Zhang, 2015e)

(1) For each node v_i, $i = 1, 2, \ldots, m$, and $\forall v_j \in S_i = \{v_k | d_{ik} = 1\}$, calculate mean similarity between v_i and $\forall v_j \in S_i$, and mean similarity of $\forall v_j \in S_i$,

$$i\_\text{mean} = \text{mean } r_{ij}, \quad v_j \in S_i$$

$$i\_\text{adj\_mean} = \text{mean } r_{kj}, \quad v_k \in S_i, v_j \in S_i$$

(2) For a node v_i, $i = 1, 2, \ldots, m$, the nodes $\forall v_j \in S_i$, and the node $v_k \notin S_i$, $k = 1, 2, \ldots, m$. First, calculate the mean similarity of between v_k and $\forall v_j \in S_i$

$$k\_i\_\text{adj mean} = \text{mean } r_{kj}, \quad v_j \in S_i$$

then calculate the similarity win

$$z_{ki} = \alpha(r_{ki} - i\_\text{mean}) + (1 - \alpha)(k\_i\_adj\_\text{mean} - i\_adj\_\text{mean})$$

where $v_k \neq v_i$, $d_{ki} = 0$, and α is the importance weight of node v_i against its adjacent node set S_i in determining whether the node v_k can connect to the node v_i or not, $0 \leq \alpha \leq 1$. Reverse v_k and v_i, and repeat steps (1) and (2), calculate z_{ik}.

(3) Calculate $z^{ik} = z_{ki} n_i / (n_k + n_i) + z_{ik} n_k / (n_k + n_i)$, where n_i is the degree of node v_i, i, $k = 1, 2, \ldots, m$; $v_k \neq v_i$, $d_{ki} = 0$. The weights, $n_i / (n_k + n_i)$ and $n_k / (n_k + n_i)$, are given because the nodes of greater degree are generally more important (Barabasi and Albert 1999; Zhang and Zhan, 2011; Huang and Zhang, 2012; Li and Zhang, 2013), and the calculation results on the nodes of greater degree are more statistically significant.

Finally, for $z^{ik} \geq 0$, calculate $y^{ik} = z^{ik}/2$, to achieve the averaged similarity win, which represents an averaged similarity win of a predicted missing link against existing links of the two nodes to be connected.

(4) Rank predicted node pairs from the larger y^{ik} to small ones. The predicted links with the larger y^{ik} have higher confidence degree.

(5) Once some of the predicted links are confirmed by observations, the adjacency matrix $d = (d_{ij})$ can be revised; return to step (1) to start a new round of prediction.

The following are Matlab codes of the algorithm:

```
raw=input('Input the file name of raw data (e.g., raw.txt, raw.xls, etc. The
matrix is z=(zij)m×n, where m is total number of nodes, n is the number of
attributes ):','s');
adj=input('Input the file name of adjacency matrix or its two-array form
(e.g., adj.txt, adj.xls, etc. Adjacency matrix is d=(dij)m×m, where m is
the number of nodes in the network. dij=1, if vi and vj are adjacent, and
dij=0, if vi and vj are not adjacent; i, j=1,2,..., m; two array form of
adjacency matrix, the 1st column is from nodes and 2nd column is to
nodes.):','s');
choice=input('Input a number to choose similarity measure (1: Pearson linear
correlation; 2: Cosine measure; 3: (Negative) Euclidean distance; 4:
Contingency correlation; 5: Jaccard coefficient):');
alpha=input('Input a weight between 0 and 1 for importance of a node to be
connected to against its adjacent nodes (e.g., 0.5, etc. weight=1, means
absolute importance of a node and no function of its adjacent nodes):');
raw=load(raw); m=size(raw,1); n=size(raw,2);
adj=load(adj);
if (size(adj,2)==2)
nn=size(adj,1);
adjj=zeros(m);
for i=1:nn
adjj(adj(i,1),adj(i,2))=1;
adjj(adj(i,2),adj(i,1))=1;
end
adj=adjj;
end
r=zeros(m);
for i=1:m-1
for j=i+1:m
ix=raw(i,:); jx=raw(j,:);
if (choice==1)
str='Pearson correlation';
ixbar=mean(ix);
jxbar=mean(jx);
aa=sum((ix-ixbar).*(jx-jxbar));
bb=sum((ix-ixbar).^2);
cc=sum((jx-jxbar).^2);
r(i,j)=aa/sqrt(bb*cc);
```

```
end
if (choice==2)
str='Cosine measure';
aa=sum(ix.*jx);
bb=sum(ix.^2);
cc=sum(jx.^2);
r(i,j)=aa/sqrt(bb*cc);
end
if (choice==3)
str='(Negative) Euclidean distance';
r(i,j)=-sqrt(sum((ix-jx).^2));
end
if (choice==4)
str='Contingency correlation';
xx=[ix;jx];
pn=1;
tt(1)=xx(1);
for kk=1:max(size(xx))
jj=0;
for ii=1:pn
if (xx(kk)~=tt(ii)) jj=jj+1; end;
end
if (jj==pn) pn=pn+1;tt(pn)=xx(kk); end;
end
for kk=1:pn
for jj=1:pn
temp(kk,jj)=0;
for ii=1:max(size(ix))
if ((ix(ii)==tt(kk)) & (jx(ii)==tt(jj))) temp(kk,jj)=temp(kk,jj)+1; end; end
end; end
for kk=1:pn
pp=0;
for jj=1:pn pp=pp+temp(kk,jj); end
ni(kk)=pp;
end
for kk=1:pn
pp=0;
for jj=1:pn pp=pp+temp(jj,kk); end
nj(kk)=pp;
end
summ=0;
for kk=1:pn
summ=summ+ni(kk);
end;
xsquare=0;
for kk=1:pn
for jj=1:pn
if (ni(kk)==0 | nj(jj)==0) continue; end
xsquare=xsquare+temp(kk,jj)*temp(kk,jj)/(ni(kk)*nj(jj));
end; end
```

```
xsquare=summ*(xsquare-1);
r(i,j)=2*sqrt(xsquare/(summ*(pn-1)))-1;
end
if (choice==5)
str='Jaccard coefficient';
bb=sum((ix==0) & (jx~=0));
cc=sum((ix~=0) & (jx==0));
dd=sum((ix~=0) & (jx~=0));
r(i,j)=(dd-(cc+bb))/(dd+cc+bb);
end
r(j,i)=r(i,j);
end; end
fprintf('\nPredicted potential links, similarity win, and similarity\n')
disp('Node Node Similarity win Similarity')
nn=0;
res=zeros(m*m,3);
ilinkmean=zeros(1,m); inlinkmean=zeros(1,m);
num=zeros(1,m); nu=zeros(1,m);
for i=1:m
nu(i)=0;
ilinkmean(i)=0;
for j=1:m
if (i==j) continue; end;
if (adj(i,j)~=0)
nu(i)=nu(i)+1;
num(nu(i))=j;
ilinkmean(i)=ilinkmean(i)+r(i,j);
end; end;
ilinkmean(i)=ilinkmean(i)/nu(i);
inlinkmean(i)=0;
if (nu(i)>1)
for j=1:nu(i)-1
for k=j+1:nu(i)
if (k==j) continue; end
inlinkmean(i)=inlinkmean(i)+r(num(j),num(k));
end; end;
inlinkmean(i)=inlinkmean(i)/((nu(i)^2-nu(i))/2); end
if (nu(i)==1)
for j=1:m
if (adj(i,j)~=0) inlinkmean(i)=r(i,j); end
end; end
jinlinkmean=zeros(1,m);
for j=1:m
if ((j==i) | (sum(j==num)==1) | (adj(i,j)~=0)) continue; end
jinlinkmean(j)=0;
for k=1:nu(i)
jinlinkmean(j)=jinlinkmean(j)+r(j,num(k));
end
jinlinkmean(j)=jinlinkmean(j)/nu(i);
z=alpha*(r(i,j)-ilinkmean(j))+(1-alpha)*(jinlinkmean(j)-inlinkmean(j));
```

```
nn=nn+1; res(nn,1)=i; res(nn,2)=j; res(nn,3)=z;
end; end
ress=zeros(m*m,4);
mm=0;
for i=1:m-1
for j=i+1:m
for k=1:nn
if ((res(k,1)==i) & (res(k,2)==j)) mm=mm+1; ress(mm,4)=r(i,j); ress(mm,1)=i;
ress(mm,2)=j; ress(mm,3)=ress(mm,3)+res(k,3)*nu(i)/(nu(i)+nu(j)); end;
if ((res(k,1)==j) & (res(k,2)==i))
ress(mm,3)=ress(mm,3)+res(k,3)*nu(j)/(nu(i)+nu(j)); end
end; end; end
ress(:,3)=round(ress(:,3)/2*10000)/10000;
ress(:,4)=round(ress(:,4)*10000)/10000;
iress=zeros(mm,4);
id=0;
for i=1:mm
if (ress(i,3)>=0) id=id+1; iress(id,:)=ress(i,:); end
end
ires=sortrows(iress(1:id,:),-3);
disp([ires])
```

Use the data of 12 Chinese human populations (nodes) and 17 common HLA-DQB1 alleles (attributes) (12×17 matrix; Zhang, 2015e) An adjacency matrix (12×12 matrix; Zhang, 2015e) for the network of 12 human populations, derived from linear correlation analysis is used also. Choosing Pearson correlation measure, the results are given in Table 1.

18.2 Screening Node Attributes that Significantly Influence Node Centrality

In nature, many networks, e.g., networks of protein–protein interactions, ecological networks, some social networks, etc. are created by links between nodes that have some correlations, similarities, or complementarities in node attributes (Zhang, 2011c, 2012e, 2015c, 2016p). Node attributes determine the evolution and topological structure of such networks. Nevertheless, we usually do not know which attributes are significant or crucial to node connection. Utilization of insignificant attributes will likely increase the noise of network information. Therefore, it is necessary to screen significant node attributes from more candidate attributes. The

Table 1   Predictions of links between 12 Chinese populations under different α (Zhang, 2015e).

$\alpha = 0$ (25 predicted links)				$\alpha = 0.5$ (24 predicted links)				$\alpha = 1$ (23 predicted links)			
Node	Node	Simil. Win	Simil.	Node	Node	Simil. Win	Simil.	Node	Node	Simil. Win	Simil.
3	9	0.1952	0.2517	7	9	**0.2172**	**0.5798**	7	9	**0.2577**	**0.5798**
4	9	0.1878	0.3699	6	12	**0.2072**	**0.58**	6	12	**0.2379**	**0.58**
6	9	0.184	0.4077	5	9	0.1816	0.41	5	12	**0.1795**	**0.4577**
5	9	0.1839	0.41	6	9	0.1812	0.4077	5	9	0.1794	0.41
5	**12**	**0.1794**	**0.4577**	5	12	**0.1795**	**0.4577**	6	9	0.1784	0.4077
7	**9**	**0.1766**	**0.5798**	4	9	0.1748	0.3699	10	12	0.173	0.4036
6	**12**	**0.1765**	**0.58**	10	12	0.1629	0.4036	4	9	0.1619	0.3699
2	9	0.1642	0.345	**10**	**11**	**0.1536**	**0.4698**	10	11	**0.1566**	**0.4698**
10	12	0.1528	0.4036	3	9	0.1527	0.2517	4	12	0.1461	0.3966
10	**11**	**0.1506**	**0.4698**	7	12	0.1326	0.2787	7	12	0.1239	0.2787
8	11	0.1467	0.32	2	9	0.1318	0.345	7	11	0.1198	0.3295
7	12	0.1413	0.2787	4	12	0.1318	0.3966	8	11	0.1164	0.32
8	12	0.1384	0.183	8	11	0.1315	0.32	**4**	**11**	**0.1162**	**0.4832**
7	11	0.1367	0.3295	7	11	0.1283	0.3295	3	9	0.1101	0.2517
4	**11**	**0.1302**	**0.4832**	**4**	**11**	**0.1232**	**0.4832**	2	9	0.0995	0.345
4	12	0.1176	0.3966	8	12	0.1099	0.183	8	12	0.0813	0.183
2	**10**	**0.1159**	**0.4808**	3	12	0.095	0.2429	3	12	0.0762	0.2429
3	12	0.1138	0.2429	3	11	0.0754	0.3054	3	11	0.0441	0.3054
3	11	0.1067	0.3054	**2**	**10**	**0.0686**	**0.4808**	9	11	0.0402	0.3214
1	9	0.1046	0.3837	1	9	0.059	0.3837	**2**	**6**	**0.036**	**0.5367**
2	**6**	**0.0695**	**0.5367**	**2**	**6**	**0.0528**	**0.5367**	9	12	0.0321	0.1284
2	5	0.0662	0.3984	9	11	0.0401	0.3214	**2**	**10**	**0.0214**	**0.4808**
9	12	0.0457	0.1284	9	12	0.0389	0.1284	1	9	0.0134	0.3837
9	11	0.04	0.3214	2	5	0.0166	0.3984				
2	12	0.0342	0.004								

Notes: Node IDs from 1 to 12 represent Tibetan, Uighur, Kazak, Xingjiang Han, Taiwanese, Hong Kong, Northern Han, Shanghai Han, Hunan Han, Manchu, Buyi, and Dai. Successfully predicted missing links are in bold.

following is a method to find node attributes that significantly influence node centrality in the network (Zhang, 2016e).

Suppose there is a weighted network X with m nodes, its weighted adjacency matrix is $d = (d_{ij})_{m \times m}$. $d_{ij} = w_{ij}$, if two nodes v_i and v_j are adjacent, and $d_{ij} = 0$, if v_i and v_j are not adjacent, where w_{ij} is the weight of the link from node v_i to node v_j, $i, j = 1, 2, \ldots, m$. For unweighted network, $w_{ij} = 1$, if two nodes v_i and v_j are adjacent, $i, j = 1, 2, \ldots, m$. Adjacency matrix d is a symmetric matrix, i.e.,

$d = d'$. Known n candidate attributes (i.e., traits) for m nodes (e.g., for m proteins, the attributes can be pH, isoelectric point, or molecular weight; or for m species, they can be n alleles; etc.), the raw data matrix is $x = (x_{ij})_{n \times m}$.

First, calculate the centrality indices of nodes, degree centrality, closeness centrality, betweenness centrality, and circuit centrality, each representing one of the topological properties of a network (Shams and Khansari 2014; Khansari *et al.*, 2016; Zhang, 2016e).

Second, use the stepwise linear regression to screen node attributes. The stepwise linear regression is a multivariable regression that can screen statistically significant variables into the linear regression equation (Zhang and Fang, 1982). The full multivariable linear regression equation is (Qi *et al.*, 2016)

$$y = b_0 + b_1 x_1 + b_2 x_2 + \cdots + b_n x_n$$

where x_i is the ith attribute. Let

$$l_{ij} = l_{ji} = \sum_{k=1}^{m} x_{jk} x_{ik} - \left[\sum_{k=1}^{m} x_{jk} \sum_{k=1}^{m} x_{ik} \right] \bigg/ m$$

$$l_{iy} = \sum_{k=1}^{m} x_{ik} y_k - \left[\sum_{k=1}^{m} x_{ik} \sum_{k=1}^{m} y_k \right] \bigg/ m$$

$$i, j = 1, 2, \ldots, n$$

Correlation coefficients between the attributes and between attribute and centrality index y are

$$r_{ij} = l_{ij} / (l_{ii} l_{jj})^{0.5}$$

$$r_{iy} = l_{iy} / (l_{ii} l_{yy})^{0.5}$$

Solve the equation

$$r_{i1} b'_1 + r_{i2} b'_2 + \cdots + r_{in} b'_n = r_{iy},$$

$$i = 1, 2, \ldots, n$$

The variance contribution of each attribute is

$$v_i = r_{iy}^2 / r_{ii}^2$$

Let $v_k = \max v_i$, and calculate the statistic $F = (m - l - 1)v_k/q$, where l is the number of attributes screened into the equation, q is the square of residuals. For first screening, $q = v_k$. If $F \geq F_a$, screen the attribute x_k into the equation ($F_a = 0.1$, etc.), or else remove x_k. The correlation matrix is changed as the following

$$r_{ij} = r_{ij} - r_{ik}r_{kj}/r_{kk}, \qquad i, j \neq k$$
$$r_{kj} = r_{kj}/r_{kk}, \qquad j \neq k, \ i = k$$
$$r_{jk} = -r_{jk}/r_{kk}, \qquad j \neq k, \ i = k$$
$$r_{kk} = 1/r_{kk}, \qquad i = k, \ j = k$$

where k is the kth screened or removed attribute. Calculate $v_k(l + 1) = \max v_i(l + 1)$, and $F = (m - l - 2)v_k(l+1)/(q(l) - v_k(l+1))$. If $F(l + 1) \geq F_a$, screen the attribute x_k into the equation, and change the correlation matrix. Let $v_k = \max v_i$, where x_k is the attribute already in the equation, $F_k = (m - l - 1)v_k(l)/q(l)$, where q is the r_{yy} in the inverse matrix of correlation matrix. If $F_k \leq F_a$, remove the attribute x_k from the equation, otherwise screen into the attribute. Repeat the procedure above, until no attribute can be screened into or remove from the equation (Zhang, 2016e).

Finally, the linear regression equation is obtained as the following

$$y = \tilde{b}_0 + \tilde{b}_i x_i + \cdots + \tilde{b}_j x_j + \cdots + \tilde{b}_k x_k$$

and the attributes remaining in the equation are qualified node attributes.

The following are Matlab codes, nodeIndicesScreen.m of the algorithm:

```
raw=input('Input the file name of node-by-attribute data (e.g., raw.txt,
raw.xls, etc. The matrix is z=(zij)n*m, where n is the number of candidate
attributes, m is total number of nodes):','s');
adj=input('Input the file name of adjacency matrix of unweighted or weighted
network (e.g., adj.txt, adj.xls, etc. Adjacency matrix is d=(dij)m*m, where
m is the number of nodes in the network. dij=1 for unweighted network and
dij=wij for weighted network (wij is the weight for the link vi to vj), if
vi and vj are adjacent, and dij=0, if vi and vj are not adjacent; i,
j=1,2,..., m):','s');
```

```
fs=input('Input the F threshold value for screening attributes (e.g., 0.1,
0.05):');
choice=input('Input the type of topological property of nodes (1: Degree
centrality; 2: Closeness centrality; 3: Betweenness centrality):');
raw=load(raw); nw=size(raw,1); m=size(raw,2); n=nw+1;
x=zeros(n,m);
adj=load(adj);
newdata=zeros(nw,m); xb=zeros(1,n); sg=zeros(1,n); ds=zeros(1,n);
degr=zeros(1,m);
switch choice
   case 1
      degr=sum(adj);
   case 2
      [ss,pat,distances,paths]=Dijkstra(adj);
      for i=1:m
      degr(i)=1/sum(distances(i,:));
      end
   case 3
      [ss,pat,distances,paths]=Dijkstra(adj);
      for i=1:m
      degr(i)=pat(i)/ss;
      end
end
a=zeros(n);
x=[raw;degr];
iss='';
for i=1:n
c=0;
for j=1:m
c=c+x(i,j);
end
xb(i)=c/m;
c=0;
for j=1:m
c=c+(x(i,j)-xb(i))^2;
end
sg(i)=sqrt(c);
end
h=sg(n);
for i=1:n-1
for j=i+1:n
c=0;
for k=1:m
c=c+(x(i,k)-xb(i))*(x(j,k)-xb(j));
end
a(i,j)=c/(sg(i)*sg(j)); a(j,i)=a(i,j);
end; end
for i=1:n
xb(i)=i; sg(i)=0; a(i,i)=1;
end
```

```
l=0; s=0;
while (n>=1)
if (l==n-1) break; end
ma=0;
for i=1:n
ds(i)=xb(i);
end
for i=1:n-1
if (ds(i)==0) continue; end
if (a(i,i)<1e-05) continue; end
v1=a(i,n)*a(n,i)/a(i,i);
if (v1<ma) ma=v1; k=i; end
end
f1=ma*(m-l-2)/(a(n,n)-ma);
if (f1<=fs) break; end
xb(k)=0; sg(k)=k;
l=l+1;
for i=1:n
for j=1:n
if ((i~=k) & (j~=k)) a(i,j)=a(i,j)-a(i,k)*a(k,j)/a(k,k); end
end; end
for j=1:n
if (j~=k) a(k,j)=a(k,j)/a(k,k);  a(j,k)=-a(j,k)/a(k,k); end
end
a(k,k)=1/a(k,k);
r=sqrt(1-a(n,n));
yn=h*sqrt(a(n,n)/(m-l-1));
if (s==0) s=1; continue; end
lab=0;
while (n>=1)
ma=-1e+18;
for i=1:n
ds(i)=sg(i);
end
for i=1:n-1
if (ds(i)==0) continue; end
if (a(i,i)<1e-05) continue; end
v1=a(i,n)*a(n,i)/a(i,i);
if (v1>ma) ma=v1;k=i; end
end
f1=-ma*(m-l-1)/a(n,n);
if (f1>fs) lab=1; break; end
sg(k)=0; xb(k)=k; l=l-1;
for i=1:n
for j=1:n
if ((i~=k) & (j~=k)) a(i,j)=a(i,j)-a(i,k)*a(k,j)/a(k,k); end
end; end
for j=1:n
if (j~=k) a(k,j)=a(k,j)/a(k,k);  a(j,k)=-a(j,k)/a(k,k); end
```

```
end
a(k,k)=1/a(k,k);
r=sqrt(1-a(n,n)); yn=h*sqrt(a(n,n)/(m-l-1));
end;
if (lab==1) continue; end
end
for i=1:n-1
a(i,1)=sg(i);
end
for i=1:n
c=0;
for j=1:m
c=c+x(i,j);
end
xb(i)=c/m;
c=0;
for j=1:m
c=c+(x(i,j)-xb(i))^2;
end
sg(i)=sqrt(c);
end
h=sg(n);
c=0;
for i=1:n-1
if (a(i,1)==0) continue; end
ds(i)=a(i,n)*sg(n)/sg(i); a(i,2)=ds(i); c=c+ds(i)*xb(i);
end
s=xb(n)-c;
iss=strcat(iss,'Screened attributes: \n');
nm=0;
for i=1:n-1
if (a(i,1)==0) continue; end
if (ds(i)~=0) iss=strcat(iss,'Attribute-',num2str(i)); end
if ((ds(i+1)~=0) & (i<n-1)) iss=strcat(iss,','); end
if (ds(i)~=0)
nm=nm+1;
for j=1:m
newdata(nm,j)=x(i,j);
end; end
end
fprintf('\nNew attribute-by-node data\n')
disp([newdata(1:nm,:)])
iss=strcat(iss,'\nStepwise regression equation:\n');
iss=strcat(iss,'y=',num2str(s));
for i=1:n-1
if (a(i,1)==0) continue; end
if (ds(i)>0) e1=num2str(ds(i)); end
if (ds(i)<0) e1=num2str(abs(ds(i))); end
if (ds(i)>0) iss=strcat(iss,'+',e1,'Attribute',num2str(i)); end
```

```
if (ds(i)<0) iss=strcat(iss,'-',e1,'Attribute',num2str(i)); end
end
iss=strcat(iss,'\nCorrelation coefficient R=',num2str(r),', ','F
  value=',num2str(fs),'\n');
fprintf(iss)
```

The functions, Dijkstra.m and foundCircuit.m, used to calculate the shortest path and corresponding distance between two nodes, and to calculate the fundamental circuits in the network, can be found in Zhang (2016b) and (Zhang, 2017b) respectively.

Data of 17 common HLA-DQB1 alleles (candidate attributes) for the world's 12 human races and populations (nodes) (17×12 matrix) are from Zhang and Qi (2014) and Zhang (2016e)) In addition, an adjacency matrix (12×12; Zhang, 2016e), derived from linear correlation analysis, is used.

In Table 2, nodes 1–12 represent Tibetan, Uighur, Kazak, Xingjiang Han, Taiwanese, Hong Kong, Northern Han, Shanghai Han, Hunan Han, Manchu, Buyi, and Dai.

Using different node centralities and statistical significances, screening results of node attributes are listed in Table 3. As an

Table 2 The world's 12 human races and populations (nodes) and 17 common HLA-DQB1 alleles (candidate attributes).

Allele ID	Alleles	1	2	3	4	5	6	7	8	9	10	11	12
1	DQB1*0201	102	325	238	186	87	135	143	140	92	106	46	60
2	DQB1*0301	71	140	214	169	230	225	159	191	146	245	199	130
3	DQB1*0302	163	114	71	59	82	49	77	67	8	53	38	40
4	DQB1*0303	153	44	71	119	184	163	170	174	62	202	86	170
5	DQB1*0401	10	44	12	51	0	58	50	45	15	64	15	0
6	DQB1*0402	41	18	71	25	0	0	11	0	0	0	0	0
7	DQB1*0501	123	53	48	25	20	37	77	45	15	43	38	80
8	DQB1*0502	51	0	48	68	71	106	28	11	92	43	331	380
9	DQB1*05031	71	44	48	76	36	56	44	6	0	74	46	90
11	DQB1*05032	10	0	0	0	0	0	0	0	31	0	0	0
12	DQB1*0504	0	0	24	25	0	0	0	0	46	0	0	0
13	DQB1*0601	10	44	36	102	158	88	99	129	146	85	136	30
14	DQB1*0602	153	105	36	42	36	47	121	124	331	64	23	0
15	DQB1*0603	10	35	60	25	5	0	17	34	8	0	0	0
16	DQB1*0604	10	18	24	25	5	12	0	28	0	21	0	0
17	DQB1*06051	20	18	0	0	20	0	6	0	0	0	0	10
18	DQB1*null	0	0	0	0	0	25	0	6	7	0	42	10

Table 3 Screened node attributes when different node centralities (dependent variables) and statistical significance levels are used. "+" denotes positive influence and "−" denotes negative influence (Zhang, 2016e).

Attributes	Node centrality		Closeness centrality		Betweenness centrality		Circuit centrality	
	$F = 0.1$	$F = 0.05$	$F = 0.1$	$F = 0.05$	$F = 0.1$	$F = 0.05$	$F = 0.1$	$F = 0.05$
1 DQB1*0201	+	+			+	+		
2 DQB1*0301			−	−	−	−	−	−
3 DQB1*0302	−	−						
4 DQB1*0303	+	+	+	+	+	+	+	+
5 DQB1*0401							+	+
6 DQB1*0402			−	−			+	+
7 DQB1*0501							−	−
8 DQB1*0502	−	−	−	−				
9 DQB1*05031			−	−				
10 DQB1*05032	−	−	+	+	−	−	−	−
11 DQB1*0504	+	+			+	+	+	+
12 DQB1*0601	+	+	+	+	+	+		
13 DQB1*0602			−	−	+	+	+	+
14 DQB1*0603	+	+			+	+	+	+
15 DQB1*0604	−	−	−	−	−	−	−	−
16 DQB1*06051			−	−				
17 DQB1*null	+	+			+	+		
R	0.9999	0.9999	0.9999	0.9999	1	1	0.9998	0.9998

example, the stepwise linear regression equation for node centrality and $F = 0.1$ is

$$y = -029311 + 0.00899\, x_1 - 0.00963\, x_3 + 0.03587\, x_4 - 0.01650\, x_8$$
$$- 0.25115\, x_1 + 0.10231\, x_{11} + 0.01261\, x_{12} + 0.05268\, x_{14}$$
$$- 0.08376\, x_{15} + 0.09326\, x_{17}$$

$$R = 0.9999$$

where $x_i, i = 1, 3, 4, 8, 10, 11\ 12, 14, 15$, and 17, denotes the screened common HLA-DQB1 alleles (attributes).

Chapter 19

Node Prediction

In the sampling of statistic networks (Zhang, 2011c, 2012b), the number of new nodes will decline with increase in sample size, and it tends to be an upper asymptote as sample size tends to infinity (Zhang, 2017a, 2017n). For example, the number of new species decreases with an increase of sample size in the studies of ecological networks (Zhang and Schoenly, 1999). In most cases, our sampling is incomplete, i.e., the samples we have taken are limited. Thus, the exact number of nodes of a statistic network is unknown. We need to find some methods to estimate node richness in statistic networks. So far, a lot of methods on link prediction have been proposed. But there are few studies on node richness estimation. As a case study, Zhang (2017a) used some nonparametric methods (Burnham and Overton, 1978, 1979; Chao, 1984; Chao and Lee, 1992; Colwell and Coddington, 1994; Zhang and Schoenly, 1999b) to estimate node richness in the network. This chapter describes this research below.

19.1 Methods

Six nonparametric estimators are used to estimate node richness of statistic networks (Colwell and Coddington, 1994; Zhang and Schoenly, 1999b; Zhang, 2017a).

(1) Chao (1984) requires presence–absence data only (Zhang and Schoenly, 1999b)

$$S = S_{\text{obs}} + L^2/(2M)$$

where L and M are the number of nodes that occur in only one and two samples, respectively.

(2) Jackknife 1 (Burnham and Overton, 1978, 1979) is the 1st-order jackknife estimator. It can be used to estimate total number of nodes from a sample (Zhang and Schoenly, 1999b)

$$S = S_{\text{obs}} + L(n-1)/n$$

where L is the number of nodes found in only one sample, and n is the number of samples.

(3) Jackknife 2 (Burnham and Overton, 1978, 1979) is the 2nd-order jackknife estimator (Zhang and Schoenly, 1999b)

$$S = S_{\text{obs}} + L(2n-3)/n - M(n-2)^2/[n(n-1)]$$

where L, M, and n are the same as the above.

(4) Bootstrap (Smith and van Belle, 1984)

$$S = S_{\text{obs}} + \sum_{j=1}^{S_{\text{obs}}} (1 - p_j)^n$$

where p_j is the proportion of samples containing node j.

(5) Two methods of Chao and Lee (1992) are based on sample coverage (Zhang and Schoenly, 1999b)

$$S = D/C + n(1-C)/C^* g^2 \quad S = D/C + n(1-C)/C^* \beta^2$$

where

$$C = 1 - f_1/n$$

$$n = \sum i f_i \quad g^2 = \max\left\{ D/C^* \sum i(i-1) f_i/(n(n-1)) - 1, 0 \right\}$$

$$\beta^2 = \max\left\{ g^2 \left[1 + n(1-C) \sum i(i-1) f_i / \ldots (n(n-1)C) \right], 0 \right\}$$

and $D = \sum f_i$ and f_i is the number of classes that have exactly i elements in the sample.

The following are Matlab codes, nodeEst.m, of the nonparametric methods to estimate node richness in a network (Zhang, 2017a):

```
samp=input('Input the file name of sampling data (e.g., raw.xls, etc.
Sampling data matrix is s=(sij)m*n, where m is the number of nodes (species,
or objects, etc.) in the network, n is the number of samples): ','s');
    ss=xlsread(samp);
    m=size(ss,1); n=size(ss,2);
    dat=ss';
    Sobs=m;
    v=sum(dat~=0);
    L=sum(v==1); M=sum(v==2);
    Chao=Sobs+L∧2/(2*M)
    Jackknife1=Sobs+L*(n-1)/n
    Jackknife2=Sobs+L*(2*n-3)/n-M*(n-2)∧2/(n*(n-1))
    Bootstrap=Sobs+sum((1-v/n).∧n)
    for i=1:Sobs
    ps(i)=0;
    for j=1:n
    if (dat(j,i)~=0) ps(i)=ps(i)+1; end
    end; end
    for i=1:n
    w(i)=0;
    for j=1:Sobs
    if (ps(j)==i) w(i)=w(i)+1; end
    end; end
    gam1=0; gam2=0;
    d=sum(w);
    sp1=0; sp2=0;
    for i=1:n
    sp1=sp1+i*(i-1)*w(i);
    sp2=sp2+i*w(i);
    end
    cbar=1-w(1)/sp2;
    sp3=d/cbar*sp1/(sp2*(sp2-1))-1;
    if (sp3>0) gam1=sp3; else gam1=0; end
    sp4=gam1*(1+sp2*(1-cbar)*sp1/(sp2*(sp2-1)*cbar));
    if (sp4>0) gam2=sp4; else gam2=0; end
    if (cbar==0)
    ChaoLee1=Sobs
    ChaoLee2=Sobs
    else
    ChaoLee1=d/cbar+sp2*(1-cbar)/cbar*gam1
    ChaoLee2=d/cbar+sp2*(1-cbar)/cbar*gam2
    end
```

19.2 Application Example

Application data are from our field sampling on communities of arthropods and weeds around Pearl River delta and Zhuhai Campus of SYS University in 2008. Arthropods data for different taxa and areas are represented by data set names gz, fampea, spepea, and weed data for different taxa and areas are represented by data set names weedspepea, weedspezhu, weedfampea.

Table 1 lists estimated taxa richness in six arthropod and weed communities.

Table 1   Estimation of taxa richness for various ecological networks.

	gz	fampea	spepea	weedspepea	weedspezhu	weedfampea
Chao	72	143	149	80	—	41
Jackknife 1	59	71	148	75	53	33
Jackknife 2	69	83	163	85	59	39
Bootstrap	50	63	131	65	49	28
Chao & Lee 1	74	71	176	92	53	42
Chao & Lee 2	101	77	218	113	55	56
Mean	71	85	164	85	54	40

Part 6

Network Construction

Chapter 20

Construction of Biological Networks

The construction of biological networks is strongly related to link prediction. In a sense, link prediction is a part of network construction. In this chapter, the methods for both integral construction of complete networks and link prediction of incomplete networks, at biochemical and ecological levels, are discussed.

20.1 Prediction Methods of Protein–Protein Interactions

Predicting protein–protein interactions is a great challenge. Such simulation and calculation are much faster and less costly than most of the experimental methods. In the past years, a variety of algorithms have been used to predict whether proteins are interacted. These methods can be summed up into four categories (Tu, 2006): (1) the methods based on genomic information; (2) the methods based on evolutionary relationships; (3) the methods based on protein sequences of the *ab initio* prediction method, and (4) the methods based on three-dimensional structure of proteins. In comparing the different methods, the main problem is that the various methods use different databases, and the reliability of the data in these databases is quite inconsistent.

20.1.1   Phylogenetic profile

The phylogenetic profile method is based on the assumption that functionally related genes are expected to exist simultaneously or not in a fully sequenced genome. The pattern for presence or absence is called phylogenetic profile. If the sequences of two genes are not homologous, but they have the similar or the same phylogenetic profiles, they can be inferred as functionally relevant (Gaasterland and Ragan, 1998; Psilegrini *et al.*, 1999). This method provides a way for annotations of unknown functional proteins. The limitation is that it cannot determine whether the function-related proteins are physically in direct contact; it can only annotate the function of nonessential functions, and its accuracy depends on the number of genomes to be sequenced and the reliability of the phylogenetic profile method used.

20.1.2   Gene neighborhood

The gene neighborhood method is based on the fact that, in the bacterial genome, functionally related genes are closely linked to a particular region to form an operon (Tamames *et al.*, 1997). The adjacency between these genes is conservative in species evolution. It can be used as the indicator of the functional relationship between gene products (Dandekar *et al.*, 1998; Overbeek, 1999). This method appears to be applicable only to the early evolution of simple microbes.

20.1.3   Gene fusion event

The gene fusion event method is based on the assumption that two or more interacting proteins of one species are fused in another species into a polypeptide chain due to genetic fusion events during species evolution. Gene fusion event can serve as an indicator of function correlation or interaction of proteins. The limitation of this method is that it cannot be judged whether the fused proteins are physically in direct contact with each other. The mechanism of gene fusion events may be complex and diverse (Marcot *et al.*, 1999).

20.1.4 *Mirror tree*

The idea of the mirror tree method is that the function-dependent proteins or the domains in the same protein are constrained by functional constraints and that their evolutionary processes should be consistent (i.e., coevolution; Fryxell, 1996; Pages *et al.*, 1997). Constructing and comparing their phylogenetic trees, if their tree topology is found to be similar, the trees are called mirror trees. Furthermore, it can be inferred that the functions of the tree genes are related. Goh *et al.* (2000) introduced the linear correlation coefficient to quantify the similarity of trees. In essence, this method is consistent with the above phylogenetic profile method. The limitation of this method is that the required sequences of the two proteins should be found in multiple identical species and that multiple sequence alignments can be obtained by multisequence alignment.

20.1.5 *Correlated mutation*

In proteins that are physically in direct contact with each other, such as proteins in the same structural complex, the accumulated residue change of a protein in the evolutionary process is compensated by the corresponding change in another protein. This phenomenon is called correlated mutation (Gobel *et al.*, 1994), which can be thought of as a mirror tree at the amino acid level (also called in-silico two-hybrid method). This method provides an early theoretical basis for *ab initio* prediction based on the three-dimensional structure of proteins (Olmea *et al.*, 1999).

20.1.6 *Correlated evolutionary rate*

The principle of the correlated evolutionary rate is that the evolutionary rate of a protein is determined by the number of interactions between the protein and other proteins, and it yields the negative correlation. That is, the more the number of interactions, the lower the evolutionary rate. The evolutionary rate of a protein is not determined by the importance of the protein.

20.1.7   *Interologs*

The interologs method is based on the principle that interacting proteins are conservative during species evolution and that we can use the protein network constructed in a species to predict the protein–protein interactions in another species (Walhout *et al.*, 2000).

20.1.8   *Prediction based on primary structure of proteins*

The method was first proposed by Bock and Gough (2001), and it does not require genetic or evolutionary information and requires only the information of a single protein sequence. This method extracts the sequence data of the interacting proteins from the DIP database and the data is trained by the support vector machine (SVM) method based on sequence information of the protein pair, including the physical and chemical properties of the amino acid residues, the charges and the hydrophobic properties, etc.

20.1.9   *Prediction based on three-dimensional structure of proteins*

Homologous structural complexes is the method that uses three-dimensional structure of proteins to predict protein–protein interactions. The principle of this method is, in the protein complex with known three-dimensional structure, the proteins in the same family interact in the same way (Gawwad *et al.*, 2013a, 2013b; Cemanovic *et al.*, 2014; Sutkovic and Gawwad, 2014). The probability of the interaction of the two proteins should be proportional to probability of the interaction between the domains (Deng *et al.*, 2002).

20.2   Gene Coexpression Network Analysis

Gene coexpression network analysis is one of the most important methods for constructing gene networks (Stuart *et al.*, 2003; Ideker and Krogan, 2012; Chen, 2016). Coexpression of genes refers to the fact that the two genes are simultaneously up-regulated or down-regulated in different biological samples (e.g., different tissues of the human body or yeast in different media). The gene coexpression network is a gene network constructed from the similarity between expression data of genes. Two genes with similar expression profiles are connected to form a link in the network. The similarity between gene expression data can be used to analyze the possible interaction between gene products so as to understand the interaction between genes and find the core genes. Here, the core genes are important hubs and play the critical role in network modules.

In general, researchers use Pearson correlation coefficient to determine whether the two genes have a coexpression relationship. That is, the expression levels of the genes A and B are measured in n biological samples. With the logarithmic of expression of the gene A as the X-axis (log A), the logarithmic of expression of the gene B as the Y-axis (log B), we get n data points, and the Pearson correlation coefficient is further obtained. We can manually set a threshold for Pearson correlation coefficient. Any pair of genes, A and B, whose Pearson correlation coefficient exceeds the threshold, has a link between them. By doing so, a gene coexpression network is eventually constructed.

Gene coexpression network analysis has several drawbacks: (1) it is known that complex networks have scale-free features (Barabasi and Oltvai, 2004), but by using this method the network constructed from very similar biological samples are often not scale-free; (2) the constructed network is unweighted and cannot reflect the continuous relationship of coexpression; moreover, it cannot reflect the level of coexpression (Langfelder and Horvath, 2008). In order to solve these problems, the researchers recently developed a method called WGCNA (Weight Gene Coexpression Network Analysis) (Langfelder and Horvath, 2008). The original idea of this method is the topological overlap matrix (Ravasz *et al.*, 2002). The simplified algorithm is as follows: For each of two given nodes, calculate all neighbors of the node, and calculate the number of neighbors shared by the two nodes. The correlation between two nodes is defined by the ratio of the shared nodes to all neighboring nodes. The algorithm of WGCNA is as follows (Langfelder and Horvath, 2008): (1) Obtain the logarithmic absolute expression of all the genes; (2) calculate the

Pearson correlation coefficient r_{ij} between gene i and gene j; (3) set a parameter β, and modify the correlation coefficient between the two genes as r_{ij}^{β}; (4) calculate the connectivity of gene i, $a_i = \Sigma_{j=1}^{n} r_{ij}^{\beta}$, and achieve the distribution of a_i of all genes; (5) adjust β so that the distribution of a_i is a power-law distribution (Zhang and Horvath, 2005); (6) define the topological overlap matrix between genes i and j, $T_{ij} = (t_{ij})$

$$t_{ij} = (l_{ij} + a_{ij})/(\min(k_i, k_j) + 1 - a_{ij}), \quad i \neq j$$
$$t_{ij} = 1, \quad i = j$$

where $l_{ij} = \Sigma_u a_{iu} a_{uj}$, $k_i = \Sigma_u a_{iu}$; (7) cluster all T_{ij} using cluster by average linkage algorithm; (8) define $r_{ij} = 1 - d_{ij}$ as the correlation of coexpression of the two genes, where d_{ij} is between-node distance in the cluster analysis. Finally, $R = (r_{ij})$ is the correlation matrix of weight gene coexpression network.

20.3   Classification-based Machine Learning

Gene coexpression network analysis needs to collect samples under a variety of conditions and development phases, which is a disadvantage for the study of specific development phase or condition. Machine learning is another option for studying gene relevance. Machine learning can be divided into supervised and unsupervised methods depending on whether the training set is used. Supervised methods use the existing knowledge to generate the prediction model. This prediction model can further help us to identify undefined variables.

Classification is a major theme of machine learning, also known as pattern recognition or supervised learning. In the past years, more and more computationally flexible classifiers have been proposed, such as classification trees, SVMs, and some regularized versions of more classical classifiers. Some of the more widely used classification algorithms are outlined below.

(1) Decision tree

Decision tree is one of the main techniques for classification and prediction (Zhang, 2012e, 2016r). Decision tree learning is a instance-based induction process. It deduces some classification rules, represented by a decision tree, from a set of unordered and irregular instances. We construct a decision tree in order to find the relationship between attributes and categories, and further use it to predict the categories of unknown records. It uses the top-down recursive method to compare nodes in terms of their attributes and determine the next link from the node according to the different attribute values. The conclusion is drawn at the leaf nodes of the decision tree.

(2) Bayesian classification

The Bayesian classification is algorithms that use Bayesian theorem to conduct classification (Zhang, 2010, 2012e, 2016r). These algorithms are mainly used to estimate the possibility that an unknown class of sample belongs to each category, and then select one of the most likely categories as the final category of the sample. Bayesian theorem itself requires a strong prerequisite for conditional independence, but this assumption is often not true in the actual situations, and therefore its classification accuracy will not be ensured. Thus, many improved algorithms of Bayesian classification are proposed, such as the Tree Augmented Naive Bayes (TAN) algorithm, which is based on the Bayesian network structure and increases the association between attribute pairs.

(3) Artificial Neural Networks

Artificial Neural Networks (ANNs) are mathematical models that are mathematically similar to the neural network of the human brain (Bian and Zhang, 2000; Yan and Zhang, 2000; Zhang, 2010; Zhang and Barrion, 2006; Zhang et al., 2007; Zhang et al., 2008a, 2008b; Zhang and Zhang, 2008; Zhang and Wei, 2009). Hundreds of ANNs have been proposed, including BP, RBF, Hopfield, Boltzmann machine, Hamming network, self-organizing networks, etc., (Zhang, 2010).

(4) k-Nearest Neighbors

k-Nearest Neighbors (KNN) is an instance-based classification method. The method is proposed to find out the k training samples closest to the unknown sample x, and classify x into the category that most of the k training samples belong to. It is a lazy learning method that stores samples until they are to be classified. The computational cost of KNN is huge if the sample set is complex.

(5) Support Vector Machine

SVM is a statistical learning method (Bian and Zhang, 2000; Yan and Zhang, 2000). Based on the criterion of structural risk minimization, it constructs the optimal classification superplanar, with maximized classification interval, to improve the generalization ability of the learning machine. SVM solves such problems as nonlinear, high-dimensional, local minimization, and other issues. For the classification problem, SVM algorithm calculates the decision surface of a region according to the samples in the region, thus determining the category to which unknown samples in the region belong.

(6) Classification based on association rules

Association rule mining is an important research field in data mining. The method tries to find the rules with the form of *condset→ C*, where *condset* is the set of items (or attribute–value pairs), and C is the class label, which are called class association rules (CARS). The association classification is usually composed of two steps: (1) use the association rule mining algorithm to extract all the association rules with specified support degree and confidence degree from the training data set, and (2) use the heuristic method to extract a set of high-quality association rules from existing association rules.

(7) Ensemble Learning

Ensemble Learning refers to integration of all results from different classifiers.

20.4 Statistic Network

The statistic network is a weighted and nondeterministic network (Zhang, 2012b). In the statistic network, a connection (link, interaction) value, i.e., connection weight, represents connection strength and connection likelihood between two nodes (species, protein, etc.) and its absolute value falls in the interval (0, 1]. The connection value is expressed as a statistical measure such as correlation coefficient, association coefficient, or Jaccard coefficient. In addition, all connections of the statistic network can be statistically tested for their validity. A connection is true if the connection value is statistically significant. If all connection values of a node are not statistically significant, it is an isolated node. An isolated node has no connection to other nodes in the statistic network. Positive and negative connection values denote distinct connection types (positive or negative association or interaction). In the statistic network, two nodes with the greater connection value will show more similar trend in the change of their states. At any time, we can obtain a sample network of the statistic network. A sample network is an unweighted and deterministic network.

All connections in a deterministic network are actually those connections that occur with connection values 1 or −1. Therefore in a sense, the deterministic network is a special case of statistic network.

To calculate between-node connection values, we need to find a statistic measure. The most useful measure is correlation coefficient (Zhang, 2007, 2011c, 2015c; Zhang and Li, 2015b). The statistically significant correlations may represent the true (direct or indirect) association or interaction between two nodes. A positive correlation (association, interaction) means the two taxa (species, family, etc.) tend to jointly occur in a sample, and a negative one means the two taxa tend to exclusively occur in a sample (Zhang, 2012b).

Various correlation measures, such as the following can be used in the construction of statistic networks (Zhang, 2012b): (1) Linear (Pearson) correlation. The statistical significance of linear correlation can be tested using t-test. (2) Partial linear correlation. Partial (pure) linear correlation is based on linear correlation, which reflects between-node direct interaction (Zhang, 2011c). The statistic significance of partial linear correlation can be tested using t-test. Using partial correlation will yield a statistic network with all connections as direct association (interaction). (3) Spearman rank correlation. Spearman rank correlation (Spearman, 1904; Schoenly and Zhang, 1999) is a correlation measure to denote a weak linear (quasi-linear) relationship. In addition, other correlation measures can be used (Zhang, 2012e, 2015d, 2016r; Zhang and Li, 2015a).

The networks derived from linear and partial linear correlation measures are linear networks, and the networks derived from Spearman correlation, which is a quasi-linear correlation measure, are a kind of quasi-linear networks. Statistically significant correlations, i.e., node pairs with significant linear or quasi-linear dependency are included in two kinds of networks respectively.

In a linear network, the states of two connected nodes will show a linear dependent relationship, whereas for a quasi-linear network the states of two linked nodes will show a quasi-linear dependent relationship (Zhang, 2012b). Nodes that never follow linear and quasi-linear relationships are excluded from the two networks. In a sense, nodes in the network are relatively predictable, but isolated nodes are hard to be predicted (Zhang, 2011c).

A network can be a linear network, quasi-linear network, or nonlinear network. However, a network that changes in a local domain (a short time, or a small extent, etc.) can be approximated with a linear network, i.e., in the local domain, all between-node changes can be treated as the linear ones (Zhang and Li, 2015b).

As mentioned above, using partial correlation will result in a network with all links as direct interactions, but Pearson linear correlation and Spearman correlation can be used to create a network with links as indirect interactions.

Many ecological networks and molecular networks are statistic networks. For example, many cancer networks are statistic networks, and thus their networks may change as time and environmental condition. It will increase the difficulty of prevention and treatment of these cancers. Due to such a property, the network or interaction derived from one experiment only is a sample network/occasional interaction but not a complete statistic network/deterministic interaction.

20.5 Homogeneity Test of Samples and Sampling Completeness

In the sampling of network construction, we need to record the nodes quickly (species, families, etc.) (Zhang, 2011b). To examine sampling completeness, a yield–effort curve may be drawn, which plots the cumulative number of taxa caught or observed (y-axis) against the cumulative effort of sampling (x-axis) (Cohen, 1978; Dickerson and Robinson, 1985; Cohen *et al.*, 1993; Zhang and Schoenly, 1999a). If sampling stops while the yield–effort curve is still rapidly increasing, then the network derived from this sampling is most likely incomplete. If sampling ceases when the slope of the yield–effort curve reaches zero or close to zero, the sampling is probably complete. In addition, how to ensure sample homogeneity is also necessary in sampling studies. Bias from sample order can be corrected by bootstrap procedure. But variation in curve shape due to environmental heterogeneity remains a likely significant source of sampling error (Zhang and Schoenly, 1999a). Coleman *et al.* (1982) presented a statistical method to test whether individuals among the samples (of definable size) obey the random placement hypothesis, which assumes a lack of correlation in the location of individuals (Zhang and Schoenly, 1999a). The method can be used to test sample homogeneity and examine sampling completeness.

20.5.1 *Coleman's method*

According to Coleman's random placement hypothesis (Coleman *et al.*, 1982), consider a collection C of N individuals from S taxa, with n_i individuals in C belonging to the ith taxon, and suppose that each member of C occurs in one of k nonoverlapping samples that have areas $a_1, a_2, \ldots, a_k$. The number of taxa, s, in a given region is a random variable whose magnitude depends on the area a of the region, and the relative area is defined as $\alpha = a / \sum a_i$. The mean number of taxa, s, and the variance σ^2 are calculated as follows

$$s(\alpha) = S - \sum (1 - \alpha)^{n_i}$$

$$\sigma^2(\alpha) = \sum (1 - \alpha)^{n_i} - \sum (1 - \alpha)^{2n_i}$$

The method to test sample homogeneity is to compare the observed mean taxa richness *vs.* the sample size with the expected taxa richness *vs.* the sample size curve (Zhang and Schoenly, 1999a; Zhang, 2011b). If 95% of the plotted points (means) of the observed curve fall two standard deviations outside the expected curve, the observed samples are considered to be statistically more heterogeneous in taxa composition (at the 0.05 level) than sampling error (alone) can account for (Coleman *et al.*, 1982). Therefore, we can conclude that these samples are more heterogeneous in taxonomic composition than is expected under the random placement hypothesis.

Bootstrap procedures are used to produce the taxa richness *vs.* the sample size curves (Zhang, 2011b). The curves plot the cumulative number of taxa, defined as the sum of the number of taxa in the previous sample(s) and the number of taxa in the present sample that were not observed in any previous sample. For the first sample, the cumulative number of taxa is defined to be equal number of taxa found in this sample.

If random placement hypothesis is met, the samples are homogeneous, otherwise they are heterogeneous. If the difference of the number of taxa between the last two (cumulative) sample sizes is less than desired percent threshold, most of the taxa are considered to be recorded and the sample size is enough.

The algorithm is implemented as a Matlab program, SampHomoTest.m. In sampling data file, the rows are taxa and the columns are samples.

```
raw=input('Input the excel file name of taxa-by-sample data (e.g., raw.xls,
    etc.):','s');
sim=input('Input the maximum number of simulations (e.g., 100):');
pro=input('Input sampling completeness threshold (e.g., 0.01, 0.05):');
sp=xlsread(raw);
spp=sp';
col=size(sp,2); nspp=size(sp,1);
n=zeros(1,nspp);
m=zeros(1,col);
pool=col-1;
tg=0;
fprintf('SSze MnObsS ExpS ExpStd ExpS-2SD ExpS+2SD ObsS-2SD Obs+2SD\n');
for i=1:nspp
n(i)=sum(spp(:,i));
end
while (col>0)
sobm=0;
su=0;
al=(col-pool)/col;
sum1=0;
sum2=0;
for i=1:nspp
```

```
sum1=sum1+(1-al)^n(i);
sum2=sum2+(1-al)^(2*n(i));
end
sexp=nspp-sum1;
std=sqrt(sum1-sum2);
for k=1:sim
cols=zeros(1,col);
temp=randperm(col);
cols(1:col-pool)=temp(1:col-pool);
sobs=0;
for i=1:nspp
np=0;
for j=1:col-pool
np=np+spp(cols(j),i);
end
if (np~=0) sobs=sobs+1; end
end
sobm=sobm+sobs;
su=su+sobs^2;
end
m(col-pool)=sobm/sim;
sx=(su-sobm^2/sim)/(sim-1);
if (sx<0) sx=0; end
obl=sobm/sim-2*sqrt(sx);
obu=sobm/sim+2*sqrt(sx);
ml=sexp-2*std;
mu=sexp+2*std;
if (((sobm/sim)>=ml) & ((sobm/sim)<=mu)) tg=tg+1; end
fprintf([num2str(col-pool) ' ' num2str(sobm/sim) ' ' num2str(sexp) ' '
num2str(std) ' ' num2str(ml) ' ' num2str(mu) ' ' num2str(obl) ' '
num2str(obu) '\n']);
if ((col-pool)>=col) break; end
pool=pool-1;
end
fprintf('\n');
if ((tg*1.0/col)<0.95)
fprintf([num2str(col-tg) ' of the ' num2str(col) ' data points ('
num2str((col-tg)/col*100)' percent) fell outside the RPH confidence
interval.\n']);
fprintf('Consequently, these sampling data are a poor fit to the Random
Placement Hypothesis. The samples are heterogeneous and the environment is
heterogeneous.\n');
else
fprintf([num2str(col-tg)' of the ' num2str(col) ' data points ('
num2str((col-tg)/col*100)' percent) fell outside the RPH confidence
interval.\n']);
fprintf('Consequently, these sampling data are a good fit to the Random
Placement Hypothesis. The samples are homogeneous and the environment is
homogeneous.\n');
end
```

Table 1   Test results for a data set (Zhang, 2011b).

Sample size	Mean observed number of taxa (ONT)	Expected number of taxa (ENT)	Standard deviation of expected number of taxa	Lower limit of ENT	Upper limit of ENT	Lower limit of ONT	Upper limit of ONT
1	5.504	6.799	1.311	4.175	9.423	2.067	8.94
2	7.71	8.561	1.464	5.632	11.49	4.253	11.166
3	9.318	9.812	1.534	6.743	12.882	5.906	12.729
4	10.534	10.81	1.557	7.694	13.926	7.252	13.815
5	11.536	11.637	1.555	8.526	14.748	8.218	14.853
6	12.322	12.338	1.538	9.261	15.416	9.227	15.416
7	12.818	12.942	1.514	9.913	15.972	9.79	15.845
8	13.486	13.469	1.486	10.496	16.442	10.45	16.521
9	13.982	13.933	1.456	11.02	16.847	11.061	16.902
10	14.416	14.346	1.426	11.493	17.199	11.581	17.25
11	14.815	14.716	1.395	11.924	17.508	12.123	17.506
12	15.011	15.051	1.365	12.319	17.782	12.307	17.714
13	15.39	15.355	1.336	12.682	18.028	12.643	18.136
14	15.634	15.634	1.307	13.019	18.248	12.904	18.363
15	15.961	15.89	1.278	13.333	18.447	13.326	18.595
16	16.181	16.128	1.249	13.628	18.628	13.525	18.838
17	16.366	16.35	1.221	13.907	18.793	13.672	19.059
18	16.637	16.557	1.192	14.172	18.943	14.075	19.198
19	16.814	16.753	1.163	14.425	19.08	14.304	19.323
20	16.969	16.937	1.133	14.669	19.204	14.513	19.424
21	17.118	17.112	1.103	14.906	19.318	14.58	19.655
22	17.293	17.278	1.071	15.136	19.42	14.884	19.701
23	17.441	17.437	1.037	15.362	19.512	15.089	19.792
24	17.671	17.59	1.001	15.586	19.593	15.442	19.899
25	17.761	17.736	0.964	15.808	19.664	15.549	19.972
26	17.876	17.878	0.923	16.031	19.724	15.813	19.938
27	18.05	18.015	0.879	16.257	19.773	15.996	20.103
28	18.142	18.148	0.83	16.487	19.809	16.089	20.194
29	18.294	18.277	0.776	16.724	19.831	16.462	20.125
30	18.444	18.404	0.716	16.971	19.837	16.795	20.092
31	18.51	18.527	0.647	17.232	19.823	16.949	20.07
32	18.661	18.649	0.567	17.515	19.783	17.327	19.994
33	18.748	18.767	0.467	17.832	19.703	17.597	19.898
34	18.898	18.884	0.334	18.215	19.553	18.139	19.656
35	19.0	19.0	0.0	19.0	19.0	19.0	19.0

```
fprintf('\n');
if (((m(col)-m(col-1))/m(col))<=pro) fprintf('Most of taxa have been
   recorded and the sample size is thus enough. \n'); end
if (((m(col)-m(col-1))/m(col))>pro) fprintf('Some taxa have not yet been
   recorded and more samples are needed. \n'); end
```

20.5.2 *Application*

Use a set of arthropod data investigated in rice fields of Guangzhou, China (35 samples; 19 families; Zhou, 2007), choose 1000 randomizations and set the sampling completeness as 0.01 (the difference of the number of taxa between the last two (cumulative) sample sizes is less than 1%) (Zhang, 2011b). The results show that the sampling is complete and samples are environmentally homogeneous (all observed points fell inside the confidence interval) and the sampling is complete for the four arthropod communities (difference is around 0.5%; Table 1).

20.6 Network Construction Based on Power-Law or Exponential-Law Distribution of Node Degree

Power-law and exponential-law distribution, in particular power-law distribution, is the most popular form of degree distribution of biological networks. The following algorithm produces an expected frequency distribution according to the given parameter of power-law or exponential-law distribution function, and then creates an adjacency matrix in which the practical frequency distribution of node degrees coincides with the expected frequency distribution (Zhang, 2016i).

The power-law and exponential-law distribution functions are as follows (Goemann *et al.*, 2011; Zhang, 2011c, 2012e)

$$p(x) = x^{-\lambda}$$
$$p(x) = e^{-\lambda x}$$

where λ is a constant. Suppose there are v nodes in the network being built. First, given the permitted error of degree distribution, *er* (e.g., 0.05), and the maximum number of simulations, *sim* (e.g.,

$v*5$). The fitting error of degree distribution is defined as

$$\text{error} = \sum_{i=1}^{m} |f_i - ff_i| / \sum_{i=1}^{m} f_i$$

where f_i and ff_i are the expected and practical frequencies of nodes in the degree interval $[in_i, in_{i+1})$ $(i = 1, 2, \ldots, m-1)$.

The adjacency matrix of the network being constructed is $d = (d_{ij}), i, j = 1, 2, \ldots, v$, where $d_{ij} = d_{ji}, d_{ii} = 0$, and if $d_{ij} = 1$ or $d_{ji} = 1$, there is a connection between nodes i and j. There are $v(v-1)/2$ unknown variables $d_{ij}, i, 1, 2, \ldots, v - 1; j > i$. However, there are only v known conditions (degrees of v nodes), $s_i, i = 1, 2, \ldots, v$. Therefore, there are, in general, multiple solutions for $d = (d_{ij})$. Thus, use the enumerating/meriting method to find the $d = (d_{ij})$ that meets the error requirement.

The procedure of the algorithm is as follows

(1) Let $fa = 1.2$, and $fac = fa$.
(2) Find degree intervals $[in_i, in_{i+1}), i = 1, 2, \ldots, m-1$, where

$$in_1 = 1;$$
$$in_{i+1} = in_i^* fac, \quad (\text{Power} - \text{law distribution})$$
$$in_{i+1} = in_i^* fac^* 2, \quad (\text{Exponential} - \text{law distribution})$$
$$i = 1, 2, \ldots, m-1$$
$$in_m = v, \quad \text{if } in_m > v$$

(3) Let class $= m - 1$. Calculate the probability of degree distribution, i.e., the probability of a node's degree falling in the interval $[in_i, in_{i+1})$

$$p_i = in_i^{-\lambda} - in_{i+1}^{-\lambda} \quad (\text{Power} - \text{law distribution})$$
$$p_i = e^{-\lambda in_i} - e^{-\lambda in_{i+1}} \quad (\text{Exponential} - \text{law distribution})$$

and the corresponding frequencies $f_i = p_i v$, $i = 1, 2, \ldots,$ class $- 1$. Calculate

$$su = \sum_{i=1}^{\text{class}-1} f_i$$

and let $f_{\text{class}} = v - su$.

(4) Let initial error, $miner = 10^{10}$, and simulation times, $tot = 1$.

(5) Let

$$s = \sum_{i=1}^{\text{class}} f_i$$

(6) Calculate node degrees

$$s_k = \text{floor}\left(\sum_{i=1}^{\text{class}}\sum_{j=1}^{f_i}(in_{i+1} - in_i) * \text{rand} + in_i\right)$$

where $floor(x)$ denotes the integer part of x, $rand$ is a random value in $(0,1)$, $k = 1, 2, \ldots$.

(7) Rearrange v pairs of (node, degree), from greater to smaller in terms of node degrees

$$v_i, s_i$$
$$v_j, s_j$$
$$\ldots$$
$$v_q, s_q$$

In this step, for each node v_i with s_i expected connections, randomly create $v_i - m_i$ connections to other nodes, and each $v_i - m_i$ nodes also has a connection, where m_i is the number of connections already created by previous nodes.

(8) Produce the candidate adjacency matrix, $d = (d_{ij})$, calculate the practical frequency distribution, $ff_i, i = 1, 2, \ldots$, and the error.

(9) If there is at least one node degree, e.g., $s_r = 0$, return (6); or else, go to (10).

(10) If the calculated $error < miner$, let $minerr = error$, $dd = dffd_i = ff_i$, go to (11), or else go to (12).

(11) If $minerr \leq er$, go to (14).

(12) If $tot \geq sim$, go to (13), or else let $tot = tot + 1$, return (5).

(13) Let $fac = fac + 0.3$, return (2). If $fac > v$, go to (14).

(14) Print adjacency matrix (dd), expected and practical frequency distributions (f_i and $ffd_i, i = 1, 2, \ldots, m$), and node degrees.

The following are Matlab codes of the algorithm, netConstr.m:

```
v=input('Input the number of nodes in the network:');
typedis=input('Input the type of frequency distribution of degrees (1:
Power-law distri (F(x)=1-x^(-c)); 2: Exponential-law distri
(F(x)=1-e^(-cx))):');
c=input('Input the parametrical value (c) of frequency distribution of node
  degrees:');
disp('e.g., c=1.5 for power-law distri (mostly falls in (0, 2]), and c=0.2
for exponential-law distri. For power-law distri, the following formula can
be used to obtain a suitable c, c=1.6347-0.1401m+0.0019v+0.0038m^2,
r^2=0.83, p=0.0006<0.01, where m is the mean of node degrees, v is the
number of nodes in the network.')
er=0.05;
sim=v*5;
fa=1.2; % It can be replaced with, e.g.,
fa=22.6712-34.4086*c+0.0764*v-0.0423*c*v+13.0632*c^2
for fac=fa:0.3:v
in=zeros(1,v);
if (typedis==1)
i=1;
in(1)=1;
while (v>0)
i=i+1;
in(i)=in(i-1)*fac;
if (in(i)>v) in(i)=v; break; end
end; end
if (typedis==2)
i=1;
in(1)=1;
while (v>0)
i=i+1;
in(i)=in(i-1)*fac*2;
if (in(i)>v) in(i)=v; break; end
end; end
deg=in(1:i)
class=i-1;
ff=zeros(1,class);
p=zeros(1,class);
su=0;
for i=1:class-1
p(i)=netConstrDistr(typedis,c,in(i),in(i+1));
ff(i)=p(i)*v;
su=su+ff(i);
end
ff(class)=v-su;
id=0;
minerr=1e+10;
tot=1;
while (v>0)
[adj0,fff0,error0]=netGen(class,deg,ff);
```

```
if (error0<minerr)
adj=adj0; fff=fff0; minerr=error0;
if (minerr<=er) id=1; break; end
end
if (tot<=sim) break; end;
tot=tot+1;
end
if (id==1) break; end
end
fprintf('\nAdjacency matrix of the generated network\n')
disp([adj])
fprintf('\n')
fprintf('\nIntervals of node degrees\n')
for i=1:class
fprintf(['['num2str(deg(i))','num2str(deg(i+1))')'])
end
fprintf('\n\nExpected frequency distribution of node degrees of the
   generated network\n')
disp([ff])
fprintf('\n')
fprintf('Practical frequency distribution of node degrees of the generated
   network\n')
disp([fff])
fprintf(['Fitting error of expected and practical distribution of node
   degrees of the generated network:' num2str(minerr) '\n'])
fprintf(['\nNode degrees of the generated network\n' num2str(sum(adj))
   '\n'])
fprintf(['\nMean of node degrees of the generated network:'
num2str(mean(sum(adj))) '\n\n\n'])
```

The functions, netConstrDistr.m and netGen.m, are as follows

```
function pab=netConstrDistr(typedis,lamda,a,b)
pab=0;
if (typedis==1) pab=a^(-lamda)-b^(-lamda); end
if (typedis==2) pab=exp(-lamda*a)-exp(-lamda*b); end

function [adj,pracff,error]=netGen(m,in,expff)
%pracff[]: degree distri of produced network; er: expected chi-square of
practical and expected degree distri
%in[]: degree threshold pairs vector; expff[]: frequency; sim: simulation
   times
%[in[i],in(i+1)): expff(i), i=1, 2, ..., m; m is the vector dimension
s=sum(expff);
while (m>0)
adj=zeros(s);
pracff=zeros(1,m);
f=zeros(1,s);
p=zeros(1,s);
w=zeros(1,s);
k=0; u=0;
```

```
for i=1:m
for j=1:expff(i)
k=k+1;
f(k)=floor((in(i+1)-in(i))*rand()+in(i));
u=u+f(k);
end; end
for i=1:s
p(i)=i;
end
for i=1:s-1
k=i;
for j=i:s-1
if (f(j+1)>f(k)) k=j+1; end
end
l=p(i); p(i)=p(k); p(k)=l;
u=f(i); f(i)=f(k); f(k)=u;
end
for i=1:s
lab=0;
vv=0;
for j=1:i-1
if (adj(j,i)==1)
adj(i,j)=1;
vv=vv+1;
if (vv==f(i)) lab=1; break; end
end; end
if (lab==1) continue; end
cc=0;
for j=1:s
w(j)=j;
end
while (f(i)~=0)
cs=floor((s-cc)*rand()+1);
if (w(cs)==i) continue; end
if ((adj(w(cs),i)==0) & (w(cs)<i)) continue; end
adj(i,w(cs))=1;
if (cs<s-cc)
for j=cs+1:s-cc
w(j-1)=w(j);
end; end
cc=cc+1;
if (cc<=(f(i)-vv)) break; end
end; end
for i=1:s
for j=1:s
if (adj(j,i)~=adj(i,j)) adj(j,i)=1; adj(i,j)=1; end
end; end
for i=1:s
p(i)=0;
for j=1:s
```

```
if (adj(i,j)==1) p(i)=p(i)+1; end
end; end
for i=1:m
pracff(i)=0;
for j=1:s
if ((p(j)>=in(i)) & (p(j)<in(i+1))) pracff(i)=pracff(i)+1; end
end; end
error=sum(abs(expff-pracff))/sum(expff);
isonodes=sum(sum(adj)==0);
if (isonodes==0) break; end
end
```

Part 7
Pharmacological and Toxicological Networks

Chapter 21

Network Pharmacology and Toxicology

21.1 Network Pharmacology

Due to lack of drug efficacy and unexpected toxicity, the failure cases in clinical trials account for more than half (Kola and Landis, 2004). The main cause was strongly attributed to the incorrect guiding ideology for drug design in traditional pharmacology that is based on the view of a drug–a target–a disease (Hopkins, 2007, 2008; Sun *et al.*, 2014; Zhang, 2016m).

Complex diseases, like cancer and diabetes, etc., are not usually attributed to mutation or dysfunction of a single molecule, but are usually caused by the dysfunction of its whole regulation network. In a network, although a single molecule changes insignificantly, this will collectively result in substantial changes in the whole signaling path. For example, research has found that the 10% increase of a single molecule expression level in metabolic pathways can lead to 100% change of final metabolite production. The research on cancer genome project revealed that the vast majority of mutations exist in a few samples only, and to find the same genetic mutation is almost impossible. At the level of network, cancer-related mutations will mostly appear in the genes of specific signaling pathways. Thus, for diagnosis and treatment of cancer and other complex diseases, the target is certainly not a single gene but may be a specific pathway or network. Thus, the analysis of molecular mechanisms of diseases based on biological networks is imperative. For this reason, Hopkins (2007, 2008) proposed the concept of network pharmacology. Network pharmacology helps to understand drug's pharmacological mechanism in the network perspective. In a sense, it is also a branch of network biology (Zhang, 2016m; Budovsky and Fraifeld, 2012; Huang and Zhang, 2012; Zeitoun *et al.*, 2012; Li and Zhang, 2013; Iqbal *et al.*, 2014; Shams and Khansari, 2014; Jesmin *et al.*, 2016).

Network pharmacology is an interdisciplinary science based on pharmacology, network biology, systems biology, bioinformatics, computational science, and other related scientific disciplines. In particular, it is a network-based science, just like other new proposed sciences (Zhang, 2016k). Network pharmacology aims to understand the network interactions between a living organism and drugs that affect normal or abnormal biochemical function. It aims to exploit the pharmacological mechanism of drug action in the biological network and helps to find drug targets and enhance the drug's efficacy. The scope of network pharmacology covers but is not limited to: (1) theories, algorithms, models, and software of network pharmacology; (2) network construction and interactions prediction; (3) theories and methods on dynamics, optimization, and control of pharmacological networks (here generally refer to disease network, disease–disease, disease–drug, drug–drug, drug–target network, network targets–disease, drug targets–disease network, etc.); (4) network analysis of pharmacological networks, including flow (flux) balance analysis, topological analysis, network stability, etc.; (5) various pharmacological networks and interactions; (6) factors that affect drug metabolism; (7) network approach for searching targets and discovering medicines (including medicinal plants, etc.); (8) big data analytics of network pharmacology, etc. (Zhang, 2016m).

21.1.1 *Theoretical fundamentals*

21.1.1.1 *Scientific foundation*

(1) Pharmacology

Pharmacology is the branch of medicine and biology which focuses on drug action. A drug can be broadly defined as any man-made, natural, or endogenous molecule which exerts a biochemical and/or physiological effect on the cell, tissue, organ, or organism (Vallance and Smart, 2006; Wikipedia, 2016b). It aims to study the interactions between a living organism and chemicals that affect normal or abnormal biochemical function.

There are a variety of branches of pharmacology, including clinical pharmacology, neuropharmacology, psychopharmacology, theoretical pharmacology, behavioral pharmacology, environmental pharmacology, biochemical and molecular pharmacology, cardiovascular pharmacology, gastrointestinal pharmacology, respiratory tract pharmacology, urogenital pharmacology, etc.

(2) Network biology

Network biology was first proposed by Barabasi and Otlvai in 2004. Since then, Zhang (2011d, 2012e) further defined the scope of network biology from cellular level to ecosystems and social networks. Network biology focuses on (both dynamic and static) nodes (molecules, metabolites, cells, etc.) and between-node interactions in biological networks (pathways, ecosystems, etc.). It covers theories, algorithms, and programs of network analysis; innovations and applications of biological networks; dynamics, optimization, and control of biological networks; ecological networks, food webs, and natural equilibrium; coevolution, coextinction, and biodiversity conservation; metabolic networks, protein-protein interaction networks, biochemical reaction networks, gene networks, transcriptional regulatory networks, cell cycle networks, phylogenetic networks, and network motifs; physiological networks; network regulation of metabolic processes, human diseases, and ecological systems; social networks, and epidemiological networks, etc. In recent years, the theory and methodology of network biology have been established (Jiang and Zhang, 2015a, 2015b; Zhang, 2011–2017; Zhang and Li, 2015a, 2015b). Numerous papers on biological networks of human diseases have also been published (Tacutu *et al.*, 2011; Budovsky and Fraifeld, 2012; Huang and Zhang, 2012; Zeitoun *et al.*, 2012; Li and Zhang, 2013; Iqbal *et al.*, 2014; Shams and Khansari, 2014; Zhang and Li, 2015a, 2015b; Jain, 2016a,

2016b; Jesmin *et al.*, 2016; Habib *et al.*, 2016, 2017; de Araujo *et al.*, 2017; Narad *et al.*, 2017).

(3) Systems biology

Hood (1998) first proposed the scientific discipline systems biology and defined it as the science that studies all components and their interactions in biological systems. In the view of systems biology, the organism is a complex system containing many interactions between components (genes, proteins, mRNAs, small molecular metabolites, etc.) at multiple layers (cell, tissue, organ, and whole body) of the organism. Biological systems have such properties as emergency, complexity, and robustness (Hood, 1998, 2002, 2003; Schrattenholz *et al.*, 2010; Zhang, 2012e, 2016n). Systems biology aims to exploit all components and their interactions under certain conditions (e.g., various genetic and environmental conditions) and to predict biological functions, phenotypes, and their behaviors. According to systems biology, the drug target should be extended from the single molecule into molecular combination, a signal transduction the pathway, or even a few of the pathways (Frantz, 2005; Schrattenholz *et al.*, 2010; Zhang, 2016m).

21.1.1.2 *Basic principles*

There are at least two fundamental principles in the ideology of network pharmacology (Zhang, 2016m): (1) Schilling *et al.* (1999) argued that throughout natural selection, the cellular metabolic activities always maintain a balance when no significant perturbation occurs, or regulatorily minimize the systematic bias from resting status. From the perspective of network biology, a biological network (human body) at steady state/natural equilibrium state is at the healthy state, i.e., it is a stable network with a specific topological structure and certain network properties. If the network equilibrium is disrupted or damaged, it will change to a pathological or disease state, i.e., an unstable network with different topological structure and network properties. A drug for disease treatment is to restore the biological network to the balance/equilibrium state, or reduce the degree of balance being destroyed (Yildirim *et al.*, 2007;

Janga and Tzakos, 2009). According to Le Chatelier's principle, if the balance (health state) of a system (network) has experienced a change (i.e., to a disease state), the role of an effective drug is to drive the balance to the direction that will weaken such change. (2) Due to biological redundancy (e.g., the redundancy of metabolites/molecules/reactions/interactions, etc.), which is the result of natural evolution, as an alternative solution, the role of an effective drug is to induce a somewhat different biological network that can guarantee the operation of basic functionality of the healthy biological network, if the normal balance is not easily achieved (Zhang, 2016m).

Network pharmacology is proposed based on the theory of network science and biological balance, and it thus provides new ideas for drug discovery, as well as for understanding the mechanism of drug functioning. In the perspective of network pharmacology, we should try to perturb the pathogenic network using the drug rather than search for pathogenic genes only (Barabasi and Otlvai, 2004).

21.1.2 *Methodology*

On the basis of high-throughput -omics data, network database retrievals and other biological information, network pharmacology stresses the construction of pharmacological networks, topological analysis of pharmacological networks, network flow analysis, structural optimization and optimal control of pharmacological networks, etc. Other experiment and observation based methods are also included (Li and Zhang, 2013).

21.1.2.1 *Data source*

There are two sources of fundamental data for research of network pharmacology: public databases and experimental verification. First, we can use public databases, i.e., the existing public data and published data, to construct network models of the specific disease and drug target, to predict the drug target (Budovsky and Fraifeld, 2012), further, to construct drug–target–disease network and analyze pharmacological mechanism of the drug, and finally to validate the mechanism through experiments (Zhou *et al.*, 2012). Second, we may use -omics technologies and high-throughput technologies to investigate the interactions between the drug and network model, to construct and analyze drug–target–disease network based on the generated data, and to analyze pharmacological mechanism of drug actions (Zhang, 2016m).

21.1.2.2 *Big data analytics*

Big data is the data set that is so large or complex that conventional data processing techniques are inadequate. Challenges include analysis, capture, data curation, search, sharing, storage, transfer, visualization, querying, and information privacy (Wikipedia, 2016c).

Big data analytics is the process of examining big data to uncover hidden patterns, unknown correlations, and other useful information. With big data analytics, e.g., high-performance data mining, predictive analytics, text mining, forecasting, and optimization, we can analyze huge volumes of data that conventional analytics can not handle. In addition, machine learning techniques are ideally suited to addressing big data needs (Zhang, 2010; Zhang and Qi, 2014; SAS, 2016). Many issues in network pharmacology, network construction, interactions prediction, etc., are also expected to be addressed by using big data analytics.

21.1.2.3 *Network construction and interactions prediction*

For a disease and the corresponding drug, the pharmacological network is the base for further pharmaceutical studies. Nevertheless, most pharmacological networks are unknown or imperfect. Thus, how to construct a pharmacological network is a prerequisite for such studies. Among them, the networks of disease-related protein interactions are the most important. The most used methods to find such interactions and construct pharmacological networks include phylogenetic profile (Gaasterland and Ragan, 1998; Pellegrini *et al.*, 1999), gene neighborhood (Dandekar *et al.*, 1998), gene fusion event, mirror tree (Fryxell, 1996), correlated mutation (Gobel *et al.*, 1994), correlated evolutionary rate (Fraser *et al.*, 2002), prediction based

on primary structure (Bock and Gough, 2001), homologues structural complexes (Aloy and Russell, 2003), Gene Coexpression Network Analysis (Stuart *et al.*, 2003; Ideker and Krogan, 2012; Chen,
2016), etc. Among them, phylogenetic profile method is considered
to be particularly useful for construction of networks and prediction of large-scale interactions. Tu (2006) used Pearson correlation
between proteins, which is based on phylogenetic profile method, to
construct the networks of small-cell lung cancer and non-small-cell
lung cancer and predict potential interactions. Zhang (2011c, 2012e,
2012b, 2015c, 2015d) has proposed a series of correlation methods
to construct networks. Pearson correlation measure will lead to a
false result (Zhang and Li, 2015b). Thus, Zhang (2015c, 2015d) used
partial linear correlation and proposed some partial correlation measures, and used them to jointly predict interactions. Moreover, there
are a lot of other studies on construction and prediction of biological
networks (Goh *et al.*, 2000; Pazos and Valencia 2001; Guimera and
Sales-Pardo, 2009).

Based on an incomplete network, we may predict missing interactions (links) (Clauset *et al.*, 2008; Guimera and Sales-Pardo, 2009;
Barzel and Barabási, 2013; Lu *et al.*, 2015; Zhang, 2015c, 2015e,
2016a, 2016e; Zhang and Li, 2015b).

Network evolution based (Zhang, 2015b), node similarity based
(Zhang, 2015e), and correlation based (Zhang, 2007, 2011c, 2012b,
2012e, 2015c, 2015e, 2016a, 2016e; Zhang and Li, 2015b) methods
are expected to be the most promising in the future.

21.1.2.4 *Network analysis*

Network analysis covers a variety of scientific areas and methods
(Zhang, 2012a). The main contents of network analysis used in network pharmacology are described as follows (Zhang, 2016l, 2016m,
2016o).

(1) Attribute analysis

Attribute analysis aims to screen node attributes according to their
contribution to topological structure of the network (Zhang, 2016p).

(2) Topological analysis

Topological analysis of networks mainly includes the following contents

Find trees in the network: DFS algorithm, Minty's algorithm, etc., (Minty, 1965; Zhang, 2012e, 2016h).

Find circuits (closed paths, loops) (Paton, 1969; Zhang, 2012e, 2017b).

Find the maximal flow: Ford–Fulkerson algorithm (Ford and Fulkerson, 1956; Zhang, 2012e).

Find the shortest path: Dijkstra algorithm, Floyd algorithm (Dijkstra, 1959; Zhang, 2012e, 2016p).

Find the shortest tree: Kruskal algorithm (Zhang, 2012e, 2017d).

Calculate network connectedness (connectivity), blocks, cut vertices, and bridges (Zhang, 2012e, 2016g, 2016j).

Calculate node centrality (Zhang, 2012e; Shams and Khansari, 2014; Jesmin *et al.*, 2016).

Find modules, mosaics, and subnetworks (Bascompte, 2009; Zhang, 2016c; Zhang and Li, 2016).

Analyze degree distribution (Huang and Zhang, 2012; Zhang, 2012e; Zhang and Zhan, 2011; Rahman *et al.*, 2013).

In addition to the methods above, other statistical methods, e.g., PCA, canonical correlation analysis (Zhang, 2017i)), etc., are also useful in network analysis.

(3) Network structure and stability

Stability of biological networks has been studied in the past (Din, 2014). These studies have been focused on ecosystems and the methods can be used in the pharmacological studies. Pinnegar *et al.* (2005) used a detailed Ecopath with Ecosim (EwE) model to test the impacts of food web aggregation and the removal of weak linkages. They found that aggregation of a 41-compartment food web to 27 and 16 compartment systems greatly affected system properties (e.g., connectance, system omnivory, and ascendancy) and influenced dynamic stability (Zhang, 2012e, 2016m).

The most advanced theory is that there is a relationship between network connectance and different types of ecosystem stability. Some models suggest that lower connectance involves higher local (May, 1973; Pimm *et al.*, 1991; Chen and Cohen, 2001) and global (Cohen *et al.*, 1990; Chen and Cohen, 2001) stability, i.e., the system recovers faster after a disturbance. Another theory suggests that a food web with higher connectance has more numerous reassembly pathways and recovers faster from perturbation (Law and Blackford, 1992).

(4) Flow (flux) balance analysis

Network flow is determined by topological structure and properties of the network (Borgatti, 2005). Flow balance analysis aims to analyze network flows at the steady state. Differential equations and other equations are usually used to describe network dynamics (Chen *et al.*, 2010; Schellenberger *et al.*, 2011). For example, Jain *et al.* (2011) used some mathematical models to decipher balance between cell survival and cell death using insulin.

Some standardized indices and matrices can be used in flow balance analysis (Latham, 2006; Fath *et al.*, 2007; Zhang, 2012e). They include Average Mutual Information (AMI) (Rutledge *et al.*, 1976) Ascendency index (Ulanowicz, 1983, 1997) compartmentalization index (Pimm and Lawton, 1980), and constraint efficiency (Latham and Scully, 2002). Zorach and Ulanowicz (2003) presented effective measures (effective connectivity, effective flows, effective nodes, effective rules) for weighted networks. Fath and Patten (1999a) developed a measure to calculate the evenness of flow in a network. Higashi and Patten (1986, 1989) and Fath and Patten (1999b) presented an index for describing the dominance of indirect effects.

(5) Network models

Network models are the foundation to understand interactions within complex networks (Zhang, 2016m). Various random graph models produce network structures that may be used in comparison to real complex networks (Wikipedia, 2016a). Some network models have been developed for food webs (Zhang, 2012e), such as cascade model (Cohen *et al.*, 1990), niche model (Williams and Martinez, 2000), multitrophicassembly model (Pimm, 1980, Lockwood *et al.*, 1997), MaxEnt models (Williams, 2010), Ecopath model

(Polovina, 1984; Christensen and Pauly, 1992; Libralato *et al.*, 2006), etc. Ecosim is the dynamic program of the EwE (Walters *et al.*, 1997, 2000), which allows a dynamic representation of the system variables, like biomasses, predation, and production. They can be revised and improved to fit pharmacological networks. In addition, some dynamic evolution models are also network models (Zhang, 2015b, 2016f).

(6) Network dynamics, evolution and control

Theoretically, Ferrarini (2011–2015) have proposed a series of thoughts and methods on the dynamics, controllability, and dynamic control of biological networks. Zhang (2015b) proposed a generalized network evolution model and self-organization theory on community assembly, in which the model is a series of differential (difference) equations with different number with time. Moreover, Zhang (2016f) developed a random network-based, node attraction-facilitated network evolution method. The two dynamic models are useful to study the network evolution and dynamics and to predict interactions.

A network can be optimized to search for an optimal search plan, and achieve a topological structure so that the network possesses relative stability (Zhang, 2012e). The dynamic control of network means to change topological structure and key parameters of the network stage by stage so that the goal function of entire network achieves the optimum or suboptimum (Zhang, 2012e). Mathematical tools, like dynamic programming, decision-making analysis, game theory, etc., can be used to address these problems.

Luo (2007) studied a constraint optimization on flux balance model in order to study cellular metabolic network. Kim *et al.* (2015) proposed a dynamic network model of Rho GTPases signal network and developed a Boolean network model used to analyze various states and emergency reconstruction of Rho GTPases signal network. In order to reveal Epithelial–Mesenchymal Transition (EMT) in the process of cancer metastasis from the dynamics view, Tanaka and Ogishima (2015) defined the state space of a gene regulation network as all possible activation patterns for the network and introduced panorama into the state space, which showed a relatively stable distribution of three steady states or phases.

21.1.2.5 *Network visualization*

Network visualization aims to present users with the static/dynamic two- or three-dimensional illustrations and images of biological networks (Zhang, 2016m). There are a variety of such network softwares for doing it (Zhang, 2012e), for example, ABNNSim (Schoenharl, 2005), Topographica (Bednar *et al.*, 2004), Pajek, NetDraw, NetLogo (Resnick, 1994), netGenerator (Zhang, 2012a, 2012e), Repast (Macal and North, 2005), Startlogo (Resnick, 1994), etc.

21.1.3 *Application*

Network pharmacology helps to better understand the influence of behaviors of cells and organs on functional phenotypes, to understand the mechanism of drug functioning, and to provide theoretical basis and technical support for drug design and for rational clinical use of drugs (Zhang, 2016m). Furthermore, it helps to predict and explain drug interactions and optimize the use of drugs, to find the factors that affect drug efficacy and safety, and to quickly discover biomarkers and drug targets (Zhou *et al.*, 2012).

21.1.3.1 *Application scope*

In summary, network pharmacology can be used in the following aspects (Zhang, 2016m)

(1) Mine pharmacological mechanism of drug action and guide clinical application

Network pharmacology helps to not only understand the network-wide mechanism of drug functioning, but also guide drug's clinical use. Because the current drugs mostly point to a single target, thus the change of a single amino acid in the target may cause drug resistance. As a result, drug resistance has become a common phenomenon. In fact, many effective antibiotics work by acting on multiple targets at the same time rather than a single target (Lange *et al.*, 2007; Shi *et al.*, 2014).

(2) Discover and confirm drug targets

Traditional research on single drug target focuses on the single molecule, and the resultant drug usually fails to work due to some pharmacokinetic or toxic effects. Network pharmacology features with multiple targets, i.e., multiple molecules or even multiple sub-networks are used to adjust the biological network. The discovery and confirmation of drug targets will be accelerated due to the strong predictability that bases on the whole network (Apsel *et al.*, 2008; Zheng *et al.*, 2013).

(3) Guide research and design of new drugs

Many approved drugs have proved to be worse than expected in terms of their selectivity (Campillos, 2008). Some drugs of multiple targets are clinically successful, especially double or multiple enzyme inhibitors (Lange *et al.*, 2007). Some medicine for treatment of tumors can be combined with multiple kinases (Frantz, 2005). Moreover, drug–target networks have scale-free properties (Campillos, 2008; Janga and Tzakos, 2009), a single protein will produce compensation mechanism for dysfunction, and thus the single target strategy fails to work.

21.1.3.2 *Application cases*

(1) Use network method to search potential therapeutic targets in medicinal plants

Budovsky and Fraifeld (2012) used network method to search potential therapeutic targets in medicinal plants. In this study, they used a list of the plants growing in the Judea region and surveyed scientific literature and ethno-pharmacological data to identify medicinal applications of the local vegetation. The validated and potential human targets of the major compounds found in the medicinal plants were downloaded from the STITCH database (one of the largest repositories of chemical–protein interactions). Meanwhile, the NetAge database was applied to data mine association of the medicinal plants targets with the major human age-related diseases and associated processes. And protein–protein interaction data for the targets was extracted from the BioGRID database. With the focus on their medicinal applications nearly 1,300 plants growing in this region were identified, and 25% of them have medicinal applications and were analyzed. It revealed that screening for chemical–protein interactions, together with the network-based analysis of potential targets, may facilitate discovery and therapeutic applications of medicinal plants.

(2) Relational networks and pharmacological mechanisms of medicinal attributes and functions of Chinese herbal medicines

Zhang (2017h, 2017i, 2017j, 2017k) established a database, CHM-DATA, with 1,127 Chinese herbal medicines mainly having recorded chemical composition, involving 7 taste attributes, 5 medicinal properties, 1 toxicity attribute, 22 chemical composition categories, 12 meridians and collaterals (Gui Jing), and 78 medicinal functions (Gong Xiao). The data were used to calculate point correlations between these 125 attributes. Totally four relational networks, i.e., the networks for medicinal attributes and functions, for chemical composition and meridians and collaterals, for meridians and collaterals and medicinal functions, and for meridians and collaterals were constructed based on the significant point correlations. Network analysis indicated that the former three are scale-free complex networks and the last one tends to be a random network (Zhang, 2017k). Node degrees of the four networks follow power-law distribution.

For the relational network of chemical composition, meridians, and collaterals, the skewness of degree distribution, aggregation index, variation coefficient, and entropy are 0.3956, 1.1963, 1.803, and 3.2121, respectively. The network is thus considered to be a complex network. Binomial distribution $p = 0.4118$, $\chi^2 = 40.9913$; Poisson distribution $\lambda = 4$, $\chi^2 = 52.2854$; the network is thus considered to be a nonrandom network. Exponential distribution $\lambda = 0.25$, $\chi^2 = 9.9663$, and it is considered to be a network with exponential degree distribution. Power-law distribution $\alpha = 4.0365$, $x_{min} = 4$, and Kolmogorov–Smirnov goodness-of-fit statistic $D = 0.2289$. Therefore, the degree distribution follows power-law distribution and the density function is $p(x) = x^{-4.0365}$, $x \geq 4$. Overall, the relational network for meridians/collaterals and functions is a scale-free complex network. The network has 3 connected components with 2 components with 1 node (attribute; isolate node) respectively.

21.2 Network Toxicology

Network toxicology aims to understand toxicological mechanism of harmful substances in the network perspective (Zhang, 2016o). It helps to analyze network interactions between living organisms and toxicants that affect normal or abnormal biological functions. The effects of toxicants are influenced by network properties, network structure, network dynamics, etc. Network toxicology aims to understand toxicological mechanism of harmful substances in the network perspective. The scope of network toxicology covers but is not limited to: (1) theories, algorithms, and software of network toxicology; (2) mechanisms and rules of flow and diffusion of toxicants in the network; (3) network analysis of toxicological networks; (4) various toxicological networks and interactions; (5) factors that influence chemical toxicity, including the dosage, the route of exposure, the species, age, sex, and environment; (6) toxicity assessment, i.e., identify adverse effects of a substance; (7) control of toxicants, etc. Details of scientific foundation and methodology of network toxicology can be found in Zhang (2016o).

21.2.1 *Common signaling pathway responses to environmental toxicants*

Environmental toxicant exposure may induce diseases by alternating cellular signaling pathways. How to understand the effects of environmental toxicant exposure at systems biology level is an interesting topic. Rager and Fry (2013) reviewed critical studies that evaluated six important toxicants, arsenic, benzene, cadmium, chromium, cigarette smoke, and formaldehyde, and given biological pathways altered by exposure to arsenic, benzene, cadmium, chromium, cigarette smoke, and/or formaldehyde. It was found that 12 pathways are commonly altered by at least two toxicants, AP-1/c-Jun, EGFR, IL1, MAPK, MYC, natural killer cell, NFκB, Nrf2, p53, TCR, TGF-β, and TNF-α signaling (Table 1).

Table 1 Pathways that are commonly altered by environmental exposures (Rager and Fry, 2013).

Pathway altered by at least two toxicants	Associated toxicants
AP-1/c-Jun signaling	arsenic, cadmium
EGFR signaling	cigarette smoke, formaldehyde
IL1 signaling	arsenic, cadmium
MAPK signaling	arsenic, benzene, cadmium, cigarette smoke
MYC signaling	arsenic, cadmium
Natural killer cell signaling	chromium, cigarette smoke
NFκB signaling	arsenic, cadmium, formaldehyde
Nrf2 signaling	arsenic, cadmium, cigarette smoke
p53 signaling	arsenic, benzene, cadmium, chromium, cigarette smoke
TCR signaling	arsenic, benzene, chromium
TGF-β signaling	cigarette smoke, formaldehyde
TNF-α signaling	arsenic, cadmium

Rager and Fry (2013) hypothesized that common signaling pathways likely overlap and respond as a common sensor to environmental exposures. They overlaid the main pathway/transcriptional regulators of each pathway onto a global interaction network to test the hypothesis. Networks containing these pathway mediators were constructed based on connectivity, using Ingenuity Pathway Analysis. With the proteins that directly interact with the toxicant-responsive pathway signals, a highly significant ($p < 10^{-2}$) network, environmental toxicant signalosome (Fig. 1), was constructed, which illustrates likely trans-pathway interactions that commonly respond to environmental exposures.

21.2.2 *Dynamic model of insecticide resistance*

Maintaining the certain dosage/concentration of an insecticide in the environment, the insecticide susceptible subpopulation of an insect species will evolve to the insecticide resistant subpopulation. The rates of reproductive transition both from susceptible subpopulation to resistant subpopulation and from resistant subpopulation

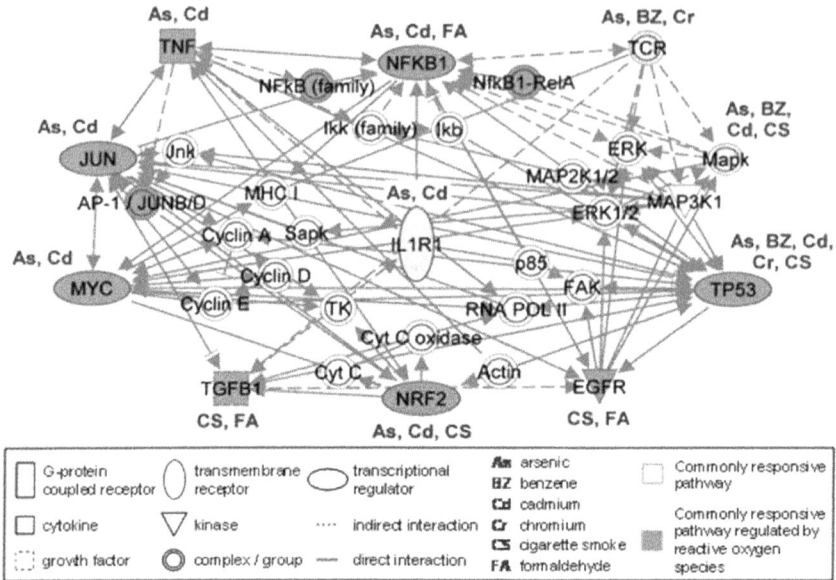

Fig. 1 Environmental Toxicant Signalosome (Rager and Fry, 2013).

to susceptible subpopulation depend on the insecticide pressure. Therefore, we have the following dynamic model of insecticide resistance, which is composed of two differential equations and an algebraic equation

$$dx/dt = r_1(c,t)x - f_1(c,t)x + g_1(c,t)y - \alpha(c,t)xy$$
$$dy/dt = r_2(c,t)y - g_2(c,t)y + f_2(c,t)x - \beta(c,t)xy$$
$$c = u(t)$$

where t: time (year, generation, etc.); $x(c,t)$: resistant subpopulation at t; $y(c,t)$: susceptible subpopulation at t; c: dosage/concentration of the insecticide. $r_1(c,t)$: c-dependent growth rate (fitness) function of resistant subpopulation, $r_1(c,t) \geq 0$; $\alpha(c,t)$: competition coefficient of susceptible subpopulation to resistant subpopulation, $\alpha(c,t) \geq 0$; $r_2(c,t)$: c-dependent growth rate (fitness) function of susceptible subpopulation, $r_2(c,t) \geq 0$. $\beta(c,t)$: competition coefficient of resistant subpopulation to susceptible subpopulation, $\beta(c,t) \geq 0$;

$f_1(c,t)$: reduction rate of resistant subpopulation due to the reproduction transition from resistant to susceptible insects, $f_1(c,t) \geq 0$; $f_2(c,t)$: growth rate of susceptible subpopulation due to the reproduction transition from resistant to susceptible insects, $f_2(c,t) \geq 0$. $f_2(c,t) = f_1(c,t)$, if there is no subpopulation loss. $g_2(c,t)$: reduction rate of susceptible subpopulation due to transition from susceptible to resistant insects under insecticide pressure, $g_2(c,t) \geq 0$; $g_1(c,t)$: growth rate of resistant subpopulation due to transition from susceptible to resistant insects under insecticide pressure, $g_1(c,t) \geq 0$. $g_1(c,t) = g_2(c,t)$, if there is no subpopulation loss. $f_1(c,t)$ and $f_2(c,t)$ are decrement functions of c. The functions, $g_1(c,t)$ and $g_2(c,t)$ are increment functions of c.

Four typical outcomes, dependent upon c, are expected in a certain period: (1) resistant subpopulation (x) persists and susceptible subpopulation (y) disappears; (2) resistant subpopulation disappears and susceptible subpopulation persists; (3) both resistant and susceptible subpopulations disappear, and (4) resistant and susceptible subpopulations stabilize at their relative equilibrium states.

The resistance strength can be measured by μ

$$\mu(c,t) = dx/dy = (r_1(c,t)x - f_1(c,t)x + g_1(c,t)y$$
$$- \alpha(c,t)xy)/(r_2(c,t)y - g_2(c,t)y + f_2(c,t)x$$
$$- \beta(c,t)xy)$$

Resistance strength increases with $\mu(c,t)$.

21.2.3 *Joint action of multiple insecticides*

It is generally believed that when several insecticides act jointly on the pest body, one of three kinds of joint actions of synergism, additive effect, and antagonism will occur. Synergism refers to the interaction of insecticides such that the total effect is greater than the sum of the individual effects, called synergy also, which can be explained by the fact that the toxicity of an insecticide is enhanced by another pesticide. Additive effect is an action in which two insecticides used in combination produce a total effect the same as the sum of the individual effects. In this combination, if in proportion, use an insecticide instead of another pesticide, the toxicity of the combination does not change. Antagonism is the action that the total effect is less than the sum of the individual effects, that is, the toxic effects are reduced by the interaction of two or more insecticides. To some extent, the site competition is the direct reason of antagonism. Synergism and antagonism are closely related to biological networks of insects.

The following are some indices and methods for detecting joint action of multiple insecticides.

(1) Toxic Unit

The toxic unit of the mixture of two insecticides is used to evaluate/analyze the combined toxic effects and mechanisms of insecticides in different proportions. It is calculated as follows

$$T_u = \text{CE}_1/\text{LC}_{50-1} + \text{CE}_2/\text{LC}_{50-2}$$

T_u is toxic unit of mixed insecticide, CE_i is concentration of the insecticide i to be tested (e.g., mg/L), and LC_{50-i} is 50% lethal concentration of the insecticide i in trial (e.g., mg/L), $i = 1, 2$.

With T_u and probit value of mortality rate as the dependent and independent variables, respectively, develop the regression equation and obtain the T_u (i.e., LC_{50}) of 50% mortality rate and its confidence interval. If LC_{50} falls outside of the confidence interval, the two insecticides show synergism ($\text{LC}_{50} < 1$) or antagonism ($\text{LC}_{50} > 1$).

(2) Concentration Addition and Independent Action

The formulae of Concentration Addition (CA) and Independent Action (IA) are as follows

$$\text{EC}_{x\text{mix}} = 1 \left/ \sum_{i=1}^{n} (p_i/\text{EC}_{xi}) \right.$$

$$x\% = 1 - \cap_{i=1}^{n}(1 - F_i(p_i^*\text{EC}_{x\text{mix}}))$$

where $\text{EC}_{x\text{mix}}$ is the concentration of the mixture to produce $x\%$ effect. EC_{xi} is the concentration at which the effect of the ith insecticide alone is the same as the total effect $x\%$ of the mixture. p_i is the ratio of concentration of ith insecticide against concentration of the mixture. F_i is the fitting function of concentration–effect curve for ith insecticide, with mortality rate as dependent variable and the concentration of the mixture as independent variable. If the experimental observation of toxicity of a mixture deviates from the predicted toxicity of the two models above, the two insecticides are considered to have toxic interactions (synergism or antagonism).

(3) Joint toxicity of insecticides

The mathematical models of Zhang and Gu (1998) on mixture infection of multispecies pathogens can be expanded to describe joint action of multiple insecticides.

For an insect individual or population, the potential sites that can be acted by insecticide molecules or active units are limited. The sites with overlapped action increase with insecticide molecules or active units and growth of the sites being acted will decrease. It can be described by the Poisson distribution (Zhang and Gu, 1998).

Suppose the probability of being acted for all sites is the same and insecticide molecules or active units act on sites randomly, the mean number of being acted for a site is assumed to be m, and thus the probability of being acted n times for a site is

$$p_n = e^{-m} m^n / n!, \quad n = 0, 1, 2, \ldots, \tag{1}$$

The probability with nonaction is thus e^{-m}, and $y = 1 - e^{-m}$, is the probability being acted (i.e., the mortality rate of insect population).

Suppose two insecticides jointly act on insect body. Their respective action strength is not affected by the other. Two insecticide molecules or active units will produce various effects, including synergism, additive effect, antagonism, etc. Assume the mean numbers of being acted for a site for respective given dosages/concentrations (x_1, x_2) of two insecticides, A and B, are m and n, respectively. If the joint action is the additive effect, we have

$$c = m + n \tag{2}$$

where c is the mean number of being acted for a site for a given dosage/concentration of the mixed pesticide. Known

$$m = -\ln(1 - y_1)$$
$$n = -\ln(1 - y_2) \tag{3}$$
$$c = -\ln(1 - y)$$

we have

$$\ln(1 - y) = \ln(1 - y_1) + \ln(1 - y_2) \tag{4}$$

i.e., $1 - y = (1 - y_1)(1 - y_2)$, or

$$y = 1 - (1 - y_1)(1 - y_2) \tag{5}$$

where y is the mortality rate of the mixed pesticide of two insecticides, A and B.

m and n are determined of insecticide concentration or dosage x_1, x_2, respectively. Known the relationships

$$y_1 = h(x_1)$$
$$y_2 = g(x_2) \tag{6}$$

according to eq. (3), we have the following relationships between (m, n) and (x_1, x_2)

$$m = -\ln(1 - y_1) = -\ln(1 - h(x_1))$$
$$n = -\ln(1 - y_2) = -\ln(1 - g(x_2)) \tag{7}$$

Let's set the concentration or dosage of two insecticides as x_1 and x_2, respectively. According to eq. (2), the concentration or dosage x_{mix} of the mixed pesticide is

$$x_{\text{mix}} = x_1 + x_2 \tag{8}$$

Suppose the observed mortality rates of the two insecticides and mixed pesticide, corresponding to the concentration or dosage x_1, x_2, and x_{mix}, respectively, are y_1, y_2, and y'. Calculate y using eq. (5). The joint action is synergistic if $y' > y$, and it is antagonistic if $y' < y$.

Assume the relationship between y' and x_{mix} is known as

$$y' = f(x_{\text{mix}}) \tag{9}$$

then we have

$$c = -\ln(1 - y') = -\ln(1 - f(x_{\text{mix}})) \tag{10}$$

Using the relationships (7) and (10) in (2), or relationships (6) and (9) in (5), the theoretical mortality rate with additive effect will be

$$y = 1 - (1 - h(x_1))(1 - g(x_2)) \tag{11}$$

The joint action is synergistic if $f(x_{\text{mix}}) > y$ (i.e., $c > m + n$), and it is antagonistic if $f(x_{\text{mix}}) < y$ (i.e., $c < m + n$).

As an alternative method, if we are interested in specified mortality rate, the concentration or dosage x_1 and x_2 can be derived from eq. (6) as

$$\begin{aligned} x_1 &= k(y_1) \\ x_2 &= p(y_2) \end{aligned} \tag{12}$$

where y_1 and y_2 are given mortality rates of two insecticides, A and B, respectively.

In order to achieve more reliability on the conclusions above, we may repeat experiments to achieve the confidential interval ($p = 0.05$) of y'

$$y' \pm 2s_{y'} \tag{13}$$

where $s_{y'}$ is the standard deviation of y'. The joint action is synergistic, if $y' - 2s_{y'} > y$, and it is antagonistic, if $y' + 2s_{y'} < y$.

Part 8
Ecological Networks

Chapter 22

Food Webs

In the past years, network analysis became the mainstream of ecological theory. A variety of ecological problems, e.g., the fundamental structure of ecological networks, the robustness of network to external disturbance, coevolution and coextinction, community assembly, succession, etc., can be addressed by network analysis (Zhang, 2012e, 2016n). Without a deep insight on ecological network structure, it will be impossible to assess network robustness under species extinction, habitat destruction, and other human activities (Bascompte, 2009; Zhang, 2016n, 2016q). To analyze the relationship between structure and functionality of network, three issues should be addressed (Bascompte, 2009): (1) network analysis must consider various interaction types. Various types of interactions jointly determine the stability of a network (Bastolla *et al.*, 2009); (2) the model must consider how population dynamics influences network structure and how network structure influences population dynamics, and (3) species invasion and climate change, etc., should also be considered.

In an ecological network, if the set of some species and the set of interactions are hyper-represented in the network, i.e., between-vertex interactions are hyper-represented based on the equivalent random network (Milo *et al.*, 2002), then they can be referred to as mosaics (Zhang, 2012e). Mosaics are modules for assembling a network. A small mosaic is usually a three-layer food chain (Bascompte, 2009; e.g., Predator–Prey–Resource, or Omnivore–Prey–Resource, etc.). A large network is usually organized from various small mosaics. The method for organizing mosaics to a large network may probably influence the stability of the entire network (May, 1972).

22.1   Fundamentals of Food Webs

In this chapter, we focus on the most fundamental ecological networks, food webs (further details can be found in Zhang, 2012e). A food web is a set of species connected by trophic relationships. It is an ecological network made of interactive species. Biodiversity, ecosystem structure and function, etc., can be represented and described by food webs.

A food web represents the trophic hierarchy and interrelation of various species in the ecosystem. In the food web, all species occupying the same trophic position make up a trophic level. For example, all plants in the food web constitute a trophic level, called the first or "primary producers", all herbivores comprise the second or "primary consumer" trophic level, and all carnivorous animals constitute the third or "secondary consumer" trophic level. In addition, if there are more advanced carnivores that eat other carnivores, they will constitute an even higher trophic level (Zhang, 2012e). By between-species trophic interaction in a community, the food web describes and quantifies the complexity of the ecosystem (May, 1972; Cohen, 1978; Pimm, 1982; Montoya *et al.*, 2006; Pascual and Dunne, 2006). Food webs are key topics of ecology (Bascompte, 2009).

There are two types of food webs, i.e., the one with autotrophic species as base species and the one with scavenger animals as base species (Gonenc *et al.*, 2007). The complexity and trophic levels of food web determine the stability, resilience, and robustness of the community (Zhang, 2016n). An ecosystem resists the extinction of species if its food web is complex enough. The species loss in food web would to some extent be detrimental the stability of ecosystem.

Food webs have long been the center of ecological studies. They began with text and table expression and thereafter linear and spatial expression. With the emergence of a large number of algorithms and software, the studies of network structure are now becoming the focus of food webs. These algorithms and software have been used to explore the ecosystem stability and robustness (Zhang, 2016n). For example, they are used to study degree distribution, connectance, and network size (Dunne *et al.*, 2002). Odum (1983) held that the community stability could be measured by energy path in the food web. MacArthur (1955) argued that stability may be increased by the increase of links in the food web. Pimm *et al.* (1991) discussed the effects of different models of food webs on ecosystem structure, stability, and robustness. Byron and Tennenhouse (2015) determined similarities in structure among food webs and found that there are pairs of systems that are highly similar in structure once appropriately normalized for size, makeup, and geographical location, and that a majority of food webs have similar structural components when compared with random food webs. Jiang and Zhang (2015a, 2015b) quantified the relative importance of species in Carpinteria Salt Marsh (CSM) food web by analyzing five centrality indices. The results showed that there are large differences in rankings species in terms of different centrality indices. Jiang *et al.* (2015) used Pajek to analyze the topological properties of four full arthropod food webs in South China. The results showed that predators are significantly more abundant than preys, and the proportion of predators to preys (3.07) is significantly higher than previously reported by Cohen in 1977 (1.33). In the food webs, the number of top species is the largest, accounting for about 50% of the total. The number of intermediate–intermediate links is far greater than the other three links. The average degree of paddy arthropod food webs is 6.0, 6.04, 5.74, and 7.75, respectively. In addition, it was found that the mechanism of evolution and population size will affect food web topology (Rossberg *et al.*, 2005). Further, habitat destruction and climate change are likely to cause the extinction of key species. Once a key species is extinct, the robustness of food web will be profoundly affected.

In recent years ecologists have analyzed the model dynamics of large food webs. There were two components in the models. First, they included the interaction network with dozens of species which spanned several trophic levels (Pascual and Dunne, 2006; Montoya *et al.*, 2006); Second, they also considered practical body weight and metabolic rate (Brose *et al.*, 2006). By including these factors in the models, the role of body weight ratio in preserving network's properties and the influence of removing some species on species richness were studied (Berlow *et al.*, 2009). In general, the larger a food web, the more precise the conclusions.

Mutualism (e.g., plants pollinators, plants seed Dispersers) is also considered in recent years' analysis on food web mosaics. These mutualistic food webs show some characteristics: (1) Heterogeneity. Many species interact with a few species. The number of interactions of these few species is usually more than other species. (2) Nesting effect. (3) Between-species interactions are weak and asymmetric. Bastolla *et al.* (2009) demonstrated that species richness benefits from the mutualistic network structure. For a given number of interactions, the nest structure of mutualistic network will maximize the number of coexisting species (Zhang, 2012e).

A central question in food web theory is how static structures emerge from dynamics of interactive species (Cohen *et al.*, 1993). A possible way to address this question is to perform community assembly experiments (Zhang, 2014c, 2015b). Most literature on dynamic assembly of multispecies ecosystems deal with competitive communities (Case, 1990; Morton *et al.*, 1996) or randomly wired ecosystems without trophic structure (May, 1973; Pimm, 1991).

It was found that deletion of the most connected species that are typical from the food webs with skewed degree distributions might result in coextinction of many other species by direct or indirect effects (Pimm, 1980; Dunne *et al.*, 2002; Montoya and Sole, 2002). Coextinction tends to occur in homologous species with similar genetic relationship (e.g., the same genera), which may result in a nonrandom pruning (Rezende *et al.*, 2007). Extinction of trophic levels occurs faster than coextinct species (Petchey *et al.*, 2008).

The studies of food web properties have provided useful information for understanding ecosystem organization and its relationship with ecological stability (Pimm, 1991; Pimm *et al.*, 1991; Morin and Lawler, 1996; McCann 2000). Most have centered on the scale-invariant (in terms of number of species) nature of food web patterns. For example, some network models predicted that food webs are robust to random species extinction but dependent upon some species of better connections. If these species are lost, then the network will abruptly collapse (Dunne *et al.*, 2002; Memmott *et al.*, 2004; Montoya *et al.*, 2006).

Most food webs have been determined by field sampling (Zhang, 2012e). However, sampling effort will exert a little effect on the topological properties (Martinez, 1991, 1994; Martinez *et al.*, 1999; Montoya and Sole, 2002, 2003). It was found that connectance is robust under different sampling efforts (Montoya and Sole, 2003). The same occurs for link distribution frequencies (Montoya and Sole, 2002).

22.1.1 *Weak interactions*

Montoya and Sole (2003) found that most interactions in complex food webs are weak, confirming some previous theoretical and empirical findings (Paine, 1992; Raffaelli and Hall, 1996; Berlow *et al.*, 1999; McCann 2000). It is important for community stability and species coexistence (May, 1973; Laska and Wooton, 1998; McCann *et al.*, 1998; McCann, 2000; Zhang, 2011c, 2012e): (1) weak interactions yield negative covariances between resources, which promotes community-level stability; (2) negative covariances make species that interact weakly dampen the destabilizing potential of strong interactions (McCann, 2000).

22.1.2 *Degree distribution*

For a food web, suppose S is the number of species, L is the number of actual links, connectance C is defined as L divided by the maximum possible number of links S^2. Some research found that food webs with different size had a roughly constant value of C from 0.1 to 0.15 (Martinez, 1992; Warren, 1994). Sometimes C varied in food webs as S increases from 0.27 to 0.02 (Montoya and Sole, 2003). It was found that the inverse hyperbolic relationship between S and C is based on the EcoWeb database (Cohen and Briand, 1984; Cohen *et al.*, 1990).

According to Montoya and Sole (2003), C was scale-variant and related to S according to a power law $S = C^{-a}$ with an exponent $a \approx -1/2$. The constant connectance hypothesis (CCH), i.e., $C \approx 0.14$ despite changes in S, reported in some previous studies (Pimm *et al.*, 1991; Havens 1992; Martinez, 1992) did not hold for the 12 food webs investigated. L increased with S in a different manner from that predicted by both the link-species scaling law (*LSSL*) and the *CCH*. Assuming the simplest relationship between L and S, $L = aS^b$, the *LSSL* holds that b must be close to 1 and on average the number of links per species in a food web is constant and scale invariant at roughly 2, i.e., $L \approx 2S$ (Cohen *et al.*, 1990; Martinez 1992). However, the *CCH* holds that L increases approximately as the square of S with $a < 1$ (a is the connectance C) and therefore $L = CS^2$ ($C = 0.14$; Martinez 1991, 1992). Other studies rejecting *LSSL* indicated values of $b \approx 1.5$ (Sugihara *et al.*, 1989; Schoenly *et al.*, 1991; Havens, 1992; Martinez, 1994).

Montoya and Sole (2003) examined differences in degree distributions across some complex food webs and found that larger food webs have skewed degree distributions that strongly depart from those expected from random wiring (Zhang, 2011c), whereas food webs with fewer species have more homogeneous degree distribution frequencies. This topological property highlights the importance of the position of species within food webs for their stability. It was found that in a species-rich network, most species had very few connections (many specialists) and only a few species were highly connected (fewer generalist preys and predators) (Montoya and Sole, 2003; Zhang, 2011c). The number of connections fluctuated around mean linkage density (L/S) when species richness was low. It was found that simulated networks always yielded Poisson degree distributions, which was independent of S.

To exploit how degree distribution affects community responses under species removal is an important topic in food web theory (Sole and Montoya, 2001; Dunne *et al.*, 2002; Montoya and Sole, 2002). According to Montoya and Sole (2003), the shape of the degree distribution of network of trophic interactions is highly dependent upon the species richness (Yodzis, 1980; Sugihara *et al.*, 1989; Martinez,

1992, 1994; Martinez *et al.*, 1999). The topological property might result from assembly processes and can be partially reconstructed by multitrophic assembly models in which dynamics are dominated by between-species weak interaction strengths.

22.1.3 *Network structure and system stability*

Pinnegar *et al.* (2005) used a detailed Ecopath with Ecosim (EwE) model to test the impacts of food web aggregation and the removal of weak linkages. They found that aggregation of a 41-compartment food web to 27- and 16-compartment systems greatly affected system properties (e.g., connectance, system omnivore, and ascendancy) and influenced dynamic stability. Highly aggregated webs recovered more quickly following disturbances compared to the original disaggregated model.

Food webs with skewed degree distributions showed two behaviors (Montoya and Sole, 2003): they exhibited high homeostasis when species were removed at random from the community, but were very fragile when removals targeted generalist or most connected species. Food webs with Poisson degree distributions were highly fragile to both types of removals (random or directed).

Increased S implies a more complex distribution where few species play key roles in community persistence. Thus, stochastic environmental fluctuations might less affect species-rich communities, but human perturbations might have larger effects. This is based on structural stability and does not consider dynamic effects of species deletion.

Many theories or hypotheses were proposed to explain diverse patterns of food web connectance (Warren, 1994). The most developed theory is that there is a relationship between C and different types of ecosystem stability. Some models suggest that lower connectance involves higher local (May, 1973; Pimm, 1991; Chen and Cohen, 2001) and global (Cohen *et al.*, 1990; Chen and Cohen, 2001) stability, that is, the system recovers faster after a disturbance. Conversely, another theory suggests that a food web with higher connectance has more numerous reassembly pathways and can thus recover faster from perturbation (Law and Blackford, 1992).

22.2   Food Web Models

Some simple models are important to describe nonrandom food web structure, which include cascade model (Cohen *et al.*, 1990), niche model (Williams and Martinez, 2000), hierarchical model (Cattin *et al.*, 2004), and others. These static models are based on one-dimensional classification and ordering of species or species hierarchy. Using these models, some properties of food webs can be primarily approached (Jiang and Zhang, 2015b; Zhang *et al.*, 2014a).

22.2.1   *Multitrophic assembly model*

The multitrophic assembly model was first developed by Pimm (1980). An extended version of Pimm's multitrophic assembly model was presented later (Lockwood *et al.*, 1997). In the extended model, different types of functional responses (i.e., how prey consumption by predators vary with prey density) in predator–prey dynamics are introduced: (1) Linear functional response (Holling type I), where interaction strength is a per capita consumption rate of prey per predator and prey consumption increases linearly with prey abundance, ultimately reaching a maximum for high prey densities, and (2) prey-dependent functional responses, where interaction strength shows the maximal attack rate of predators on prey and increases in a decelerated (Holling type II) or sigmoid (Holling type III) fashion, ultimately reaching the maximal attack rate. Lotka–Volterra equations are used to model population dynamics (Montoya and Sole, 2003). All model features are predefined in a matrix that represented the regional species pool. Link distribution frequencies in species pools are constructed randomly (Bollobás, 1985). At each t_i time iteration, one species is randomly chosen to put into the community from the species pool at a population density of 0.001. A species was considered extinct if its density was below that value (Lockwood *et al.*, 1997).

22.2.2   *Cascade model*

The cascade model was first proposed by Cohen and Newman (1985). All species are arranged as a cascade structure so that the top species can feed on the lower species. The adjacent matrix $A_{s \times s}$, built from the food web, is thus a strict upper triangular matrix ($a_{ij} = 0$, $i > j$), i.e., it is a strict trophic hierarchy or cascade (Zhang, 2012e). Therefore, the food web has no cycle. The species with ID No.1 are only fed by other species and they do not feed on any other species, etc. The species with ID No. S may feed on any other species. Furthermore, this model assumes that there is a positive real value c ($c = 2CS^2/(S-1)$, and C is the connectance, C= L/S^2), so that for $S \geq c$, all elements a_{ij}, $i \leq j$, follow the Bernoulli distribution with the parameter $p = c/S$. Thus, some properties of the food web can be derived (Zhang *et al.*, 2014a).

Suppose $p = c/S$, $q = 1 - p$, and $c \geq 0$, $S \geq c$. Topological structure of a food web can be derived by (Cohen and Newman, 1985)

$$E(T) = E(B) = (1 - q^S)/p$$
$$E(I) = S[1 - 2(1 - q^S)/c + q^{S-1}]$$
$$E(L) = pS(S - 1)/2 = c(S - 1)/2$$
$$E(L_{BI}) = E(L_{IT}) = (S - 1)(1 + q^{S-1}) - (1 + q)(1 - q^{S-1})/p$$
$$E(L_{BT}) = (1 - q^{S-1})/p - (S - 1)q^{S-1}$$
$$E(L_{II}) = pS(S - 1)/2 - (S - 1)(2 + q^{S-1}) + (1 - q^{S-1})(1 + 2q)/p$$

From the first three formulae, we can calculate the proportions of the number of top species T, number of intermediate species I, and number of basal species B, divided by total number of species. And from the last three formulae, we may calculate the proportions of number of top species — basal species links BT, number of top species — intermediate species links IT, number of intermediate species — intermediate species links II, and number of intermediate species — basal species links BI, divided by total number of species.

Zhang *et al.* (2014a) used the cascade model above to analyze the effect of parasitism on food webs collected from CSM. It was demonstrated that median and mean of generality and vulnerability for predator–prey and parasite–host subwebs are greater than reported previously. Inclusion of parasites significantly increases the mean generality and vulnerability of the full food web. The fitting goodness of cascade model on the predator–prey subweb without parasites is lower than that on the full CSM food web.

The cascade model of Cohen *et al.* (1990) is used to theoretically generate a food web. It assigns each species a random value drawn uniformly from the interval $[0, 1]$, and each species has the probability $P = 2CS/(S-1)$ of consuming only species with the values less than its own. It helps to explain species richness among trophic levels but underestimates interspecific trophic similarity and overestimates length and number of food chains in larger webs (Zhang, 2012e).

The following are Matlab codes, cascadeModel.m, for the cascade model:

```
%Cascade model of food web generation (Cohen et al., 1990).v: number of
 %species. d[1-v][1-v]: food web matrix, where if species j feeds on
species i,
%the element (i,j) is 1, or else 0.
   v=input('Input the number of species:');
   c=input('Input directed connectance (e.g., 0.3):');
   d=zeros(v);
   b=zeros(1,v);
   x=zeros(1,v);
   p=2*v*c/(v-1);
   for i=1:v
   x(i)=rand();
   end
   for i=1:v
   k=0;
   for j=1:v
   if (x(j)<x(i))
   k=k+1; b(k)=j;
   end; end
   if (rand()<=p)
   for r=1:k
   for j=1:v
   if (b(r)==j) d(i,j)=1; end
   end; end; end; end
   fprintf(['Food web generation (Cohen et al., 1990). There are'
num2str(v)'
species in the food web.\n\n'])
   fprintf('Food web matrix (if species j feeds on species i, the element
(i,j) is 1,
or else 0):\n')
   disp([d'])
   s=0;
```

```
fprintf('\n\nNumber of predators per species (in-degree):\n');
for i=1:v
k=0;
for j=1:v
k=k+d(j,i); s=s+d(j,i);
end;
fprintf([num2str(k) ' '])
end
fprintf('\n\nNumber of preys per species (out-degree):\n');
for i=1:v
k=0;
for j=1:v
k=k+d(i,j);
end
fprintf([num2str(k) ' '])
end
fprintf('\n\nTotal degree of every species:\n')
for i=1:v
k=0;
for j=1:v
k=k+d(i,j);
end
for j=1:v
k=k+d(j,i);
end
fprintf([num2str(k) ' '])
end
fprintf(['\n\nThere are ' num2str(s) ' links.\n']);
fprintf(['\n\nDirected connectance (C)=' num2str(s/v^2)]);
disp([d])
```

22.2.3   Niche model

22.2.3.1   Niche model

In the niche model (Williams and Martinez, 2000), each of S species is assigned a "niche value" parameter (n_i), drawn uniformly from the interval $[0,1]$. Species i consumes all species falling in a range (r_i) in which the center (c_i) is uniformly drawn from $[r_i/2, n_i]$. It helps to produce loops and cannibalism by increasing up to half of r_i to include values greater than n_i. The value of r_i is determined by using a beta function to randomly draw values from $[0,1]$ whose expected value is $2C$ and then multiplying the value by n_i (expected $E(n_i) = 0.5$) to obtain desired C. A beta distribution with $\alpha = 1$

is used. It has the form $f(x|1, \beta) = \beta(1-x)^{\beta-1}$, and the expectation $E(x) = 1/(1+\beta)$. In this case, $x = 1 - (1-y)^{1/\beta}$ is a random variable from the beta distribution if y is a uniform random variable and b is chosen to obtain the desired expected value. The fundamental generality of species i is measured by r_i. The number of species falling within r_i represents realized generality. Sometimes food webs generated by models contain completely disconnected species (isolated species) or trophically identical species. All such species should be eliminated and replaced. The species with the smallest n_i has $r_i = 0$ so that every food web has at least one basal species (Zhang, 2012e).

22.2.3.2 *Variant model*

The niche model assigns each species a randomly drawn "niche value". The species are then constrained to consume all prey species within one range of values whose randomly chosen center is less than the consumer's niche value. A variant of niche model was thus needed (Williams and Martinez, 2000).

In the variant of niche model, there are two steps in the construction of food web. First, randomly assign species $i = 1, 2, \ldots, s$, to the niche nutrients in the interval $[0,1]$. Second, randomly assign species i into the interval $[0,1]$; all species within the interval are considered to be preys of species i. The food web matrix is thus constructed.

Trophic similarity of a pair of species (s_{ij}) is the number of predators and prey shared in common divided by the pair's total number of predators and prey. Mean maximum similarity of a web can be calculated by averaging all species' largest similarity index (Williams and Martinez, 2000).

The following are Matlab codes, nicheModel.m, for the variant of niche model:

```
%Niche model of food web generation (Williams and Martinez, 2000).v: number
%of species. d(1-v,1-v): food web matrix, where if species j feeds on
species i,
%the element (i,j) is 1, or else 0.*/
   v=input('Input the number of species: ');
   %nicheModel
```

```
d=zeros(v+1);
x=zeros(1,v+1);
for i=1:v
x(i)=rand();
end
max=1000;
for i=1:v
s=0;
for k=1:max
s=s+rand();
end
meann=s/max;
beta=1/meann-1;
s=0;
for k=1:max
xx=1-(1-rand())^(1/beta);
s=s+xx;
end
meann=s/max;
r=meann*x(i);
c=rand();
c=x(i)*c+r/2*(1-c);
a=c-r/2;
b=c+r/2;
for j=1:v
if (i==j) continue; end
if ((x(j)>=a) & (x(j)<=b))
d(i,j)=1;
end; end; end
fprintf('Food web generation (Williams and Martinez,2000).\n')
bool=niche(v,d);
%nicheModelVariant;
for i=1:v
x(i)=rand();
end
for i=1:v
a=x(i)*rand();
b=x(i)*rand();
if (b<a)
c=a; a=b; b=c;
end
for j=1:v
if (i==j) continue; end
if ((x(j)>=a) & (x(j)<=b))
d(i,j)=1; end
end; end
fprintf('Variant model of food web generation (Williams and Martinez,
2000).\n');
bool=niche(v,d);
```

The following is the Matlab function, niche.m, used in the model above:

```
function bool=niche(v,d)
bool=0;
%print
fprintf(['There are ' num2str(v) ' species in the food web.\n'])
fprintf('Food web matrix (if species j feeds on species i, the element
(i,j) is 1,
or else 0): \n')
disp([d'])
fprintf('\n')
s=0;
fprintf('Number of predators per species (in-degree):\n')
for i=1:v
k=0;
for j=1:v
k=k+d(j,i);
s=s+d(j,i);
end
fprintf([num2str(k) ' '])
end
fprintf('\n');
fprintf('Number of preys per species (out-degree):\n');
for i=1:v
k=0;
for j=1:v
k=k+d(i,j);
end
fprintf([num2str(k) ' '])
end
fprintf('\n');
fprintf('Total degree of every species:\n');
for i=1:v
k=0;
for j=1:v
k=k+d(i,j);
end
for j=1:v
k=k+d(j,i);
end
fprintf([num2str(k) ' '])
end
fprintf('\n');
fprintf(['There are ' num2str(s) ' links.\n'])
c=s/v^2;
fprintf(['Directed connectance (C)=' num2str(c)])
fprintf('\n');
%speciestype
a=zeros(1,v+1);
```

```
b=zeros(1,v+1);
for i=1:v
a(i)=0; b(i)=0;
for j=1:v
a(i)=a(i)+d(j,i);
b(i)=b(i)+d(i,j);
end; end
s=0;
fprintf('Top species:\n');
for i=1:v
if (a(i)==0)
fprintf([num2str(i) ' '])
s=s+1;
end; end
fprintf('\n')
fprintf(['Total number: ' num2str(s)])
fprintf('\n');
s=0;
fprintf('Intermediate species:\n');
for i=1:v
if ((a(i)~=0) & (b(i)~=0))
fprintf([num2str(i) ' '])
s=s+1;
end; end
fprintf('\n');
fprintf(['Total number: ' num2str(s)])
fprintf('\n');
s=0;
fprintf('Basal species:\n');
for i=1:v
if (b(i)==0)
fprintf([num2str(i) ' '])
s=s+1;
end; end
fprintf('\n');
fprintf(['Total number: ' num2str(s)]);
fprintf('\n');
%tropsimil
ss=zeros(v+1);
mxsim=0; summ=0;
for i=1:v
max=0;
for j=1:v
if (i==j)
ss(i,j)=1;
continue; end
c=0; cc=0;
for k=1:v
if ((d(i,k)==1) & (d(j,k)==1)) c=c+1; end
if ((d(i,k)==1) | (d(j,k)==1)) cc=cc+1; end
```

```
end
for k=1:v
if ((d(k,i)==1) & (d(k,j)==1)) c=c+1; end
if ((d(k,i)==1) | (d(k,j)==1)) cc=cc+1; end
end
ss(i,j)=c*1.0/cc;
if (ss(i,j)>max) max=ss(i,j); end
end
summ=summ+max;
end
mxsim=summ/v;
fprintf('\nTrophic similarity of paired species:\n');
disp([ss])
fprintf('\n\n');
fprintf(['Maximum similarity of the web: ' num2str(mxsim)])
disp([d])
fprintf('\n\n');
bool=1;
```

Niche model performs well in predicting degree distribution of food webs in particular the distribution of total degree and outde-gree. Moreover, it helps to develop a general function for degree dis-tribution.

In the niche model, species are randomly assigned to an interval. This is reasonable in some cases. However, the assignment is not accordant with true situations if between-species trophic similarity is high. In these situations species will not randomly but intensively distributed over the interval [0,1], which results in the increase of species with high links compared with the prediction of niche model. Diverse species aggregate with similar trophic requirements, and are thus different from the predicted by niche model (Zhang, 2012e).

22.2.3.3 *MaxEnt model*

A model, MaxEnt, which is based on the maximum-entropy princi-ple, can be used to predict degree distribution of ecological networks (Williams 2010). It is considered as a niche model also.

The MaxEnt model includes prior knowledge of the number of basal species B and does not attempt to predict the fraction of basal species. Similarly, two consumer distributions are considered, the "all-species consumer distribution" and the "restricted consumer dis-tribution". The "all-species consumer distribution" is defined as the distribution of the number of consumers of each species, including the top species, which have no consumers. The "restricted consumer distribution" is defined as the distribution of the number of consumers of the resource species, includes prior knowledge of the number of top species T, and does not attempt to predict the fraction of top species (Williams 2010).

In the all-species consumer distribution, the resource species or consumers of each species can range from 0 to S (S: number of species; L: links), and the mean number of links per species is L/S. In the restricted resource distribution, the number of links per consumer varies within 1 and S, and the mean number of links per consumer is $L/(S-B)$. In the restricted consumer distribution, the number of links from each resource varies within 1 and S, and the mean number of links from each resource is $L/(S-T)$.

First, a discrete distribution should be found based on some values, for example $\{x_1, \ldots, x_n\}$. The mean μ (i.e., indegree or outdegree of a node) maximizes $H = -\sum_i p_i \ln p_i$, and satisfies some conditions. The maximum entropy distribution is

$$p_i = P(X = x_i) = Ce^{\lambda xi}, \quad \text{for } i = 1, \ldots, n.$$

Williams (2010) used the model $p_i = P(X = x_i) = x_i^\lambda$, i.e., using power-law distribution but not exponential distribution. Set these conditions, $\sum_i P_i = 1$ and $\sum_i X_i P_i = \mu$. C and λ can thus be obtained by using Lagrange multipliers. Williams (2010) assumes that the number of consumers and resources are independent on each node of the food web. For T top species without consumers, the number of links is from maximum entropy resource distribution. Similarly, For B basal species, the number of links is from maximum entropy consumer distribution. And For $S-B-T$ intermediate species, the number of links is the sum of the numbers from consumer and resource distributions (Zhang, 2012e).

In addition, Simpson index can also be used as the entropy index

$$H = 1 - \sum p_i^2$$

As a supplement, some other degree distributions, such as Bararasi (2009)

$$f(r) = r^{-\beta}$$

and Zhang (2011c)

$$f(r) = \alpha(r-a)^{\beta}(b-r)^{\eta}, \quad a \le r \le b$$
$$f(r) = 0, \quad r > a \quad \text{or} \quad r < b$$
$$\alpha > 0, \quad \beta \ge 0, \quad \eta \ge 0$$

can be used, where α: scale parameter; β, γ: shape parameters; a, b: position parameters.

The following are Matlab codes, maxEnt.m, for maximum entropy algorithm:

```
%Maximum entropy algorithm for network generation. In the basal
%species connection number file (maxent1) and top species connection
%number file (maxent2), the numbers of basal and top species connections
%of the network are given respectively.
    str1=input('Input the file name of basal species connection number (store
different number of connections, e.g., 3,5,7,6): ','s');
    str2=input('Input the file name of top species connection number (store
different number of connections,e.g., 5,7,10,8,6,3,1): ','s');
    ff1=load(str1); dim=size(ff1); nf1=dim(1);
    ff2=load(str2); dim=size(ff2); nf2=dim(1);
    miuin=input('Input the expected mean number of basal species
connections:');
    miuout=input('Input the expected mean number of top species
connections:');
    typedis=input('Input the type of frequency distribution of degrees (1:
exponential distri (p(x)=ce^(dx)); 2: power law distri (p(x)=x^(-c)); 3:
Zhang distri (p(x)=a(x-b)^c*(d-x)^e): ');
    typeent=input('Input the type of entropy expression (1: Shannon-Wiener
index; 2: Simpson index): ');
    vtb=input('Input the number of total species: ');
    vt=input('Input the number of top species: ');
    vb=input('Input the number of basal species: ');
    x=zeros(1,10);
    xi=zeros(1,nf1+nf2+1);
    p=zeros(1,nf1+nf2+1);
    u=zeros(1,nf1+nf2+2);
    t=zeros(1,vt+1);
    b=zeros(1,vb+1);
    k=0;
    nf=nf1;
    for i=1:nf
```

```
xi(i)=ff1(i);
end
%fprintf('Degree distribution of basal species:\n')
if (typedis==1)
x(1)=rand();
x(2)=-1+rand();
end
if (typedis==2) x(1)=1+rand(); end
if (typedis==3)
x(1)=rand();
x(2)=rand();
x(3)=2+rand();
x(4)=xi(nf)+rand();
x(5)=2+rand();
end;
x=fminsearch(@(x) maxEntObj(typedis,typeent,nf,miuin,x,xi),x');
for i=1:nf
if (typedis==1) p(i)=x(1)*exp(x(2)*xi(i)); end
if (typedis==2) p(i)=xi(i)^(-x(1)); end
if (typedis==3) p(i)=x(1)*(xi(i)-x(2))^x(3)*(x(4)-xi(i))^x(5); end
end
u(1)=p(1);
for i=2:nf
u(i)=u(i-1)+p(i);
end
for i=1:vb
ran=rand()*u(nf);
for j=1:nf
if (j==1) th=0; end
if (j>1) th=u(j-1); end
if ((ran>=th) & (ran<u(j)))
k=j;
break;
end; end
b(i)=round(xi(k));
end
nf=nf2;
for i=1:nf
xi(i)=ff2(i);
end
fprintf('Degree distribution of basal and top species:\n')
if (typedis==1)
x(1)=rand();
x(2)=-1+rand();
end
if (typedis==2) x(1)=1+rand(); end
if (typedis==3)
x(1)=rand();
x(2)=rand();
x(3)=2+rand();
x(4)=xi(nf)+rand();
```

```
x(5)=2+rand();
end;
x=fminsearch(@(x) maxEntObj(typedis,typeent,nf,miuout,x,xi),x');
for i=1:nf
if (typedis==1) p(i)=x(1)*exp(x(2)*xi(i)); end
if (typedis==2) p(i)=xi(i)^(-x(1)); end
if (typedis==3) p(i)=x(1)*(xi(i)-x(2))^x(3)*(x(4)-xi(i))^x(5); end
end
u(1)=p(1);
for i=2:nf
u(i)=u(i-1)+p(i);
end
for i=1:vt
ran=rand()*u(nf);
for j=2:nf
if ((ran>=u(j-1)) & (ran<u(j)))
k=j;
break;
end; end
t(i)=round(xi(k));
end
mis=0; mos=0;
for i=1:vt
mis=mis+t(i);
end
for i=1:vb
mos=mos+b(i);
end
fprintf('Food web generation (Williams, 2010; Bararbas, 2009; Zhang,
2011)\n')
fprintf(['Indegree of ' num2str(vb) ' basal species: '])
for i=1:vb
fprintf([num2str(b(i)) ','])
end
fprintf('\n');
fprintf(['Total degree of ' num2str(vtb-vt-vb) ' intermediate species: '
num2str(mos+mis)])
fprintf(['\nOutdegree of ' num2str(vt) ' top species: '])
for i=1:vt
fprintf([num2str(t(i)) ','])
end
fprintf('\n')
```

The following are the Matlab functions, maxEntObj.m and maxEntObjFun.m, used in the model:

```
function f=maxEntObj(typedis,typeent,nf,miu,x,xi)
f=0;
p=zeros(1,nf+1);
for i=1:nf
```

```
if (typedis==1) p(i)=x(1)*exp(x(2)*xi(i)); end
if (typedis==2) p(i)=xi(i)^(-x(1)); end
if (typedis==3) p(i)=x(1)*(xi(i)-x(2))^x(3)*(x(4)-xi(i))^x(5); end
end
switch typeent
case 1
f=maxEntObjFun(3,nf,p,xi)+abs(maxEntObjFun(1,nf,p,xi)-1)+abs(maxEntObj
Fun(2,nf,p,xi)-miu);
case 2
f=maxEntObjFun(4,nf,p,xi)-1+abs(maxEntObjFun(1,nf,p,xi)-1)+abs(maxEntO
bjFun(2,nf,p,xi)-miu);
end
function f=maxEntObjFun(term,nf,p,xi)
f=0;
for j=1:nf
switch term
  case 1
      f=f+p(j);
  case 2
      f=f+p(j)*xi(j);
  case 3
      f=f+log(p(j))*p(j);
  case 4
      f=f+p(j)^2;
end; end
```

Resource distributions of consumers and various trophic levels for 51 food webs have been analyzed and clarified (Cohen *et al.*, 1990). Degree distributions of the 51 empirical food webs and the corresponding prediction of maximum entropy model are consistent. It demonstrates that it is not necessary to consider the detailed ecological process in many food webs in order to predict the (undirected) degree distributions of consumers and resources.

In addition to the models above, Ecopath (Polovina, 1984; Christensen and Pauly, 1992; Walters *et al.*, 1997, 2000; Pauly *et al.*, 2000; Christensen and Walters, 2004) and Ecosim (Walters *et al.*, 1997, 2000; Libralato *et al.*, 2006) are food web models also.

22.3 Arthropod Food Webs

To explore the topological properties of paddy arthropod food webs is of significance for understanding natural equilibrium of rice pests. Jiang *et al.* (2015) used degree/link analysis, chain cycle/length analysis, and keystone index to analyze the topological properties

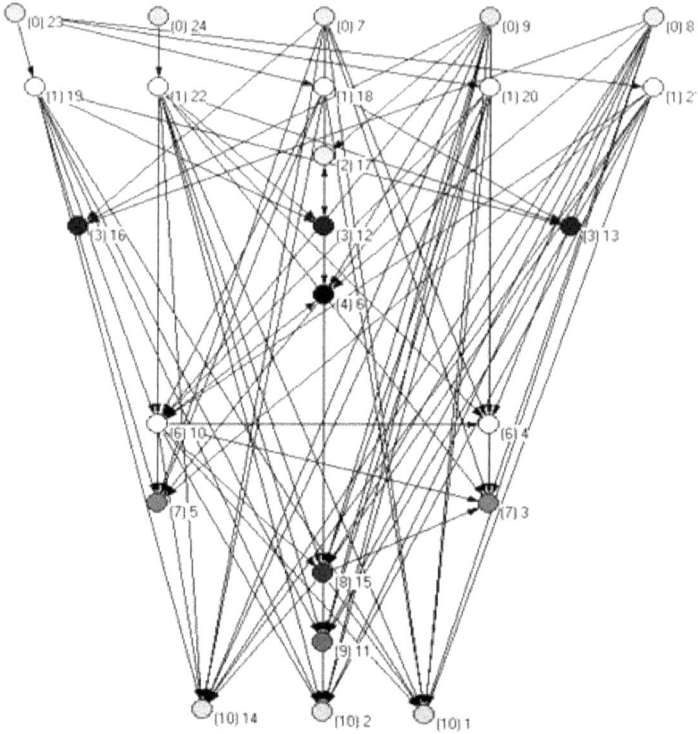

Fig. 1 Food web links of incoming degree analysis on an arthropod food web. For each species, the number in parenthesis is incoming degree and the number outside parenthesis is species ID code (Jiang *et al.*, 2015).

of four full arthropod food webs in South China (Fig. 1). The results (Tables 1 and 2) showed that predators are significantly more abundant than preys, and the proportion of predators to preys (3.07) was significantly higher than previously reported by Cohen (1.33). In the food webs, the number of top species is the largest, accounting for about 50% of the total. The number of intermediate–intermediate links is far greater than the other three links. The average degree of paddy arthropod food webs is 6.0, 6.04, 5.74, and 7.75, respectively. Average degree and link density do not change significantly with the change of the number of species, but the connectance reduces significantly. In the paddy ecosystems, the increase of species diversity does not lead to an increase proportionally in

Table 1 Species analysis of food webs (Jiang *et al.*, 2015).

Food web	Trophic level	Number of species	Total number of species	Proportion %	Species ID
FW1	T	13	26	50	1–6, 8–11, 13–15
	I	11		42.3	7, 12, 16–24
	B	2		7.7	25, 26
FW2	T	31	57	54.4	1, 3–19, 21–26, 28, 31–36
	I	24		42.11	2, 20, 27, 29, 30, 37–55
	B	2		3.49	56, 57
FW3a	T	11	23	47.83	1–6, 11–13, 15, 16
	I	10		43.48	7–10, 14, 17–21
	B	2		8.69	22, 23
FW3b	T	12	24	50	1–6, 11–14, 16, 17
	I	10		41.67	7–10, 15, 18–22
	B	2		8.33	23, 24

Table 2 Link analysis of food webs (Jiang *et al.*, 2015).

Food web	Trophic level	Number of links	Total number of links	Proportion (%)
FW1	B–I	5	78	6.41
	B–T	0		0
	I–I	21		26.92
	I–T	52		66.67
FW2	B–I	19	172	11.04
	B–T	0		0
	I–I	32		18.61
	I–T	121		70.35
FW3a	B–I	5	66	7.57
	B–T	0		0
	I–I	11		16.67
	I–T	50		75.76
FW3b	B–I	5	93	5.38
	B–T	0		0
	I–I	14		15.05
	I–T	74		79.57

the links among species. The link density and connectance of food webs of early-season rice field are less than that from late-season rice field. Cycles of all food webs cycles are 0. The maximum chain length of the basal species is 3, and the largest chain length of the top species is typically 2 or 3. Neutral insects are found to play a very important role in the paddy ecosystem. *Nilaparvata lugens* and *Sogatella furcifera* are found to be the dominant species of rice pests. *Pardosa pseudoannulata, Tetragnatha maxillosa, Pirata subparaticus, Arctosa stigmosa,* and *Clubiona corrugate* are identified as the important predatory species that may effectively control the pest population. The keystone species calculated from keystone index and network analysis are analogous, indicating either keystone index or network analysis can be used in the analysis of keystone species.

Part 9

Microscopic Networks

Chapter 23

Molecular and Cellular Networks

In this chapter, molecular networks refer to biological networks with molecules as nodes, and cellular networks are those with cells as nodes. Molecular networks cover metabolic networks, protein–protein interaction networks, biochemical reaction networks, gene networks, transcriptional regulatory networks, etc. Cellular networks include brain, tissues, organs, etc.

Most of the network analyses on molecular networks focused on degree distribution of network nodes. With regard to the parameter λ of degree distribution of molecular networks (Zhang, 2016i) in power-law distribution $p(x) = x^{-\lambda}$ and exponential law distribution $p(x) = e^{-\lambda x}$, Goemann *et al.* (2011) provided a set of values. For power-law distribution, it is 1.63 (transcription network, the number of nodes $v = 279$), 1.71 (signaling network, $v = 1571$), and 1.58 (metabolic network, $v = 1793$), respectively; for exponential-law distribution, it is 0.19 (transcription network), 0.21 (signaling network), and 0.15 (metabolic network), respectively. Zhang and Feng (2017) found that that power-law λ is 4.43 for metabolic pathway of nonalcoholic fatty liver disease (NAFLD) ($v = 61$). Martínez-Antonio (2011) demonstrated that that power-law λ is 1.11 for *Escherichia coli* transcriptional regulatory network ($v = 1531$). Huang and Zhang (2012) found that the mean is between 2.1 and 2.9 for tumor pathways ($v \approx 20 \sim 100$). Goemann *et al.* (2011) found the means are 2.35 (transcription network), 2.18 (signaling network), and 3.09 (metabolic network), respectively. However, Shams and Khansari (2014) found that the means are between 4 and 15.

Rahman *et al.* (2013) revealed that the means are between 4.68 and 10.58 for normal and cancer pathways ($v = 192 \sim 631$). Overall power-law λ is mostly around 1.5, and exponential-law λ is around 0.2; the mean of node degrees is mostly around 2.6, and some exceed 4 (Zhang, 2016i).

So far, numerous molecular networks have been reported in the past years. In this chapter, several representative molecular and cellular networks and corresponding analyses are presented.

23.1 Metabolic Pathway of Nonalcoholic Fatty Liver Disease

NAFLD is a systematic and complex disease involving various cytokines/metabolites. Zhang and Feng (2017) used the methodology of network biology to analyze network properties of NAFLD metabolic pathway (Fig. 1). Data of metabolic pathway of nonalcoholic fatty liver disease NAFLD were collected from KEGG-PATHWAY database (Kanehisa Laboratories, 2016).

In addition to power-law distribution, binomial distribution, Poisson distribution, and exponential distribution, more indices are used to determine network type. Connectedness was measured. Cut nodes, centrality indices (degree, closeness, betweenness), and sub networks/modules were calculated. Use a simple method to estimate the robustness of metabolic pathway. First, chose several cytokines/metabolites with higher node degree, i.e., crucial nodes in terms of degree centrality. Remove a cytokine/metabolite, and count the number of its downstream links and nodes.

All cytokines/metabolites in the pathway were numbered as, 1: IL6, 2: IL6R, 3: SOCS3, 4: TNF-α, 5: TNFR1, 6: NF-κB, 7: INS, 8: INSR, 9: IRS1/2, 10: PI3K, 11: Akt, 12: GSK-3, 13: LXR-α, 14: RXR, 15: SREBP-1c, 16: ChREBP, 17: L-PK, 18: LEP, 19: ObR, 20: ACDC, 21: adipoR, 22: AMPK, 23: PPAR-α, 24: Cdc42, 25: Rac1, 26: MLK3, 27: JNK1/2, 28: ASK1, 29: JNK1, 30: AP-1, 31: c-Jun, 32: ITCH, 33: IRE1α, 34: TRAF2, 35: IKKβ, 36: IL1, 37: CYP2E1, 38: ROS, 39: FasL, 40: IL8, 41: TGF-β1, 42: PERK, 43: elF2α, 44: ATF4, 45: CHOP, 46: Bim, 47: Bax, 48: Fas, 49: CASP8, 50: Bid, 51: CxI/II/III/IV, 52: CxII, 53: CxIII, 54: CxIV, 52:

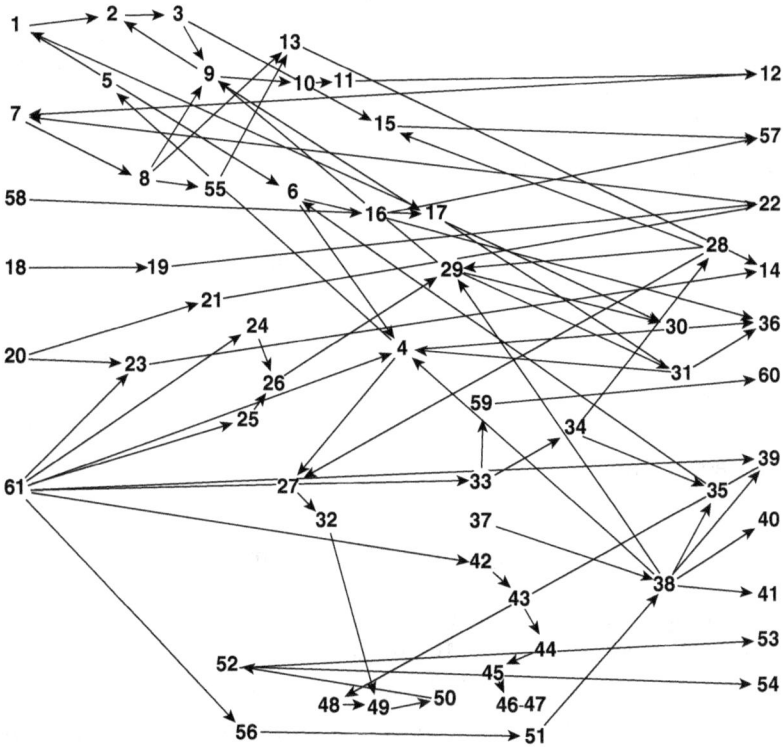

Fig. 1 Network diagram of metabolic pathway of NAFLD (Zhang and Feng, 2017).

Cytc, 53: CASP3, 54: CASP7, 55: Oxysterol, 56: Fatty Acyl-CoA, 57: Lipogenic enzymes, 58: Glucose, 59: XBP1, 60: C/EBPα, and 61: FFAs, respectively.

It was found that the metabolic pathway of NAFLD is not a typical complex network with power-law degree distribution, $p(x) = x^{-4.4275}$, $x \geq 5$. There is only one connected component in the metabolic pathway. The calculated cut cytokines/metabolites of the metabolic pathway are SREBP-1c, ChREBP, ObR, AMPK, IRE1α, ROS, PERK, elF2α, ATF4, CHOP, Bim, CASP8, Bid, CxII, lipogenic enzymes, XBP1, and FFAs. The most important cytokine/metabolite for possible network robustness is FFAs, and the seconded by TNF-α. FFAs, LEP, ACDC, CYP2E1, and glucose

Table 1　Possible robustness contribution of crucial cytokines/metabolites.

ROS		FFAs		TNF-α		NF-κB		JNK1/2	
A	B	A	B	A	B	A	B	A	B
24	28	37	43	24	27	18	18	22	26
(39%)	(34%)	(61%)	(52%)	(39%)	(44%)	(29%)	(22%)	(36%)	(32%)

Notes: A: No. cytokines/metabolites to be influenced; B: No. links to be influenced.

are the only cytokines/metabolites that affect others without influences from other cytokines/metabolites (Fig. 1). The possible contributions of some crucial cytokines/metabolites to network robustness are listed in Table 1.

23.2　*E. coli* **Transcriptional Regulatory Network**

E. coli is the most well-known bacterial model to explore the function of molecular components. Martínez-Antonio (2011) has discussed several structural and functional aspects of its transcriptional regulatory network made of transcription factors (TFs) and target genes. For each interaction in the network, there are experimental evidences about the existence of functional DNA-binding site(s) for the respective TF, where the arc start, onto the regulatory region of the TU encoding for the target gene. In this way, the network contains 1,531 genes (nodes) and 3,421 regulatory interactions (arcs, or connections, or links) (Martínez-Antonio, 2011).

Of the total nodes, 176 nodes correspond to TFs and the rest for structural and mRNAs genes. The network involves around one-third of the total genes in *E. coli*. Nevertheless, it is not easy to calculate which proportion of the total regulatory interactions in *E. coli* represents the 3,421 interactions between TFs and TGs. The network is generated by regulatory interactions from TF to TG, thus it is a directed graph with topological properties common to other biological networks. In this network, the most highly connected nodes (i.e., hubs in the network) correspond to global regulators, which have the highest output degree and they are regulating to many TG (Shen-Orr *et al.*, 2002). TFs are global regulators if they have the high number of regulated genes and (Martínez-Antonio and Collado-Vides, 2003) (1) they should regulate to a larger number of other TFs; (2) they coregulate with them; (3) their target genes should be transcribed by more than one factor;(4) the products of their TG should fall in different functional classes; (5) TFs should be active in different growth conditions: and (6) TFs commonly pertain to protein families with few paralogos.

Global TFs commonly hold the highest positions in the regulatory network and they normally auto-regulate in a dual way (they auto-activate and auto-repress themself), which guarantees the protein level of these important regulators fluctuate between certain levels but never fall to zero (Savageau, 1977; Thomas, 1973).

According to Martínez-Antonio (2011), in *E. coli* these global regulators might be divided into two groups, those controlling the global metabolism and those known as nucleoid-associated proteins (NAPs). The first group contains regulators for controlling carbon uptake (CRP), respiration mode (FNR and ArcA), and stringed response to the lack of important amino acids (Lrp) (the 1st part in Table 2; Martínez-Antonio, 2011). The second group contains FIS, H-NS, and IHF whose maximal production are associated with different points in a growing population curve (Ali Azam *et al.*, 1999). NAPs have properties of DNA-bending and bridging, these proteins are thus considered as analog regulators of gene expression (Marr *et al.*, 2008). They perform regulation by structuring the bacterial nucleoid in different forms.

The most highly regulated TU (i.e., nodes with the highest input degree) are indicated in the 2nd part of Table 2 (Martínez-Antonio, 2011). Most of these genes are encoding for products that define metabolic and adaptive capabilities including motility, response to stresses, and nitrogen metabolism. In addition, two TUs encoding for TFs are yielded from the most highly regulated ones. These genes are not the most highly conserved in bacteria but probably are more specific to *E.coli* genus, and their tight regulation may be responsible for giving the fitness required for the lifestyle distinctive of the bacterium (Martínez-Antonio, 2011).

Table 2 Hubs and highly regulated target genes in the regulatory network of *E. coli* (Martínez-Antonio, 2011).

Global regulators (HUB TFs)	Name	Number of regulated genes
CRP	Cyclic AMP receptor protein, also known as catabolite activator protein (CAP)	440 genes in 128 regulons
H-NS	Histone-like NAP	286 genes in 44 regulons
FNR	Fumarate and nitrate reductase regulatory protein	284 genes in 69 regulons
FIS	Factor for inversion stimulation	225 genes in 48 regulons
IHF	Integration host factor	223 genes in 60 regulons
ArcA	Aerobic respiration regulatory protein	160 genes in 48 regulons
Lrp	Leucine-responsive regulatory protein	97 genes in 26 regulons
Most highly regulated transcription units	TFs regulating	Encoded gene(s)
*flh*DC	9: CRP, Fur, H-NS, HdfR, IHF, LrhA, OmpR, QseB, RcsAB	Heterodimer master regulator of flagella synthesis
*sod*A	8: ArcA, CRP, FNR, Fur, IHF, MarA, Rob, SoxS	Superoxide dismutase, Alleviate oxidative stress
*nir*BCD	8: CRP, FNR, Fis, FruR, H-NS, IHF, NarL, NarP	Large and small subunits of nitrite reductase and nitrite transporter
*mic*F	8: H-NS, HU, IHF, Lrp, MarA, OmpR, Rob, SoxS	Antisense negative regulator of OmpF abundance
*gad*AX	8: ArcA, CRP, FNR, GadE, GadW, GadX, H-NS, TorR	Regulators of glutamate decarboxylase synthesis
*omp*F	7: CRP, CpxR, EnvY, Fur, IHF, Lrp, OmpR	Outer membrane porin for secretion of toxic compounds
*nrf*ABCDEFG	7: FNR, Fis, FlhDC, IHF, NarL, NarP, NsrR	Nitrite reductase, formate-dependent, cytochrome c
*glt*BDF	7: ArgR, CRP, FNR, GadE, IHF, Lrp, Nac	Large and small subunits of glutamate synthase
*dcu*B-*fum*B	7: ArcA, CRP, DcuR, FNR, Fis, Fur, NarL	C4-dicarboxylate antiporter and fumarase B
*nap*FDAGHBC-*ccm*ABCDE FGH	6: FNR, FlhDC, IscR, ModE, NarL, NarP	Proteins with predicted roles in electron transfer to periplasmic nitrate reductase
*mar*RAB	6: CRP, Fis, MarA, MarR, Rob, SoxS	Regulators of weak acids and antibiotics resistance systems

The network analysis indicated that (1) Node degrees of the undirected network are highly diverse. Generally node degrees follow power-law distribution. (2) Input degree distribution (analysis centered on target genes) is a scale-free one. (3) Output degree distribution follows the scale-free distribution. Output interactions in the whole network dominate their entire distribution in the undirected network. (d) Clustering coefficients show that in nodes with few neighbors are connected, while in nodes connected to many other nodes, a certain degree of unconnectedness exists. (4) A giant connected component of 1,476 nodes and 23 isolated clusters with less of 15 nodes each are observed in the network (Martínez-Antonio, 2011; Fig. 2).

In addition to topological properties, Martínez-Antonio (2011) discussed hierarchical organization, sensing and condition-dependent activity, operation mechanism, and evolutionary dynamics of the *E. coli* transcriptional network.

23.3 Common Disease Regulatory Network for Metabolic Disorders

Metabolic disorder causes the failure of metabolism process. Jesmin *et al.* (2016) predicted a common metabolic pathway that is shared by obesity, type-2 diabetes, hypertension, and cardiovascular diseases due to metabolic disorder. They created a protein–protein interaction network to describe the protein coexpression, coregulations, and interactions among gene and diseases. Genes associated with metabolic diseases are accumulated from different gene databases with verification and they are mined to establish gene interaction network models for expressing the molecular linkages among genes and diseases that affect disease progression. By analyzing the gene network model and PPI network, a common metabolic pathway among metabolic diseases is examined. The following are procedures of Jesmin *et al.* (2016) to create and analyze the network:

(1) **Gene integration and processed.** Genes associated with type 2 diabetes, hypertension, obesity, and cardiovascular diseases are collected from PubMed, and further verified by KEGG.

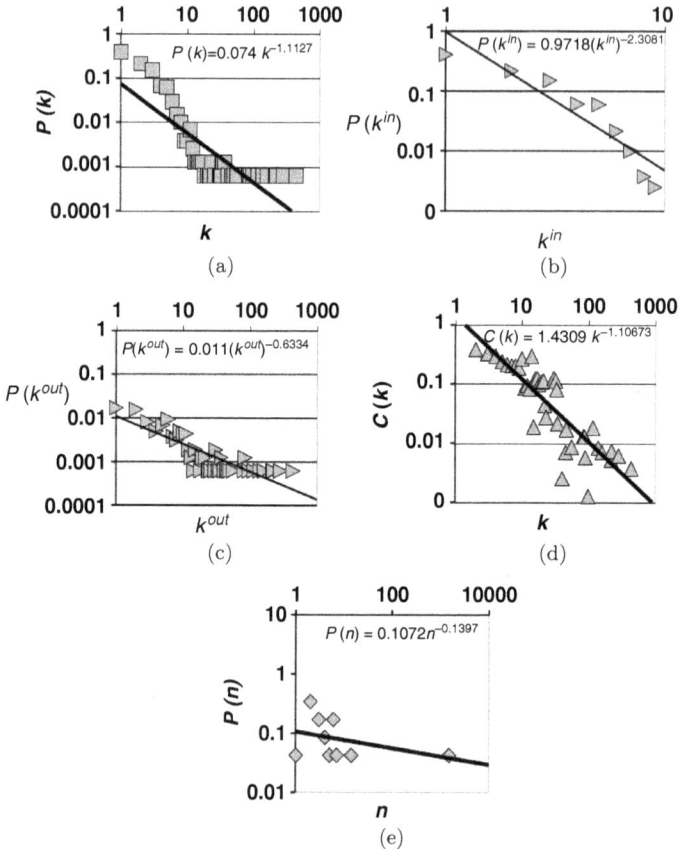

Fig. 2 Topological properties of the *E. coli* transcriptional regulatory network (Martínez-Antonio, 2011). (a) Degree distribution $P(k)$. (b) Input degree distribution $P(k^{in})$. (c) Output degree distribution $P(k^{out})$. (d) Clustering coefficient $C(k)$. (e) The size distribution of connected components in the network $P(n)$.

(2) **Gene mining.** The listed candidate genes concerning type-2 diabetes, obesity, hypertension, and cardiovascular disease are mined according to metabolic disorder by using data mining technique and stored in Unigene data warehouse of NCBI.

(3) **Gene identification according to disease.** To identify the interrelated genes among metabolic diseases, use EXPASY database which helps to find out genes not only related to diseases but also causing crashing of metabolism process of specific diseases.

(4) **Gene sorting.** Sorting is the most crucial part of the research since any tiny mistakes can remove the important gene. Sorting algorithm for taxonomy database is used to sort the identified genes internally correlated among obesity, T2D, hypertension, and cardiovascular disease. The common genes among four diseases are determined.

(5) **Gene filtering.** Gene filtering is the critical step of the research and UniHi tool is used for gene filtering. Unified Human Interactome (UniHI) is a Linkage Network filtering Technique which is applied to find out those genes that have minimum binary and also complex interaction among themselves. It is used to identify the common genes within the four target diseases. The investigated common genes are used to establish a PPI network.

(6) **PPI network creation.** PPI network helps to understand the molecular mechanism of signaling pathways of human diseases and to identify new modules of disease processes. UniHi is a reliable tool to represent PPI maps among genes. UniHi 7 includes almost 350,000 molecular interactions between genes, proteins, and drugs, as well as numerous other types of data such as gene expression and functional annotation. Finally, PPI network is created for common genes using UniHi tool (Fig. 3).

(7) **Common regulatory pathway.** The final step is the construction of the common gene regulatory pathway. The common genes among these diseases are verified with KEGG database in different biological pathways. The selected genes are cross-validated and clustered using the KEGG mining tools (Kanehisa *et al.*, 2008). From the pathway information on the selected genes, the common regulatory pathway is predicted using UniHi.

23.4 Find Molecular Pathological Events in Different Cancer Tissues

Using network analysis of cancer transcriptomes, Iqbal *et al.* (2014) presented a method to hunt for critical genes or networks helping in metastasizing cancers. Transcriptomic analysis of different cancerous

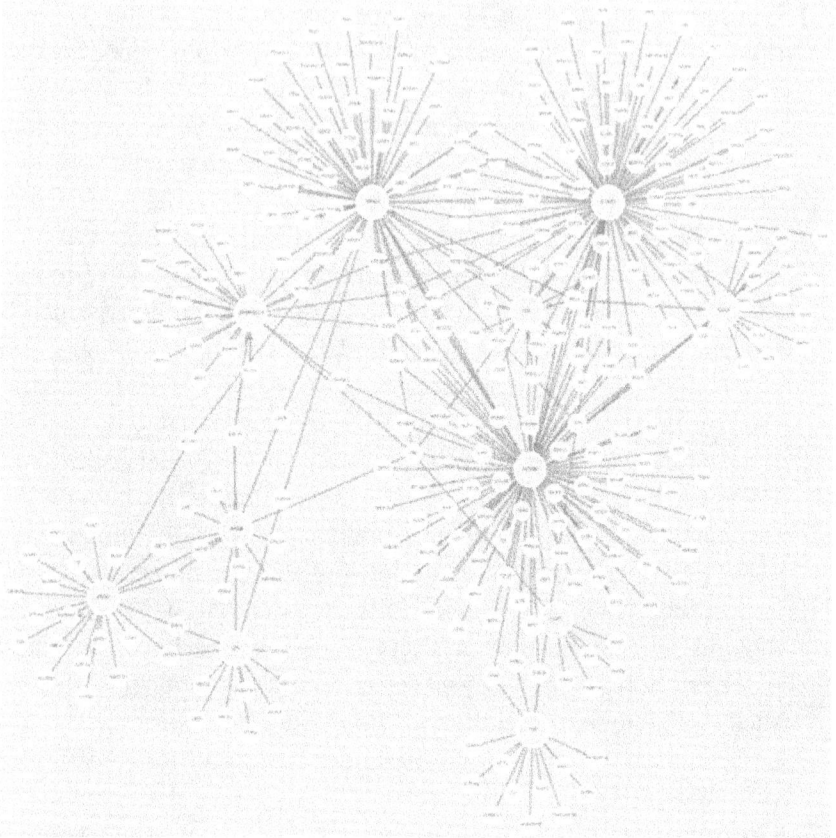

Fig. 3 PPI network among common genes represents the common regulatory pathway (Jesmin *et al.*, 2016).

tissues causing leukemia, lung, liver, spleen, colorectal, colon, breast, bladder, and kidney cancers was performed by extracting microarray expression data from online resource.

In their study, data analysis is divided into three parts: (1) Enrichment analysis was performed by using the Dynamic Impact Approach (DIA) with the help of a freely available webserver, DAVID (v.6.7) (The Database for Annotation, Visualization and Integrated Discovery (DAVID version 6.7) to identify affected pathways, biological processes, and molecular functions within the DEGs. (2) For the analysis of sequences and key regulatory networks that regulate transcriptional responses under specific environment, oPOSSUM version 2.0 Human Site Analysis and Cytoscape are used. (3) Gene Regulatory Networks are generated and analyzed by using Cytoscape (v2.8.2). Genes are mapped against their respective TF, represented by two different colors. In network analysis, 100 Gene regulatory networks between TF and genes were generated (Fig. 4).

Iqbal *et al.* (2014) found that DAVID analysis uncovered the most significantly enriched pathways in molecular functions that were up-regulated in blood, breast, bladder, colorectal, lung, spleen, prostate cancer was "Ubiquitin thioesterase activity". Transforming growth factor beta receptor activity was inhibited in all cancers except leukemia, colon, and liver cancer. oPOSSUM further revealed highly over-represented TFs, Broad-complex_3, Broad-complex_4, and Foxd3 in exception of leukemia and bladder cancer. From these findings, it is possible to target genes and networks that play a crucial role in the development of cancer. Iqbal *et al.* (2014) held that these

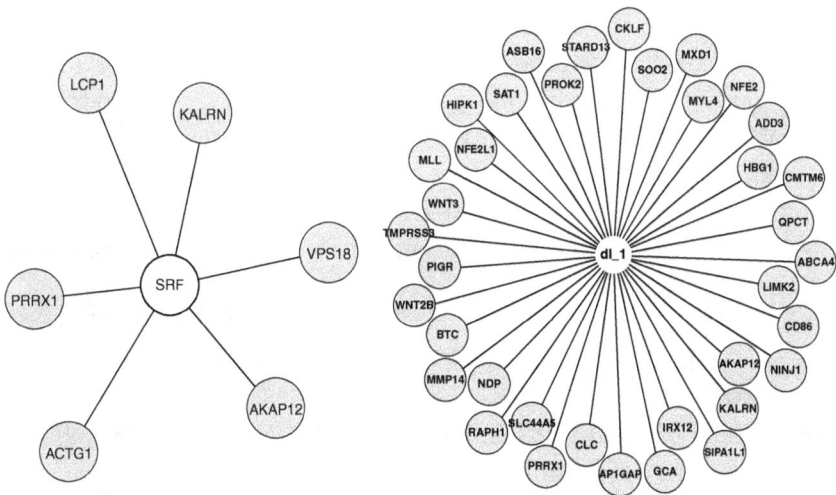

Fig. 4 Two networks of leukemia representing association of TFs SRF, dl_1, ELF5, and FOXD1 (white color) with its respective genes (grey color), generated by using Cytoscape (v. 2.0) (Iqbal *et al.*, 2014).

TFs might serve as potential candidates for the therapeutic drug targets which can impede the deadly spread of tumors.

23.5 Crucial Metabolites/Reactions in Tumor Signaling Networks

Tumor signaling pathways of the human body include mainly JAK-STAT signaling pathway, p53 signaling pathway, NF-κB signaling pathway, Ras, PI3K, and mTOR signaling pathway, Wnt NF-κB signaling pathway, BMP signaling pathway, etc. Various ligands, receptors, and signaling proteins are associating with these signaling pathways. They form a complex and directed network (Fig. 5). Although a tumor signaling pathway may be very complex, some metabolites/reactions in the network are more important for tumorigenesis than the remaining metabolites/reactions.

Using network analysis methods, Li and Zhang (2013) calculated betweenness centrality, degree, and k-core value of every metabolite/reaction in tumor signaling pathways p53, AKT, Ras, JAK-STAT, TNF and VEGF and identified crucial metabolites/reactions in these tumor signaling networks using betweenness centrality.

According to betweenness centrality, p53-P-P is the most important metabolite/reaction in p53 signaling pathway, followed by (Ac-p53-P)2, DNA damage, and ATM.

Akt is identified as the most important metabolite/reaction in AKT signaling pathway, followed by PI3K and PIP3; Ras-GTP is the most important metabolite/reaction in TNF signaling pathway, followed by MEKK1, JNKK, and Ras-GDP.

For p53 pathway (Fig. 5), the k-core value of UV, ATM (DNA damage), ART, JNK, Chk1-P, HIPK2, CSNK1, p38, PTEN (proteasome), MDM2, 14-3-3θ, DNA-PK, Akt, GSK3β, Bax, BCL2 (Gene expression), (Ac-p53-P)2, Ub-Ub-p53-MDM2, and p53-P-P is 2, and k-core value is 1 for remaining metabolites/reactions.

The k-core analysis shows that VEGF signaling pathway is the most compact network among these signaling pathways.

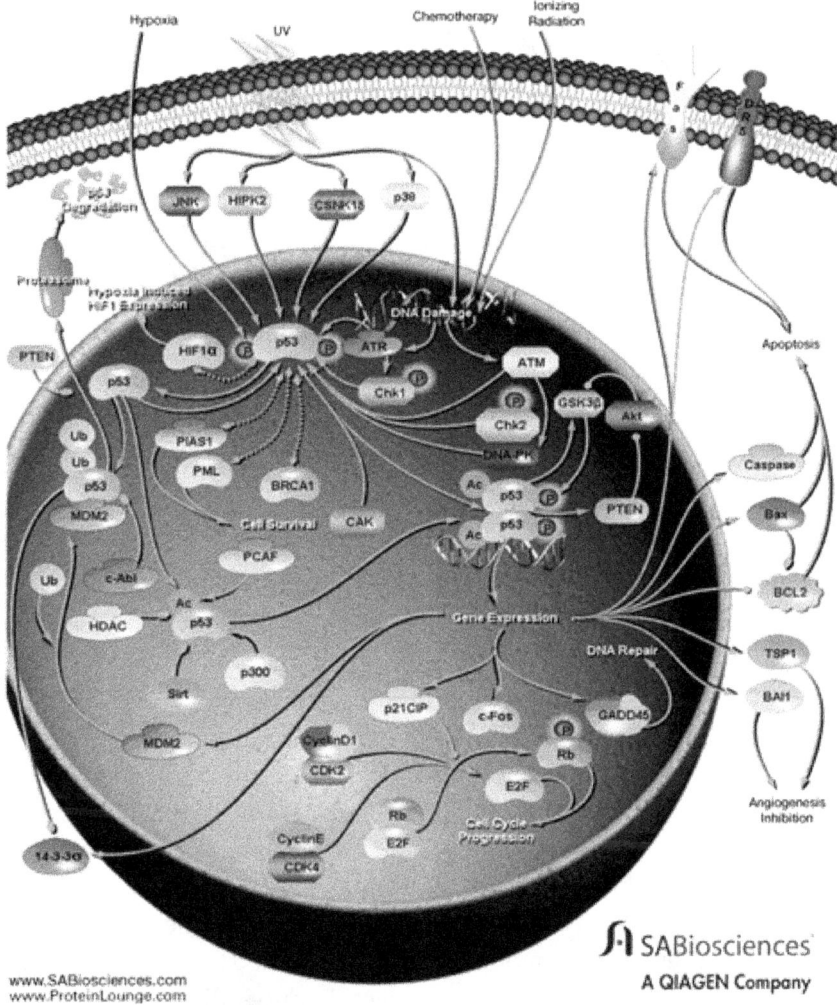

Fig. 5 P53 signaling pathway (Pathway Central, 2012).

23.6 Brain Network

Human brain is one of the most complex systems in nature. It consists of numerous neurons, neuronal clusters, and various regions that interact and coordinate with each other to form a network to function (Fig. 6). Neurons are the basic units that make up the

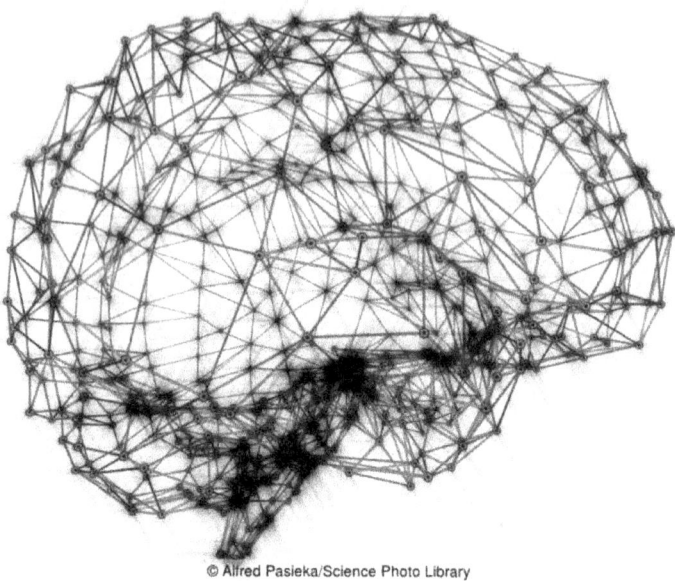

© Alfred Pasieka/Science Photo Library

Fig. 6 A neural network simulation of human brain (Pasieka A/Science Photo Library).

structure and function of nervous system. Neurons form neural loops through interactions and thus process brain information (Liang *et al.*, 2010). Even a simple brain function must be achieved by interactions between neuronal clusters, functional columns, or brain regions. At present, the human brain is generally understood as the complex system with hierarchical organization. Each hierarch is produced by interactions among many neurons at the next hierarchy, and it exhibits some of the emergence features different from the next hierarchy. Studies can be conducted at three spatial scales, microscale, mesoscale, and macroscale, which represent neurons, neuronal clusters, and brain regions, respectively. The current research and recognition on the human brain network are concentrated at the macro and systematic levels. With the development of modern brain imaging technology and statistical physics, especially the theory of complex networks, the study of brain network has been shifted to using structural magnetic resonance imaging (structural MRI), and diffusion magnetic resonance imaging (diffusion MRI), etc., to construct the structural connection network of human brain, or to using electroencephalogram (EEG), magnetoencephalography (MEG) and functional magnetic resonance imaging (fMRI), to construct the functional connection network of human brain, and then to using network analysis to reveal its topological principle, and further to understand the working mechanism within the brain (Bullmore *et al.*, 2009).

The neuronal connection network of human brain can be divided into structural brain networks (or anatomical brain networks), functional brain networks, and effective brain networks. The effective brain network is a directed network, which describes interactions or information flows between the nodes. It is actually a network composed of functional connections. The structural brain networks are mainly explored through MRI and DTI that can reflect the brain's physiological structure. The functional brain networks and the effective brain networks are mainly studied based on EEG, MEG, and fMRI that can reflect the brain's functions.

Brain network can be explored by data-driven studies and computational model-based studies. On the one hand, we can calculate the connections among predefined nodes and brain regions based on the data of structural connections and functional connections, and then construct the brain network for further analysis. On the other hand, we may explore it using specific neural computational models, which are often composed of mutually coupled oscillators, each of which is a group of differential equations with several state variables. These differential equations can characterize the dynamic behavior of certain neurons or neuronal clusters.

The dynamic relationship between the oscillators can be assigned to a random variable that satisfies a certain probability distribution or can be determined by the structural connection of the brain (Honey *et al.*, 2009).

The structural brain networks refer to the anatomical and structural connections between neurons (including electrical and chemical connections between axons and dendrites), which reflect the physiological structure of the brain and the material basis for brain's functional connectivity. The nodes of a structural brain network can be defined on different spatial scales: from single neuron to local circuits and to brain or cortical regions with specific functions, and the local neural networks on small scales can be understood as a subnetwork or a node of the brain network at the larger scale level (Zhou *et al.*, 2006). Because of the difficulty of obtaining connections in living human brain, the earlier studies on structural brain networks were conducted using animal brains. In recent years, the researchers have found that structural MRI and diffusion MRI can be used to capture the structural information of living human brain, which makes the construction of structural brain networks possible.

Functional connectivity describes the statistical relationship of between-node (the nodes that can represent brain's functional units on different scales, such as neurons, neuronal clusters, functional regions, etc.) functional signals in a period of time. Functional brain networks are undirected networks, and these relationships do not reflect the causal relationships between nodes. The functional brain network is constructed from functional signals (electrical signals, magnetic signals, signals that reflect hemodynamics or metabolism, etc.) that are based on brain/nerves. On the microscopic scale, it can be constructed from the electrical potential relationships between individual neurons. On the meroscale, it can be constructed from the local field of electrical potential, which reflects the activity of neuronal clusters. And on the macro-scale, it can be constructed from the EEG/MEG/fMRI between specific brain regions. At present, the research on functional brain networks is mostly focused on the macro-scale level.

At the network level, the dynamic functional network changes over time and exhibits obvious "small world" and corresponding architecture. For example, some parts (brain island, sensory motor cortex, medial prefrontal cortex) are highly connected areas with persistent functions, while at the same time the degree centrality changes greatly on different spatial and temporal scales. The temporal characteristics (intensity and deformation) of properties of connections and nodes of the dynamic brain network are significantly related to their structure (Liao *et al.*, 2015). It proves the property of spontaneous network dynamics of the human brain.

As briefly mentioned above, the structural network can be constructed based on the structural magnetic resonance images (morphological indicators of gray matter, such as cortical thickness, cortical area, etc.) and diffusion magnetic resonance images (white matter fiber bundles). The functional network can be constructed based on functional magnetic resonance image (the time series of activity of brain's function) and EEG/MEG signals. The most important procedures for network construction are as follows: (1) node definition: for structural, diffusion, and functional magnetic resonance data, we need to define nodes by classifying brain regions using *priori* chromatogram or by using image voxel, while for EEG/MEG data, the nodes are directly defined by recorded electrodes/channels; (2) connection definition: the network connection based on the structural magnetic resonance data is defined as the statistical relationship between morphological indicators of network nodes, and the anatomical connection based on diffusion magnetic resonance data is defined through deterministic or probabilistic fiber-tracking technology, and finally, the network connection based on functional magnetic resonance and EEG/MEG data can be defined by statistical relationship, like Pearson correlation, partial correlation, and synchronized likelihood between activity signals of nodes, and (3) construction of the structural network and functional network: the correlation matrices obtained in the step (2) are transformed to binary matrices under different thresholds, which are the desired structural and functional networks (Liang *et al.*, 2010). However, using the unweighted binary matrices leads to a significant loss of valuable information, which should be improved in the future.

Part 10
Social Networks

Chapter 24

Social Network Analysis

24.1 Definition of Social Networks

Social network analysis is a relatively advanced branch in the development of sociological research (Wasserman and Faust, 1994; Scott, 2000; Freeman, 2006). As early as the 1930s, there were many representative cases on social network analysis. Early studies put focus on friendship of people (Moreno, 1934; Rapoport and Horvath, 1961; Fararo and Sunshine, 1964). However, these studies have some common weaknesses First of all the size of these networks was generally smaller; second, the construction of the network featured a lot of subjective factors, and finally, the network data were often not complete. Due to the emergence of massive databases and the improvement of computer processing, research on social networks has begun to proceed to a large-scale, precise, and quantitative stage (Borgatti *et al.*, 2009).

24.1.1 *Types of social networks*

There are totally five types of social networks (Zhou, 2012).

(1) Social collaboration network

The nodes of the social collaboration network are those who participate in collaboration, and if there is some kind of collaboration between people, they are connected to each other. For example, as coauthors, scientists of the same paper can be seen as having collaborations In fact, many new discoveries of complex networks have arisen from the studies on collaboration networks of Hollywood film actors (Watts and Strogatz, 1998; Amaral *et al.*, 2000; Ravasz and Barabasi, 2003) and scientific collaboration networks (Newman, 2001a, 2001b, 2001c; Barabasi, *et al.*, 2002). It is important to note that research on collaboration networks have originated from social networks, but it is not limited to social networks (Zhang *et al.*, 2006).

(2) Friendship network

Some of the recent studies have focused on friendship in online dating sites, such as Facebook (Golder *et al.*, 2007). The online friendship network shows a lot of features consistent with the practical ones. For example, by analyzing the friendship information of 4.2 million users on Facebook, Golder *et al.* (2007) found that the average number of friends per user was 180, with a median of 144, which was consistent with the Dunbar's number, that is, a person can maintain up to 150 friends. Liben-Nowell *et al.* (2005) found that even for virtual online networks, users who are geographically close to IP addresses are more likely to be friends.

(3) Information network

The information network arises from transmission, sharing, and exchange of information. A typical form of information network is the relational network of senders and the receivers of information, connected by the information propagation path. To MicroBlog, for example, the users get information from the concerned object, and their own information will be automatically sent to the fans, and the later form links and a directional information network is thus produced. By analyzing the massive data from Flickr, Twitter, Youtube, etc., Zhou *et al.* (2011) found that in these representative information networks, the user's leadership, characterized by the number of fans obeyed the power-law distribution, i.e., few people have amazing influence, and most people get very few responses. Wu *et al.* (2011) analyzed the pattern and mechanism of the connection in Twitter by classifying the user's identity. If we treat commentators and forwarders in the BBS and forums as the recipients of landlord messages, an information network can be created also (Goh *et al.*, 2006).

There is a strong coupling and evolutionary relationship between the structure and function of the information network: the structural changes will affect the dynamic property of information dissemination, and some new links will be established due to the user's interest in certain information, and thus result in structural changes of the information network (Zhou, 2012).

(4) Communication network

A study on the communication networks containing fixed telephones and mobile phones found that distribution of the number of communication tools followed the powerlaw distribution (Xia *et al.*, 2005). Onnela *et al.* (2007) studied a large-scale mobile communication network containing 4.6 million nodes and found that the key to determining the connectivity of the network was weak connections. Ebel *et al.* (2002) exploited an e-mail network containing nearly 60,000 nodes and found that the network was a typical small-world and scale-free network.

(5) Social contact network

Social contact network is a network created from real physical contacts between people. Analysis of such networks is of great help in understanding the spread of epidemics (Zhou, 2012). Most of the studies on social contact networks focused on relatively closed systems, such as schools, communities, corporate boards, etc. (Borgatti *et al.*, 2009). More in-depth studies focus on sexual networks, including the characteristics of the network's own structure and its impact on various sexually transmitted diseases (Liljeros *et al.*, 2003; Doherty *et al.*, 2005). However, it is very difficult to obtain true and accurate data on sexual partners and sexual behavior, so empirical analysis is always performed in the smaller networks.

24.1.2 *Several basic theories of social networks*

(1) **Six Degrees of Separation.** Each person in the world is only six introductions away from any other person in the world.
(2) **Becken's number.** The number of brokers needed to connect with the Hollywood actor, Kevin Bacon, averaging between 2.6

and 3, which further validates the Six Degrees of Separation theory.

(3) **Dunbar' number.** Dunbar's number is a suggested cognitive limit to the number of people with whom one can maintain stable social relationships, i.e., relationships in which an individual knows who each person is and how each person relates to every other person (Wikipedia, 2017).

(4) **Weak relationships.** The connections between groups are called weak relationships. The weak relationship leads to the flow of information between different groups and plays a strong role in the message communication.

24.1.3 *Analytical framework for social network analysis*

Weerasinghe (2013) has proposed an analytical framework for social network analysis. The following is a general description on the analytical framework of static social networks.

Weerasinghe (2013) points out that existing relationships and connections among humans, animals, and micro or macroorganisms can be explored, analyzed, and interpreted using social network methodologies. Human communications through interactions on Facebook, Twitter, and blog spaces have added new dimensions, where dynamic information exchange and related text analytical methods have become prominent components in social network analytics.

The network analytical framework is illustrated in Fig. 1 The framework is divided into three stages sampling, data processing, and inference, and the corresponding methodology in each stage is data collection, data mining, and statistical analysis (Weerasinghe, 2013).

(1) Sampling frame: data collection

As illustrated by Weerasinghe (2013) in Fig. 1, network analytical frame starts with a sampling scheme. Three different types of sampling methods, including simple random sampling can be used (Kolaczyk, 2009).

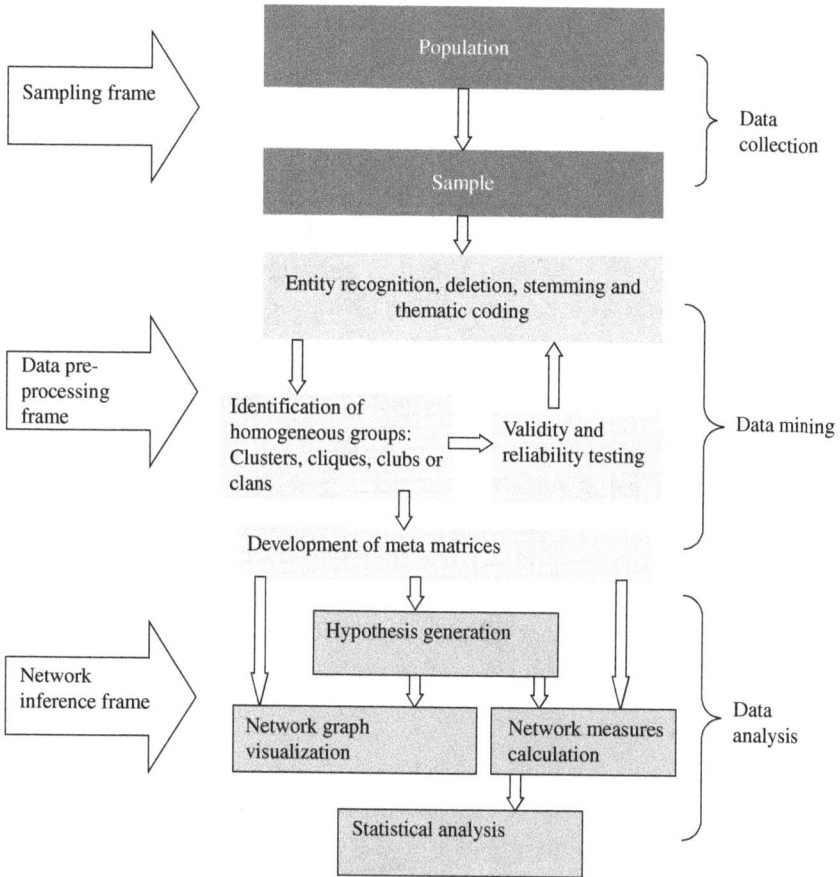

Fig. 1 Analytical framework for social network analysis (Weerasinghe (2013).

(2) Data preprocessing: data mining

Data preprocessing is done by using data mining text, analysis techniques, or software. Network analysis is applicable to different types of data. We can skip the first step and move to the next step, identification and validity and reliability testing (Fig. 1). Concepts mapping for information exchange (text data) are relatively complicated and require additional steps before numerical coding to form data matrices. Thus, this analysis is known as text analysis, or text mining, i.e., concept-driven extraction and visualization framework (Jahiruddin *et al.*, 2010). If network data are collected in terms of quantitative measurements, then the text mining step can be ignored (Weerasinghe, 2013).

(3) Data preprocessing to meta-matrix data

Data preprocessing is substantially text mining. Data mining software uses four steps of preprocessing, i.e., entity recognition, deletion, stemming, and thematic coding, done by thesaurus creation and application (Yong *et al.*, 2010). Entity recognition is carried out by deletion of unrelated contents prior to importing into text mining software.

(4) Validity and reliability testing of preprocessed data

Data mining is carried out on text coded data, etc., by using manual coding and using data or test mining software without manual coding (Weerasinghe, 2013). Validity of the concepts and attributes and categorization into clubs or cliques need to be justified using sensitivity analyses. Manual text mining or coding as well as categorizations can be used as gold standard in the sensitivity analyses of the development data set (Weerasinghe, 2013).

(5) Statistical inference

Statistical inference includes development of hypothesis exploratory data analysis, using network measures, and visual display of network graphs. Statistical modeling is also used to understand more comprehensive relationships and testing hypotheses.

(6) Exploratory data analyses: network graph visualizations and summary measures

Network exploratory analyses can be conducted using primary nodes. Edges are used to connect all nodes. Initial contacts become the primary node, and their relational contacts or clusters become secondary nodes (Weerasinghe, 2013). Network data can be presented based on classifications or characteristics of the primary nodes or initial data collection units.

(7) Statistical analyses of network data

Network data analysis includes analysis of intra- and inter-network measures, analysis of network pathways using structural equation models, and regression analysis of network effects on social, behavioral, and medical/health outcomes of individual subjects (Weerasinghe, 2013).

24.2 Network Criminology

Network criminology is the interdisciplinary science rooted in criminology, network science, sociology, and other related scientific disciplines (Zhang, 2017g). It is network analysis based and is a branch of criminology. It uses the theory and methodology of network science to analyze, predict, and control criminal patterns and behavior. Network criminology aims to investigate and understand the structure, properties, organization, function, identification, control, etc., of criminal networks (e.g., offender networks, offender–victim networks, offender–victim–area networks, etc.). The scope of network criminology covers but is not limited to (Zhang, 2017g): (1) theories, models, algorithms, and software of network criminology; (2) network analysis of criminal networks; (3) identification of criminal groups, terrorism networks, offenders, victims, and areas; (4) causes and consequences of criminal behaviors; (5) dynamics and control of criminal networks; (6) methods to prevent the development of criminal groups, terrorism networks, etc.

Criminology is one of the most important scientific foundations of network criminology. It is the scientific area on the nature, extent, management, causes, control, consequences, and prevention of criminal patterns and behavior, both at the individual and social levels (Wikipedia, 2016d).

Criminology has several branch parties, e.g., Marxist criminology, conflict criminology, biosocial criminology, critical criminology, etc. Marxist criminology states that "defiance is normal — the sense that men are now consciously involved ... in assuring their human diversity." (Sparks, 1980). Marxist criminology, conflict criminology, and critical criminology hold that most of the relationships between state and citizen are nonconsensual and criminal law is not necessarily representative of public beliefs and wishes. Criminal law is performed in the interests of the ruling or dominant class. Some theories in criminology include (Wikipedia, 2016d; Zhang, 2017g):

(1) **Sociological positivism.** Sociological positivism suggests that societal factors, e.g., poverty, membership of subcultures, low levels of education, etc., are causes of crimes (Beirne, 1987; Lochner, 2004).

(2) **Differential association.** It holds that crime is learned through association (Anderson, 1992).

(3) **Social structure theories (Hester and Eglin, 1992).** There are various viewpoints on social structure theories. Social disorganization theory suggests that neighborhoods with poverty and economic problems tend to have a high population turnover (Bursik and Robert, 1988). Further, social ecology argues that crime rates are associated with poverty, disorder, rich abandoned buildings, and other signs of community deterioration (Bursik and Robert, 1988; Morenoff *et al.*, 2001). Subcultural theory focuses on small cultural groups fragmenting away from the mainstream to generate their own values and meanings about life. Social control theories explain why people do not become criminals (Hirschi, 1969). Finally, social network analysis stresses using network analysis to explain criminal patterns and behavior.

(4) **Symbolic interactionism.** Symbolic interactionism focuses on the relationship between the powerful state, media, and conservative ruling elite and other less powerful groups.

(5) **Individual theories.** Traitor theory explores how a process of brutalization by parents or peers that usually occurs in childhood results in violent crimes in adulthood (Rhodes, 2000). Rational choice theory holds that punishment, if certain, swift, and proportionate to the crime, is a deterrent for crime, with risks outweighing possible benefits to the offender. Routine activity theory draws upon control theories and explains crime in terms of crime opportunities that occur in everyday life (Felson, 1994).

(6) Biosocial criminology is an interdisciplinary field aiming to explain crime and antisocial behavior by exploring both biological and environmental factors (Kevin and Anthony, 2011).

The methodology and more details on network criminology can be found in Zhang (2017g).

Part 11
Software

Chapter 25

Software for Network Analysis

Network analysis is generally a computation-intensive process. Usually it cannot be conducted without various programs and software. This chapter describes some software for network analysis, typically those for comprehensive network analysis, network graph drawing, self-organization, agent-based modeling, etc.

25.1 Software for Comprehensive Network Analysis

25.1.1 *NetLogo*

NetLogo is a multiagent programmable modeling environment, which provides frameworks of computer codes of models/algorithms, and suggestions for extending these models/algorithms. It provides an alternative platform for self-organization modeling (Zhang, 2016r).

25.1.1.1 *Network algorithms*

(1) Giant component model (Janson *et al.*, 1993; Wilensky, 2005)

A component of the network refers to a group of nodes which are directly or indirectly connected to each other. A giant component means that almost every node is reachable from almost every other. This model demonstrates how quickly a giant component is produced from a random network (Zhang, 2016r). In the model, the largest connected component produced by randomly connecting two nodes in the network each time step grows quickly after the average connections per node equals 1. In this model, initially we have nodes but no connections (edges) between them. At each step, two nodes are checked randomly, and an edge between them is added if they were not directly connected. All possible connections have the same probability for producing. During the model running, small chain-like components, in which all nodes in each component are directly or indirectly connected, are produced. Two components merge into one if a connection is created between nodes in two different components. The component with the most nodes is defined as the giant component.

(2) Diffusion in a directed network (Stonedahl and Wilensky, 2008a)

The model demonstrates how a quantity diffuses through a directed network. The quantity moves through nodes along established and directed edges between two nodes. Simple rules used in the model will result in various patterns regarding the topology, density, and stability of the network. For each step, each node shares some percentage of its value quantity with its neighboring nodes. A node will not share any of its value if it has no outgoing edges. It just accumulates any that its neighbors have provided through incoming edges (Zhang, 2016r).

(3) Preferential attachment (Barabasi and Albert, 1999; Barabasi, 2002)

In some of the networks, a few nodes have a lot of connections but other nodes have fewer (Zhang, 2012e, 2016r). This model demonstrates how such networks are produced. These networks are generated by a process called preferential attachment. In this process, new nodes tend to connect to existing nodes with more connections. It starts with two connected nodes. A new node is added in each step. It randomly, with a difference chances, picks an existing node to connect to. The chance is directly proportional to the number of connections of the node to be linked.

(4) Team assembly (Guimera *et al.*, 2005; Bakshy and Wilensky, 2007)

This model demonstrates how the behavior of individuals in assembling small teams for short-term projects can yield a variety of large-scale network structures over time (Bakshy and Wilensky, 2007). It is an adaptation of the team assembly model of Guimera *et al.* (2005). Team members are inexperienced newcomers or established incumbents. At each step, a new team is assembled. Each member is chosen sequentially. All members are linked to one another when a team is created. An agent and its links should be removed from the network if the agent does not participate in a new team for a prolonged period of time (Zhang, 2016r).

25.1.1.2 *Computer algorithms*

(1) Particle swarm optimization (Kennedy and Eberhart, 1995; Stonedahl and Wilensky, 2008b)

Particle swarm optimization is a search method and is usually used on multidimensional search spaces (Zhang, 2016r). This model demonstrates two-dimensional space. In the model, suppose there is an unknown function, fitness function $f(x, y)$, and we try to search x and y such that $f(x, y)$ is maximized. In particle swarm optimization, particles are placed in the search space and they move through the space based on some rules that take into account each particle's personal knowledge and swarm's knowledge. Through their movement, particles discover particularly high values for $f(x, y)$. This simple discrete model is based on the algorithm of Kennedy and Eberhart (1995). In this model, the particle swarm is trying to optimize a function that is determined by the values in the discrete cell grid. The procedure is as follows (Stonedahl and Wilensky, 2008b): each particle has a position (x_{cor}, y_{cor}) in the search space and a velocity (v_x, v_y) at which it is moving through the space. Particles have a certain amount of inertia to keep them moving in the same direction they were moving previously. In addition, they have acceleration (change in velocity), which depends on two main things: (a) each particle is attracted toward the best position that it has personally found (personal best) previously in its history, and (b) each particle is attracted toward the best position that any particle has ever found (global best) in the search space. The strength with which the particles are pulled in each of these directions is dependent upon the parameters ATTRACTION-TO-PERSONAL-BEST and ATTRACTION-TO-GLOBAL-BEST. As particles move farther away from these best positions, the attraction force becomes stronger. There is also a random factor about how much the particle is pulled toward each of these positions (Stonedahl and Wilensky, 2008b). The model runs until some particle in the swarm has found the true optimum value (i.e., 1.00).

(2) Hexagonal cellular automata (Wilensky, 2007)

In NetLogo, the model of hexagonal cellular automata runs on a two-dimensional hexagonal grid of cells. Cells are alive or dead. Only the center cell is alive at the start. The switches determine which dead cells are to become alive. At each step, one candidate dead cell goes live.

(3) Genetic algorithm (Holland, 1975; Stonedahl and Wilensky, 2008c)

It is a simple genetic algorithm in NetLogo. It works by generating a random population of solutions to a problem, evaluating those solutions and using cloning, recombination, and mutation to create new solutions of the problem (Stonedahl and Wilensky, 2008c). In this model, Stonedahl and Wilensky (2008c) use the simple "ALL-ONES" problem to demonstrate the method. The idea of the "ALL-ONES" problem is to find a string of bits that contains all ones: "111111...111". The genetic algorithm in this model includes the following procedures (Stonedahl and Wilensky, 2008c; Zhang, 2016r):

(I) Create a population of random solutions. In this step, each solution consists of a string of randomly mixed "1"s and "0"s.

(II) Evaluate each solution based on how well it solves the problem. The goodness measure of the solution is called its fitness. As described, here the goal is simply to find a solution that consists of all "1"s.

(III) Create a new generation of solutions from the old generation. Solutions with a higher fitness are more likely to be chosen as parent solutions than those with low fitness. In this step, the following cases are considered.

(a) The "tournament selection" method is used in the model, with a tournament size of 3, i.e., three solutions are drawn randomly from the old generation, and the one with the highest fitness is chosen to become a parent.

(b) Choose either one or two parents to create children. The child with one parent is a clone or copy of the parent and. With two parents, the process is the digital analog of sexual recombination, i.e., the two children inherit part of their genetic material from one parent and part from the other.

(c) Mutation will occasionally occur in the process. Some of the child's bits will be changed from "1"s to "0"s, and vice versa.

(IV) Repeat steps II and III, until a solution is found.

25.1.1.3   *Biological algorithms*

(1) Ants colony algorithm (Resnick, 1994; Wilensky, 1997a)

In this model, a colony of ants forage for food. Each ant follows a set of simple rules and the whole colony behaves in a more sophisticated way. When an ant finds a piece of food, it carries the food back to the nest, at the same time labeling its path by releasing a chemical as it moves. When other ants detect the chemical, they will follow the chemical trail toward the food. As more ants carry food to the nest, they reinforce the chemical trail (Wilensky, 1997a).

(2) Termites model (Resnick, 1994; Wilensky, 1997b)

Termites gather wood chips into piles by following a set of simple rules (Zhang, 2016r). At first, each termite wanders randomly. If it bumps into a wood chip, it picks the chip up and continues to wander randomly. When it bumps into another wood chip, it finds a nearby empty space and puts its wood chip down. Thus, the wood chips eventually end up in a single pile (Wilensky, 1997b).

(3) Tumor model (Wilensky, 1998)

The model demonstrates how a tumor grows and how it resists chemical treatment. Generally a tumor contains two types of cells, stem cells and transitory cells. A stem cell can divide asymmetrically or symmetrically during mitosis. In asymmetric mitosis, one of the two daughter cells replaces its parent, remaining a stem cell. A stem cell never dies. Meantime, the other daughter cell changes into a transitory cell that moves outward. Young transitory cells may divide and breed other transitory cells. The transitory cells stop dividing at a certain age and changes colors, eventually dying (Wilensky, 1998). In addition, a stem cell can divide symmetrically into two stem cells. In this case, the original stem cell divides symmetrically only once. The first stem cell remains static but the second stem cell moves to the right (Fig. 1). This process is called metastasis. With the progresses of disease, cells die younger and younger (Zhang, 2016r).

(4) Wolf–Sheep Predation (Wilensky and Reisman, 1999)

This model is used to explore the stability of predator–prey ecosystems. It has two main variations. In the first variation, wolves and

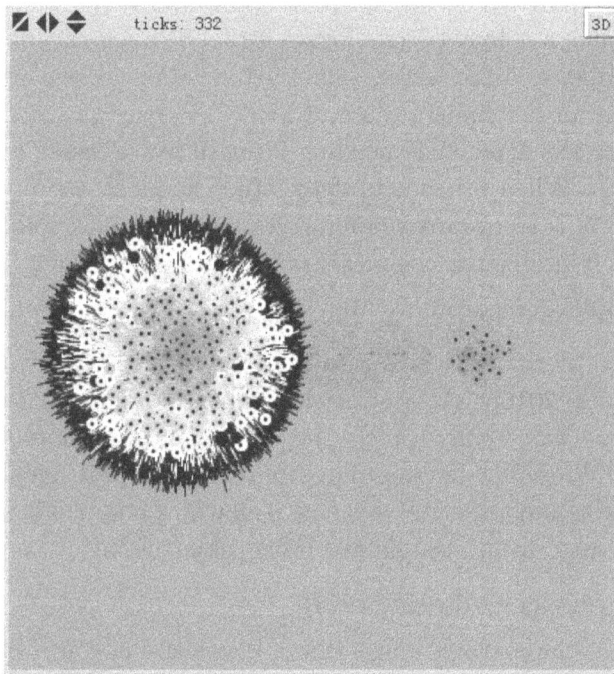

Fig. 1   Tumor model in NetLogo (Zhang, 2016r).

sheep wander randomly in the field and the wolves look for sheep to prey on. In each step wolves must eat sheep in order to replenish their energy, or they die for running out of energy. To maintain the population, each wolf/sheep has a fixed probability of reproducing at each step (Wilensky and Reisman, 1999). This variation produces a population that is ultimately unstable. The second variation adds grass, in addition to wolves and sheep. The sheep must eat grass in order to maintain their energy, or they die because they running out of energy. Once grass is eaten it will only regrow after a fixed amount of time. This variation leads to a population that is generally stable (Zhang, 2016r).

25.1.1.4 *Social algorithms*

One of the social algorithms in NetLogo is cooperation model (EACH Unit, 1997). This model is the part of the EACH unit (Evolution of Altruistic and Cooperative Habits: Learning About Complexity in Evolution; http://ccl.northwestern.edu/cm/ EACH/). It is an evolutionary model. In this model, agents (cows) compete for natural resources (grass). Cows getting more grass reproduce more often and are thus more evolutionarily successful. This model includes two types of cows, i.e., greedy and cooperative cows. It demonstrates the behavior of the two strategies when they compete against each other within an evolving population. In each step, each cow wanders around the current patch and eats a unit of grass. The greedy cows eat the grass regardless of the length of the grass. But the cooperative cows may not eat the grass below a certain height after which the grass will grow rapidly. The cooperative cows thus leave more food for the whole population at a cost of their individual well-being, and the greedy cows eat the grass regardless of their effect on the whole population (Zhang, 2016r).

25.1.1.5 *Algorithms in chemistry*

(1) Crystallization Basic (Wilensky, 2002)

A metal solidifies when it cools. The first atom solidifies at a random orientation. However, the next atom will solidify at the orientation that the already solidified atom takes. Thus, they create a crystal grain. When more atoms solidify, the grains grow (Wilensky, 2002). All atoms in each grain have the same orientation, but different grains take different orientations. Deformations occur in the crystal structure when a metal is stressed. As the stress increases, deformations pass through the crystal structure and the metal tends to bend. However, grain boundaries will prevent deformations from passing through the metal. Thus, the metal piece with fewer grain boundaries tends to be tough, while the metal piece with more grain boundaries tends to be brittle (Wilensky, 2002). This model demonstrates the formation process of grains when a metal crystallizes. In this model, a liquid metal is placed in a room with temperature much lower than that of the metal. The metal begins to solidify as the metal temperature drops down. In this process, liquid atoms are free to rotate. If a liquid atom is next to a solid atom, it will orient itself with it, or else it will rotate randomly (Wilensky, 2002).

(2) Diffusion-limited aggregation (Wilensky, 1997a, 1997b)

This is a model to demonstrate diffusion-limited aggregation. In this model, randomly moving particles aggregate together to form treelike fractal structures. The model begins with an initial seed in the center of the model view. Particles move around the view randomly, and when a particle hits a square, it sticks and a new particle is then created to continue the process (Wilensky, 1997a, 1997b).

25.1.2   *Pajek*

Pajek is used to make analysis and visualization of large scale networks. It supports abstraction by recursive factorization of a large network into smaller networks. Pajek provides some powerful visualization tools and implements some algorithms for analysis of large networks (Zhang, 2016r).

Pajek uses several data structures as: (1) Network, The main object, which includes vertices (nodes) and lines (links). (2) Permutation, Reordering of vertices. (3) Vector, A set of values of vertices. (4) Cluster, subset of vertices. (5) Partition, Shows for each vertex to which cluster the vertex belongs. (6) Hierarchy, Hierarchically ordered clusters and vertices.

Using Pajek, we can conduct simplifications and transformations like deleting loops, multiple edges, transforming arcs to edges, etc., calculate components (strong, weak, biconnected, and symmetric components), make decompositions (symmetric-acyclic, hierarchical clustering), find paths (shortest path(s), all paths between two vertices), calculate flows (maximum flow between two vertices), and make neighborhood analysis (k-neighbors). Furthermore, the following algorithms are also provided: (1) CPM (find critical paths); (2) social networks algorithms: centrality measures, hubs, and authorities; (3) measures of prestige, brokerage roles, structural holes; (4) measures of dependencies among partitions/vectors; (5) Cramer's V, Spearman rank correlation coefficient, Pearson correlation coefficient, Rajski coefficient; (6) extracting subnetwork; (7) shrinking clusters in network (generalized block modeling); (8) topological ordering, Richards's numbering, Murtagh's seriation and clumping algorithms, depth/breadth first search.

25.1.3 *Netwrk*

Netwrk is used to calculate input–output structure matrices, trophic chain and tropic aggregation, and ascendency metrics, and to identify biogeochemical cycles (Ulanowicz, 2004; Latham, 2006). Netwrk is designed to run under MS-DOS.

Netwrk and documentation is available at: http://www.cbl. umces.edu/~ulan/ntwk/network.html.

25.1.4 *EcoNetwrk*

EcoNetwrk is a Microsoft Windows-based version of Netwrk 4.2 (Ulanowicz, 2004). It is used to make the following analysis: (1) input/output analysis; (2) the determination of trophic status and the identification of an equivalent linear food chain; (3) analysis of biogeochemical cycling and the supporting flows, and (4) calculation of ecosystem indices (Zhang, 2012e, 2016r).

EcoNetwrk and documentation are available at: http://www.glerl.noaa.gov/EcoNetwrk/.

EcoNetwrk includes the following input files: (1) ASCI files. These files preserve biomasses, imports, exports, respiration, diet exchanges, consumption, production, ratios, assimilation efficiencies, egestion, diet proportions, and import proportions; (2) SCOR files. They preserve biomasses, imports, exports, respiration, and diet exchanges. Ratios and variable states are not saved; (3) CSV files. They are comma separated value files that are easily imported into Excel. They preserve the same values as SCOR files. However, Excel should only be used to change cell values. EcoNetwrk will not correctly interpret the file if some columns of an input file are deleted or rearranged.

The summary page of EcoNetwrk indicates the existing state of the network. Values of this page can be changed. Biomass and ratios page contains biomass and ratio information. The ratios listed on this page are Consumption/Biomass (C/B), Production/Biomass (P/B), Respiration/Biomass (R/B), and Respiration/Consumption (R/C). Diet proportions page summarizes the relative components of each predator's diet. Imports are included and makes the divisor Total Inputs (Consumption+Imports). The totals must sum to 1 before network analysis can proceed (Zhang, 2012e, 2016r).

25.1.5 *Ecopath*

Ecopath is used to calculate ascendency measures (Ulanowicz, 1980, 1986, 1997), a decomposition of network cycles (Ulanowicz, 1983, 1986), mixed trophic impact (Ulanowicz and Puccia, 1990), and trophic aggregation (Ulanowicz, 1995), etc.

25.1.6 *Swarm*

Swarm is the first general ABM tool based on Java. It is a software platform on the basis of agents which provides some advantages for ecological modeling, including a set of standardized object bases, task scheduling base, detection base, and some structural characteristics as inheritance, message delivery, packaging, and hierarchical structures (Zhang, 2012e, 2016r).

25.1.7 *NEURON and GENESIS*

NEURON and GENESIS can exactly simulate the electrical and chemical nature of individual neurons. However, the size of network that can be simulated is finite. GENESIS uses compartmental model to simulate biological neurons. Each section of a neuron is simulated with path equation, and the parameters in path equation include conductivity and capacity of input and output signals of cell. Similar to GENESIS, NEURON uses compartmental model to simulate biological neurons also. But the focus of NEURON is modeling rather than compartment details. NEURON included higher GUI than GENESIS (Zhang, 2012e, 2016r).

25.1.8 *CiteSpace*

CiteSpace is a freely available Java application for visualizing and analyzing trends and patterns in scientific literature. It focuses on finding critical points in the development of a field or a domain, in particular intellectual turning points and pivotal points. It is a tool for progressive knowledge domain visualization (Chen, 2004, 2006).

CiteSpace provides various functions to facilitate the understanding of network patterns and historical patterns, including identifying the fast-growth topical areas, finding citation hotspots in the land of publications, decomposing a network into clusters, automatic labeling clusters with terms from citing articles, geospatial patterns of collaboration, and unique areas of international collaboration. It supports structural and temporal analyses of various networks derived from scientific publications, including collaboration networks, author cocitation networks, and document cocitation networks. In addition, it supports networks of hybrid node types such as terms, institutions, and countries, and hybrid link types such as cocitation, cooccurrence, and directed citing links.

The primary source of input data for CiteSpace is the Web of Science. Search for a topic of interest and download the search results (including full records and references). It also provides simple interfaces for obtaining data from PubMed, arXiv, ADS, etc. Furthermore, CiteSpace can be used to generate geographic map overlays viewable in Google Earth based on the locations of authors. Details can be found at: http://cluster.cis.drexel.edu/~cchen/citespace/.

25.1.9 *Mascopt*

Mascopt is designed to provide a set of tools for network optimization problems, such as routing, grooming, survivability, or virtual network design. Mascopt helps to implement a solution to such problems by providing a data model of the network and the demands, as well as libraries to handle networks and graphs. Mascopt is Open Source and intends to use the most standard technologies as Java and XML format providing portability facilities. It provides implementation of graph data structure, several basic algorithms working on graph and input/output classes. It also provides some graphical tools to display graph results. Details can be found at: http://www-sop.inria.fr/mascotte/software/mascopt/.

25.1.10 *NEGOPY*

NEGOPY is one of the oldest network analysis programs. It is a discrete, linkage-based program for the analysis of networks. It finds cliques, liaisons, and isolates in networks that have up to 1,000 nodes and 20,000 links. NEGOPY is designed to define clusters of nodes that have more contact with one another than with nodes in other clusters (called groups) and are conceptually similar (but not identical) to the cliques. NEGOPY also sorts nodes into a number of role categories on the basis of their linkage with one another. Details can be found at: http://www.sfu.ca/personal/archives/richards/Pages/negopy.htm.

25.1.11 *NetEvo*

NetEvo is a computing framework and collection of end-user tools allowing researchers to investigate evolutionary aspects of dynamical complex networks. Details can be found at: http://netevo. sourceforge.net/.

25.1.12 Network Workbench

Network Workbench is a large-scale network analysis, modeling, and visualization toolkit for biomedical, social science, and physics research. It helps to design, evaluate, and operate a unique distributed, shared resources environment for large-scale network analysis, modeling, and visualization. Network Workbench helps to make network analysis with the most effective algorithms available. In addition, they will be able to generate, run, and validate network models to advance their understanding of the structure and dynamics of particular networks. Details can be found at: http://nwb.cns.iu.edu/ about.html.

25.1.13 UCINET

UCINET is a social network analysis program. It works in tandem with freeware program called NETDRAW for visualizing networks. NETDRAW is installed automatically with UCINET. Details can be found at: https://sites.google.com/site/ucinetsoftware/home.

25.1.14 GUESS

GUESS is an exploratory data analysis and visualization tool for graphs and networks. It contains a domain-specific embedded language Gython (an extension of Python) that supports the operators and syntactic sugar necessary for working on graph structures. An interactive interpreter binds the text typed in the interpreter to the objects being visualized for more useful integration. GUESS provides a visualization front end supporting the export of static images and dynamic movies. Details can be found at: http:// graphexploration.cond.org/.

25.1.15 Cuttlefish

Cuttlefish is a highly extensible visualization and analysis platform for various network data (Geipel 2007). It provides a versatile plug-in mechanism which allows users to extend its functionality. Furthermore, it is a platform independent open source project in SourceForge, released under GNU General Public License. GraphML and Pajek files can be loaded, as well as file formats for special visualization, or file formats for changes taking place in the network. Cuttlefish creates figures from snapshots and tex files using PSTricks or Tikz for scalable graphics. It allows browsing a large network stored in a database by applying filters to the nodes and edges displayed. In addition to ARF, other well-known layout algorithms are implemented, like Kamada–Kawai, Fruchterman–Reingold, Spring, and ISOM. Details can be found at: http://sourceforge.net/projects/cuttlefish/.

25.1.16 *CFinder*

CFinder is the software for finding and visualizing overlapping dense groups of nodes in networks. It offers a fast and efficient method for clustering data represented by large graphs, like genetic or social networks and microarray data. CFinder is also very efficient for locating the cliques of large sparse graphs. Details can be found at: http://cfinder.org/wiki/?n=Main.HomePage.

25.1.17 *JUNG*

The Java Universal Network/Graph Framework (JUNG) is a software library providing a common and extendible language for the modeling, analysis, and visualization of data that can be represented as a graph or network. It is written in Java and allows JUNG-based applications to make use of the extensive built-in capabilities of the Java API. It is designed to support a variety of representations of entities and their relations, like directed and undirected graphs, multimodal graphs, graphs with parallel edges, and hypergraphs. JUNG provides a mechanism for annotating graphs, entities, and relations with metadata. Details can be found at: http://jung.sourceforge.net/.

25.1.18 *NetMiner*

NetMiner is the software for exploratory analysis and visualization of network data. It allows users to explore network data visually and interactively, and helps to detect underlying patterns and structures of the network. Details can be found at: http://www.netminer.com/main/main-read.do.

25.2 Software for Network Layout and Others

25.2.1 *GKIN*

GKIN is a simulator and a comprehensive graphical interface where one can draw the model specification of reactions between hypothesized molecular participants in a gene regulatory and biochemical reaction network (genetic network; Arnold *et al.*, 2012). The solver is written in C++, which can run on PCs, Macintoshes, and UNIX machines. Its graphical user interface is written in Java which can run as a standalone or WebStart application. The drawing capability for rendering a network significantly enhances the ease of use of other reaction network simulators as KINSOLVER and enforces a correct semantic specification of the network.

25.2.2 *NetDraw*

NetDraw is a program for drawing social networks (Borgatti, 2011). NetDraw exhibits some advantages: (1) users can read in multiple relations on the same nodes, and switch between them (or combine them) easily; (2) if users read in valued data, they can sequentially "step" through different levels of dichotomization, selecting only strong ties, only weak ties, etc. User can choose to let the thickness of lines correspond to strength of ties; (3) the program makes it convenient to read in multiple node attributes for use in setting colors and sizes of nodes, as well as rims, labels, etc. Diagrams can be rotated, flipped, shifted, resized, and zoomed. The program makes it easy to turn on and off groups of nodes defined by a variable. (4) A few of analytical procedures are included, e.g., the identification of isolates, components, k-cores, cut-points, and bi-components (blocks) (Zhang, 2012e, 2016r).

NetDraw can read 2-mode data and automatically create a bipartite representation of it. Using the VNA file format (the VNA data format allows the user to store not only network data but also attributes of the nodes, along with information about how to display them like color, size, etc.), the program can save a network along with its spatial configuration, node colors, shapes, etc. Network diagrams can be saved as bitmaps, jpegs, windows metafiles, and enhanced metafiles. Moreover, the program can export to Pajek and Mage.

NetDraw is available at: http://www.analytictech.com/down loadnd.htm.

25.2.3 *netGenerator*

The software is used to draw directed, undirected, cyclic, and acyclic, network graphs, which is developed based on JDK 1.1.8, in which several classes are included (http://www.iaees.org/publications/ software/index.asp;BioNetAnaly; Zhang, 2012e; Fig. 2). It can be freely downloaded and run on Windows platforms.

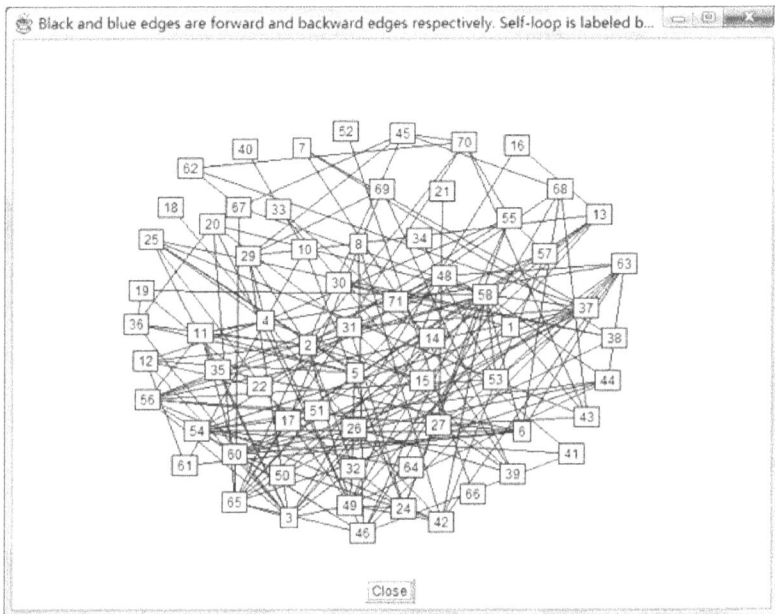

Fig. 2 Interface of the netGenerator software (Zhang, 2012e, 2016r).

25.2.4 *CCVisu*

CCVisu is a tool for force-directed graph layout. It reads the input graph from a file in Relational Standard Format (RSF). The layout of the graph is computed using standard techniques from force-directed layout. It supports several energy models, which can be selected by setting command line parameters. The weighted edge-repulsion LinLog energy model (default) is ideal for producing layouts that fulfill certain clustering criteria. The Fruchterman–Reingold energy model is available for producing layouts that fulfill certain esthetic criteria like uniform edge length. The minimizer is based on the Barnes-Hut-Algorithm. Details can be found at: https://www.sosy-lab.org/~dbeyer/CCVisu/.

25.2.5 *Otter*

Otter is a CAIDA tool used for visualizing arbitrary network data that can be expressed as a set of nodes, links or paths. It was developed to handle visualization tasks for a wide variety of Internet data, including data sets on topology, workload, performance, and routing. Details can be found at: http://www.caida.org/tools/visualization/otter/.

25.2.6 *Cytoscape*

Cytoscape is an open source software platform for visualizing complex networks and integrating them with attribute data. A lot of apps are available for various kinds of problem domains, including bioinformatics, social network analysis, and semantic web. Details can be found at: http://www.cytoscape.org/.

25.2.7 *GDToolkit*

GDToolkit (GDT) is a Graph Drawing Toolkit that can manipulate several types of graph, and automatically draw them according to many different aesthetic criteria and constraints. Details can be found at: http://www.dia.uniroma3.it/~gdt/gdt4/index.php.

25.2.8 *HyperGraph*

HyperGraph is an open source tool that provides java code to work with hyperbolic geometry and particularly with hyperbolic trees. It provides a very extensible API to visualize hyperbolic geometry, to handle graphs, and to layout hyperbolic trees. Details can be found at: http://hypergraph.sourceforge.net/.

25.2.9 *KrackPlot*

KrackPlot is a program for network visualization designed for social network analysts. There is a brief tutorial on how to use the program, and contact addresses are also given. It uses a simple screen-oriented interface. Users can drag nodes with the mouse to move them and click to add new nodes. Each node can be assigned a number of attributes, which can be highlighted using color and shape. KrackPlot provides several tools for automatic layout of the network, including algorithms based on multidimensional scaling and simulated annealing as well as circular layouts. Details can be found at: http://www.andrew.cmu.edu/user/krack/krackplot.shtml.

Part 12

Big Data Analytics

Chapter 26

Big Data Analytics for Network Biology

Big data analytics is a fast growing technology, which has been used in various areas, including network biology. Nevertheless, there are many different opinions regarding the definition of big data (Zhang *et al.*, 2013). In the general sense, big data is a collection of data that cannot be perceived, acquired, managed, processed, and serviced with traditional IT technologies and hardware and software tools for a limited period of time. In 2010, the Apache Hadoop organization defined big data as large-scale data sets that ordinary computer software cannot capture, manage, and manage in an acceptable time frame. On the basis of this definition, McKinsey & Company further defined big data. It is defined as a data set whose size is beyond the capabilities of the collection, storage, management, and analysis of typical database software. According to McKinsey's definition, it can be seen that the size of the data set is not the only standard for big data, and the growing size of the data and the inability to rely on traditional database technology to manage are important features of big data. However, International Data Corporation (IDC) held that big data technology describes a new generation of technology and architecture that, through high-speed acquisition, discovery or analysis, we can extract the economic value of a large amount of data. From this point of view, the characteristics of big data can be summarized as 4Vs, namely volume (massive size), variety (numerous modals), velocity (generated quickly), and value (the value is very remarkable but the density is very low) (Zhang *et al.*, 2013). The 4Vs definition has been widely recognized. It stresses the significance and necessity of big data — that is, mining the huge value inside big data. Therefore, the most important topic of big data is how to extract valued information from large-scale, numerous, and rapidly growing data sets.

This chapter describes the fundamentals of big data, presents a case example for extracting topics from mountainous text data using Python, and gives two case studies of big data analytics of herbal medicines.

26.1 Fundamentals of Big Data Analytics

26.1.1 *Data analysis*

The most remarkable difference between big data analytics and the traditional analysis is that the former focuses on the whole data rather than the data sample used by the latter. Big data analytics does not follow complexity and accuracy of the algorithm used but considers the whole data set to be analyzed efficiently.

26.1.1.1 *Methods for traditional data analysis*

Traditional data analysis refers to the use of appropriate statistical methods for analyzing the first-hand information and second-hand information collected, i.e., collecting and extracting valuable information from a large number of seemingly chaotic data and finding the intrinsic law, in order to maximize the functionality of the data. The following are some of the methods (Zhang *et al.*, 2013):

(1) Cluster analysis (Zhang, 2017j).
(2) Factorial analysis (Zhang, 2010).
(3) **Correlation analysis.** It can be divided into two types, functional relationship and correlation (Zhang, 2012e, 2015d, 2016r; Zhang and Li, 2015b).
(4) Regression analysis (Zhang, 2012e, 2016p).
(5) A/B test (Manyika *et al.*, 2011), also known as bucket test. By comparing the tested population, it determines which plan can

improve the target variable. Big data allows a large number of tests to be performed and analyzed to ensure that the population has sufficient size to detect meaningful differences between the control and treatment groups.

(6) **Data mining (Hand, 2007).** Data mining is an advanced technology to achieve in-depth information. It is the process of extracting information, which are unknown to people but are potentially useful, from a large number of incomplete, chaotic, vague, and random data. Data mining is mainly used to accomplish the following different tasks, corresponding to different analytical methods: classification, estimation, prediction, affinity grouping or association rules, clustering, description, and visualization. There are three categories of methods of data mining: machine learning methods, neural network methods, and database methods. Machine learning can be subdivided into: induction learning methods, case study-based methods, genetic algorithms, etc. Database methods include mainly multidimensional data analysis or on-line analytical processing (OLAP), as well as attribute-oriented induction method.

26.1.1.2 *Methods for big data analysis*

Value chain of big data can be divided into four stages: data generation, data collection, data storage, and data analysis (Zhang *et al.*, 2013). Data analysis is the last and the most important stage. It is the basis of big data application. It aims to extract useful values, provide suggestions, or support decisions (Manyika *et al.*, 2011). Traditional data analysis methods have been applied to big data analytics. Nevertheless, they are unable to meet user expectations when using them to handle complex data, such as unstructured data. So far, a variety of methods specific to big data analytics have been proposed for integration, management, and analysis of big data. Some of them include:

(1) **Bloom filter.** Bloom filter is essentially a bit array and a series of Hash functions. The principle of the Bloom filter is to use the Hash values stored in the bit array rather than the stored data themselves, which is essentially a bitmap index that uses the

Hash functions to store the compressed data. Its advantage is to have a high spatial efficiency and query rate, and the disadvantage is a certain degree of misrecognition and deletion difficulties. Bloom filter is suitable for big data applications where low error recognition rates are allowed.

(2) **Hashing.** Hashing, also known as the Hash method, is essentially a method of converting data into shorter, fixed-length values or index values. The advantage of this method is the ability to quickly read, write, and query, and the disadvantage is the difficulty to find a good Hash function.

(3) **Indexing.** Indexing is an effective method to reduce cost in disk reading/writing and to improve the rate of deletion, addition, query, etc. The drawback of the index is that it requires additional overhead to store the index file and needs to be dynamically maintained in response to updation of the data.

(4) **Trie tree.** Also known as the dictionary tree, Trie tree is a variant of the Hash tree, mostly used for quick retrieval and word frequency statistics. The idea of the Trie tree is to use the public prefix of the string to minimize the comparison of strings and improve query efficiency.

(5) **Parallel computing.** Compared to the traditional serial computing, parallel computing refers to the joint use of multiple computing resources to complete the operation. The basic idea is to decompose the problem and handle tasks by a number of independent processors in order to achieve the purpose of collaborative processing. At present, the typical parallel computing model includes message passing interface (MPI), MapReduce, Dryad, etc.

26.1.2 *Data types*

In big data analytics, the data can be divided into six types (Zhang *et al.*, 2013) as follows:

(1) **Structured data.** Structured data have always been an important research area of traditional data analysis. The

current mainstream structured data management tools, such as relational database, etc., have provided data analysis capabilities for this type of data. Thanks to the development of relational database technology, the methods for structured data analysis are more mature, and most of them are based on data mining and statistical analysis.

(2) **Text.** It is the most common unstructured data and is the usual way to store text and transfer information. Text analysis, also known as text mining, refers to the process of extracting useful information and knowledge from unstructured text. Text mining is an interdisciplinary field that involves information retrieval, machine learning, statistics, and computational linguistics, especially data mining. Most text mining systems are based on textual expression and natural language processing (NLP), focusing on the latter.

(3) **Web data.** Web data are the major source of big data. The web analytics aims to automatically retrieve, extract, and evaluate information from web documents and services in order to discover knowledge. Web analytics is based on several research areas, including databases, information retrieval, NLP and text mining. Web analytics can be divided into three related areas: web content mining, Web fabric mining, and web usage mining (Pal *et al.*, 2002).

(4) **Multimedia data.** With the development of communication technology, pictures, audio, video, and other large volumes of data, can also be quickly spread. Due to the lack of text information, the analysis has significant characteristics compared to text mining.

(5) **Social network data.** To a certain extent, social network data reflect the characteristics of human's social activities. Social network analysis can be divided into two categories: link-based structural analysis and content-based analysis (Aggarwal, 2011). Link-based structural analysis has focused on link prediction, community discovery, social network evolution, social impact analysis, and other areas.

(6) **Mobile data.** Obviously different from the web data, mobile data have such characteristics as geographical location, user's personal characteristics, and other information.

26.2 Text Mining with Python and R

Python is the most used language tool in big data analytics. Let's examine a method for extracting topics (themes) from the humongous text data using Python (Wang, 2017).

It is known that Latent Dirichlet allocation (LDA), based on unsupervised machine learning, is able to handle large-scale unstructured and unmarked data. Using the LDA method, Wang (2017) used Python to extract topics from more than 1,000 text documents. The LDA method used in the case study was topic model or topic extraction.

In the areas of machine learning and NLP, the topic model refers to a statistical model used to discover abstract topics in a series of documents. A topic can be characterized with a series of keywords. There are several ways to extract topics, of which LDA is currently the most popular method. The procedures are as follows:

(1) Install the Anaconda toolkit (details can be found at: http://www.jianshu.com/p/e4b24a734ccc).
(2) Move the text file, datascience.csv, scrapped from a network platform, to the working directory "demo".
(3) Go to the system terminal (macOS, Linux) or command prompt line (Windows), enter the working directory "demo", and execute the following command

```
pip install jieba
pip install pyldavis
```

So far, the operating environment configuration is complete.
(4) At the terminal or command prompt line, type

```
jupyter notebook
```

It can be found that Jupyter Notebook is running correctly.
(5) In Jupyter Notebook, build a Python 2 notebook, called topic-model.

(6) In order to process the table data, we use the data frame tool Pandas. First, call it

```
import pandas as pd
```

(7) Read the data file datascience.csv

```
df = pd.read_csv (''datascience.csv'')
```

(8) Perform word segmentation to extract the keywords of each document. First, let's call word segment package, jieba

```
import jieba
```

We need to deal with more than 1,000 text data, so it should be parallelly conducted. First, we need to write a function to deal with word segmentation of a single text document

```
def chinese_word_cut(mytext):
    return '' ''.join(jieba.cut(mytext))
```

Using this function, we can repeatedly call it to process all text data. However, were we can use a more efficient "apply" function

```
df [''content_cutted''] = df.content.apply (chinese_word_cut)
```

(9) Vectorization of text. Transform each keyword in the document into a feature (column), and then count the number of keywords for each document. First, call the relevant packages

```
from sklearn.feature_extraction.text import TfidfVectorizer, CountVectorizer
```

There are a variety of words in the document, and perhaps we are interested in a few words. We can extract, e.g., the most important 1,000 keywords only

```
n_features = 1000
```

The following is the procedure for keywords extraction and vectorization

```
tf_vectorizer = CountVectorizer(strip_accents = 'unicode',
max_features=n_features, stop_words='english', max_df = 0.5, min_df = 10)
tf = tf_vectorizer.fit_transform(df.content_cutted)
```

Use LDA to call the package

```
from sklearn.decomposition import LatentDirichletAllocation
```

Set the required number of topics and other parameters

```
n_topics = 5
lda = LatentDirichletAllocation(n_topics=n_topics, max_iter=50,
learning_method='online', learning_offset=50., random_state=0)
```

Now make LDA find topics in 1,000 vectorized documents

```
lda.fit(tf)
```

Each topic is described by a series of keywords. We can define the following function to display the first several keywords of each topic:

```
def print_top_words(model, feature_names, n_top_words):
for topic_idx, topic in enumerate(model.components_):
print(''Topic #%d:'' % topic_idx)
print('' ''.join([feature_names[i]
for i in topic.argsort()[:-n_top_words - 1:-1]]))
print()
```

Set the required number of keywords of each topic

```
n_top_words = 20
```

Sequentially print the keyword table of every topic

```
tf_feature_names = tf_vectorizer.get_feature_names()
print_top_words(lda, tf_feature_names, n_top_words)
```

To better exhibit the results, we can implement the following commands to visualize topics:

```
import pyLDAvis
import pyLDAvis.sklearn
pyLDAvis.enable_notebook()
pyLDAvis.sklearn.prepare(lda, tf, tf_vectorizer)
```

A dynamic graph can thus be generated. At the left side of the graph, circles represent different topics, and the size of the circle represents the number of documents that each topic contains. At the right side of the graph, the most important keywords (the highest frequency) are listed. These keywords are the most important extracted from all documents.

If we want to set the number of topics as 10, implement the following commands

```
n_topics = 10
lda = LatentDirichletAllocation(n_topics=n_topics, max_iter=50,
learning_method='online', learning_offset=50., random_state=0)
lda.fit(tf)
print_top_words(lda, tf_feature_names, n_top_words)
pyLDAvis.sklearn.prepare(lda, tf, tf_vectorizer)
```

In addition to the method above, we can use the following method to draw word cloud (https://mp.weixin.qq.com/):

```
#Load package
import jieba
from wordcloud import WordCloud
#Read the text
text = """Network biology is an emerging ..."""
#Extract keywords and frequencies
word=jieba.cut(text)
#Use default setting to draw word cloud
wordcloud = WordCloud().generate(", ".join(word))
plt.figure()
plt.imshow(wordcloud, interpolation="bilinear")
plt.axis("off")
plt.show()
```

The following R program implements the same method

```
library(jiebaR)
library(wordcloud2)
# Read the text
text="Medical statistics ..."
# Extract keywords and frequencies
mixseg=worker(type="mix")
seg=mixseg[text]
num=table(seg)
# Use default setting to draw word cloud
wordcloud2(num)
```

26.3 Big Data Analytics of Medicinal Attributes and Functions of Chinese Herbal Medicines

In China, there are more than 11,000 medicinal plants and fungi, accounting for more than 85% of traditional Chinese medicines. Zhang (2017h) explored the basic statistics and classifications of medicinal attributes and functions of Chinese herbal medicines by analyzing the large-scale data sources. It is expected to provide a scientific basis for further application (Zhang, 2017h, 2017j).

26.3.1 *Data sources*

Data for Chinese herbal medicines were collected from Pharmacopoeia People's Republic of China, Chinese Materia Medica, Compilation of National Herbal Medicines, Dictionary of Chinese Traditional Medicines, Query Platform for Promotion of Norms of Chinese Medicines, Terminology, Achievements, and Medical Information, Chinese Flora, Medicinal Plants in Southern China, Medicinal Plants in Northern China, Network of Chinese Medicine, Compendium of Materia Medica, Chinese Traditional Medicines, Baidu Encyclopedia, Practical Alias Handbook of Chinese Traditional Medicines, etc. The Pharmacopoeia People's Republic of China, Chinese Materia Medica, and Compilation of National Herbal Medicines were three of the major data sources. Collection was focused on the Chinese herbal medicines with reported chemical composition. A total of 1,127 medicines, involving about 2,000 species of medicinal plants and fungi, were collected, which account for approximately 1/5 of medicinal plants and fungi in China. Among them, medicinal plants accounted for 98.94%, and medicinal fungi accounted for 1.06%. The list included the most commonly used or important Chinese herbal medicines (Zhang, 2017h).

For the 1,127 Chinese herbal medicines, Boolean coding was made for medicine name, species, family, the above 69 medicinal attributes, and 78 medicinal functions. Finally, an interactive database that includes eight tables, CHM-DATA Version 1.0, was established (Zhang, 2017h, 2017j).

26.3.2   Basic statistics of medicinal attributes and functions for Chinese herbal medicines (Zhang, 2017h)

Some of the results for basic statistics of medicinal attributes and functions for Chinese herbal medicines are described as follows (Zhang, 2017h).

(1) Medicinal organs/tissues

Utilization percentage of each organ/tissue over all organs/tissues of Chinese herbal medicines is whole body (25%), root (19%), leaf (12%), stem/stalk (8%), fruit (7%), rhizome (7%), seed (5%), etc. All herb medicines that used the whole body accounted for 32.83% (370 medicines), and those that used root only accounted for 24.4% (275 medicines), followed by leaf (15.26%, 172 medicines), stem/stalk (10.03%, 113 medicines), fruit (9.49%, 107 medicines),

rhizome (9.05%, 102 medicines), flower/flower bud/flower inflores-
cence (6.65%, 75 medicines), seed (6.12%, 69 medicines), etc. In general, Chinese herbal medicines use the whole body, root, and leaf as
the major medicinal organs/tissues.

(2) Chemical composition categories

The chemical composition category with the maximum occurrence
percentage over all composition categories is glycosides (14.99%), followed by organic acids (10.9%), alcohols (8.2%), esters/fats (8.18%),
ketones/flavonoids (7.78%), alkaloids (6.81%), carbohydrates/starch
(6.6%), sterols (6.5%), volatile oils/ordinary oils (5.48%), olefins
(4.45%), phenols (4.22%), aldehydes (2.15%), ethers (1.94%), etc.

Glycosides (occurrence frequency or probability 0.5301; 572
medicines) is the most frequently occurring category in 1,127
Chinese herbal medicines, followed by organic acids (0.3855, 416
medicines), alcohols (0.2901, 313 medicines), esters/fats (0.2892,
312 medicines), ketones/flavonoids (0.2753, 297 medicines), alkaloids (0.241, 260 medicines), carbohydrates/starch (0.2335, 252
medicines), sterols (0.2298, 248 medicines), volatile oils/ordinary
oils (0.1937, 209 medicines), olefins (0.1576, 170 medicines), phenols (0.1492, 161 medicines), aldehydes (0.0076, 82 medicines), ethers
(0.00686, 74 medicines), etc. Overall, active ingredients of Chinese
herbal medicines mainly include glycosides, organic acids, alcohols,
esters/fats, ketones/flavonoids, alkaloids, etc.

(3) Meridians and collaterals

In all of the meridians and collaterals, liver meridians and collaterals account for 23.2%, followed by lung meridians and collaterals (18.9%), stomach meridians and collaterals (13.38%), spleen
meridians and collaterals (12.37%), kidney meridians and collaterals (10.19%), heart meridians and collaterals (9.02%), large intestine
meridians and collaterals (7.48%), urinary bladder meridians and
collaterals (3.45%), small intestine meridians and collaterals (1.43%),
gall bladder meridians and collaterals (0.37%), triple burner (0.16%),
and blood phase (0.05%).

26.3.3 Classification and network analysis of medicinal functions of Chinese herbal medicines

26.3.3.1 Principal component analysis

Principal Component Analysis (PCA) (Zhang and Fang, 1982; Qi, 2003; Vieira, 2012) is used to classify and order medicinal functions (Zhang, 2017j).

The following is the Matlab program, PCA.m, of PCA (Zhang, 2017j)

```
clear
t=input('Input threshold t of variable significance beyond the (mean+t*std)
(1-2):
');
file=input('Input the excel file name of data, e.g., pca.xls. Columns are
indices and rows are samples: ','s');
raw=xlsread(file);
z=raw';
m=size(z,1); n=size(z,2);
zbar=mean(z');
s=std(z');
for i=1:m
for j=1:n
x(i,j)=(z(i,j)-zbar(i))/s(i);
end; end
for i=1:m
for j=1:m
r(i,j)=0;
for k=1:n;
r(i,j)=r(i,j)+x(i,k)*x(j,k)/(n-1);
end; end; end
[u,p]=eig(r);
sums=sum(sum(p));
for i=1:m
p2(i)=i;
end
for i=1:m-1
k=i;
for j=i:m-1
if (p(j+1,j+1)>p(k,k)) k=j+1; end
end
i2=p2(i); p2(i)=p2(k); p2(k)=i2;
l=p(i,i); p(i,i)=p(k,k); p(k,k)=l;
end
iss='';
iss=strcat(iss,'Eigenvalues, contribution percentages, and eigenvectors for
principal components\n');
```

```
for i=1:m
thr=i;
iss=strcat(iss,num2str(p(i,i)),' ','(',num2str(p(i,i)/sums*100),'%%',')')\n');
for j=1:m
iss=strcat(iss,num2str(u(j,p2(i))),',');
end
iss=strcat(iss,'\n');
end
iss=strcat(iss,'\n');
for j=1:m
for i=1:m a(i,j)=sqrt(p(j,j))*u(i,p2(j));
end; end
iss=strcat(iss,'Principal Components\n');
for j=1:m
iss=strcat(iss,'F',num2str(j),'=');
for i=1:m
e1=num2str(i);
if (a(i,j)>0) e2=num2str(a(i,j));
elseif (a(i,j)<0) e2=num2str(abs(a(i,j)));
end
if ((a(i,j)>0) & (i~=1)) iss=strcat(iss,'+',e2,'x',e1);
elseif ((a(i,j)>0) & (i==1)) iss=strcat(iss,e2,'x',e1);
elseif (a(i,j)<0) iss=strcat(iss,'-',e2,'x',e1);
end
end
iss=strcat(iss,'\n');
end
iss=strcat(iss,'\n');
mea=mean(abs(a));
stds=std(abs(a));
thr=mea+t*stds;
for j=1:m
for i=1:m
if (abs(a(i,j))>=thr(j)) sig(j,i)=sign(a(i,j));
else sig(j,i)=0;
end
end; end
for i=1:m
iss=strcat(iss,'Significant variables of principal
component-',num2str(i),':\n');
for j=1:m
if (sig(i,j)~=0)
if (sig(i,j)<0) iss=strcat(iss,'-x',num2str(j),',');
else iss=strcat(iss,'x',num2str(j),',');
end
end
end
iss=strcat(iss,'\n');
end
fprintf(iss)
```

PCA of 78 medicinal functions ($x_1 \sim x_{78}$) indicates that the medicinal functions $x_1 \sim x_{15}$ account for 33% of the total variance, $x_1 \sim x_{50}$ account for 79%, and $x_1 \sim x_{60}$ account for 88% (Zhang, 2017j). There are no absolutely significant components and medicinal functions.

The 78 components from PCA are substantially 78 independent and comprehensive medicinal functions. Major medicinal functions for every component can be simply determined by their importance and contribution coefficients in the component.

Given different threshold t value, the new definition of medicinal functions for some of the principal components can be overall given as follows

F_1: Clear away heat — Detoxification function

F_2: Dispel endogenous damp — Dispel endogenous wind — Relieve pain function

F_3: Consolidate or warm kidney — Invigorate male impotence (Yang) or strengthen male essence function

F_4: Relieve constipation — Loosen the bowels function

F_5: Kill or expel parasites — Relieve itching function

F_6: Relieve external syndrome function

. . .

At a certain level, for example, the medicinal functions — Consolidate or warm kidney, Whet the appetite or reinforce stomach, Cool blood, Regulate or enhance energy flow (Qi), Nourish or warm spleen/stomach/Qi, Clear away heat, Detoxification, and Dispel endogenous cold — are the major attributes of the principal component F_1. Of which the medicinal functions with the same sign (+ or −) have the same direction of medicinal action (Zhang, 2017j).

26.3.3.2 *System cluster analysis*

It is called hierarchical cluster analysis also. The following is the Matlab program, SystemCluster.m, of system cluster analysis (Zhang, 2017j):

```
clear
cluster=input('Choose cluster method (1: The shortest distance; 2: The
longest distance; 3: Cluster averaging): ');
```

```
sele=input('Choose between sample distance or correlation (1: Euclidean
distance;
2: Pearson correlation; 3: Point correlation; 4: Jaccard coefficient): ');
sel=input('Choose data standardization method (1: Standard deviation; 2:
Max-min): ');
file=input('Input the excel file name of data, e.g., cluster.xls. Columns
are indices and rows are samples to be clustered: ','s');
x=xlsread(file);
m=size(x,2); n=size(x,1);
xbar=mean(x); st=std(x); maxx=max(x); minn=min(x);
for i=1:m
for j=1:n
if (sel==1) a(i,j)=(x(j,i)-xbar(i))/st(i);
elseif (sel==2) a(i,j)=(x(j,i)-minn(i))/(maxx(i)-minn(i));
end
end; end
for i=1:n-1
for j=i+1:n
if (sele==1)
r(i,j)=sqrt(sum((a(:,i)-a(:,j)).^2))/m;
elseif (sele==2)
r(i,j)=1-corr(a(:,i),a(:,j));
elseif (sele==3)
aa=sum((a(:,i)==0) & (a(:,j)==0));
bb=sum((a(:,i)==0) & (a(:,j)~=0));
cc=sum((a(:,i)~=0) & (a(:,j)==0));
dd=sum((a(:,i)~=0) & (a(:,j)~=0));
r(i,j)=1-(aa*dd-bb*cc)/sqrt((aa+bb)*(cc+dd)*(aa+cc)*(bb+dd));
elseif (sele==4)
bb=sum((a(:,i)==0) & (a(:,j)~=0));
cc=sum((a(:,i)~=0) & (a(:,j)==0));
dd=sum((a(:,i)~=0) & (a(:,j)~=0));
r(i,j)=1-(dd-(cc+bb))/(dd+cc+bb);
end
r(j,i)=r(i,j);
end; end
for j=1:n r(j,j)=0; end
r1=r;
bb1=1;
u(bb1)=0;
nu(bb1)=n;
for i=1:nu(bb1) x(bb1,i)=i; end
for i=1:nu(bb1) y(bb1,i)=1; end
while (nu(bb1)>1)
aa=1e+10;
for i=1:nu(bb1)-1
for j=i+1:nu(bb1)
if (r(i,j)<=aa) aa=r(i,j); end
end; end
aa1=0;
```

```
for i=1:nu(bb1)-1
for j=i+1:nu(bb1)
if (abs(r(i,j)-aa)<=1e-06)
aa1=aa1+1; v(aa1)=i; w(aa1)=j;
end; end; end
for i=1:nu(bb1) s(i)=0; end
nn1=0;
for i=1:aa1
if ((v(i)~=0) & (w(i)~=0))
nn1=nn1+1;
for j=1:aa1
if ((v(j)==v(i)) | (v(j)==w(i)) | (w(j)==w(i)) | (w(j)==v(i)))
s(v(j))=nn1; s(w(j))=nn1;
if (j~=i) v(j)=0; w(j)=0; end; end
end
v(i)=0; w(i)=0;
end; end
for i=1:nn1
y(bb1+1,i)=0;
for j=1:nu(bb1)
if (s(j)==i)
for k=1:n
if (x(bb1,k)==j) x(bb1+1,k)=i; end
end
y(bb1+1,i)=y(bb1+1,i)+y(bb1,j);
end; end; end
for i=1:nu(bb1)
if (s(i)==0)
nn1=nn1+1;
for k=1:n
if (x(bb1,k)==i) x(bb1+1,k)=nn1; end
end
y(bb1+1,nn1)=y(bb1,i); end
end;
bb1=bb1+1;
u(bb1)=aa;
nu(bb1)=nn1;
for i=1:nu(bb1)-1
for j=i+1:nu(bb1)
if (cluster==1) r(i,j)=1e+10;
elseif (cluster==2) r(i,j)=-1e+10;
elseif (cluster==3) r(i,j)=0;
end
nk=0;
for k=1:n
if (x(bb1,k)==i)
nk=nk+1;
nkk=0;
for kk=1:n
if (x(bb1,kk)==j)
```

```
nkk=nkk+1;
if (cluster==1)
if (r1(k,kk)<r(i,j)) r(i,j)=r1(k,kk); end
elseif (cluster==2)
if (r1(k,kk)>r(i,j)) r(i,j)=r1(k,kk); end
 elseif (cluster==3)
r(i,j)=r(i,j)+r1(k,kk)^2;
end
end; end; end; end
if (cluster==3)
r(i,j)=sqrt(r(i,j)/(nk*nkk));
end
r(j,i)=r(i,j);
end; end; end;
for k=1:n
y(bb1,k)=1; end
for i=bb1-1:-1:1
rr=0;
for j=1:nu(i+1)
ww=0;
for k=1:n
if (y(i+1,k)==j) ww=ww+1; v(ww)=k; end
end
vv=0;
for ii=1:ww
ee=0;
for jj=ii-1:-1:1
if (x(i,v(ii))==x(i,v(jj))) y(i,v(ii))=y(i,v(jj)); break; end
ee=ee+1;
end
if (ee==ii-1) vv=vv+1; y(i,v(ii))=rr+vv; end
end
rr=rr+vv;
end; end
for k=1:bb1
rs(k)=1-u(k);
end;
s=1; i=0;
while (n>0)
ss=1;
for j=s+1:bb1
if (rs(j)==rs(s)) ss=ss+1; end;
end
s=s+ss;
i=i+1;
la(i)=s-1;
if (s>=bb1) break; end
end
bb1=i;
```

```
yy=zeros(n);
for k=1:bb1
for i=1:nu(la(k))
for j=1:n
if (y(la(k),j)==i) yy(k,j)=i; end;
end; end; end
for k=1:bb1
rss(k)=rs(la(k)); uu(k)=u(la(k)); nuu(k)=nu(la(k));
end
iss='';
for k=1:bb1
iss=strcat(iss,'\n','r=',num2str(rss(k)),'\n');
for i=1:nuu(k)
iss=strcat(iss,'(');
for j=1:n
if (yy(k,j)==i) iss=strcat(iss,num2str(j),','); end
end
iss=strcat(iss,')');
end; end
fprintf(iss)
```

Choose distance measure as point correlation $(1-r)$, use max–min normalization and class averaging method. We can determine belonged class of any medicinal function at different hierarchy. At a certain level, e.g., $r = -0.00025741$, 78 medicinal functions can be classified to the following eight categories (Zhang, 2017j):

Category 1:
Clean liver/relax liver/consolidate liver/bright eyes/eliminate eye strain, thick or dark hair, consolidate or warm kidney, invigorate male impotence (Yang)/strengthen male essence, strengthen bones and muscles, antidiarrheal, tonify blood, prevent miscarriage or abortion, decrease internal heat, relieve rheumatism or lubricate the joints, dry dampness, suppress perspiration, antimalarial;

Category 2:
Benefit gall bladder/cure jaundice, induce diuresis/treat strangurt, repel foulness, strengthen heart or treat heartburn, relieve restlessness/calm the nerves/alleviate mental depression or arrest convulsion, antidiabetic;

Category 3:
Remove lung-heat/nourish lung, eliminate or relieve phlegm, antiasthma, eliminate or relieve cough, promote secretion of saliva or

body fluids, relieve sore throat, resolve food stagnation, relieve constipation, loosen the bowels, moisten dryness, nourish essential fluid (Yin), inhibit or break energy flow (Qi), antiaging, Quench one's thirst;

Category 4:
Eliminate or relieve stuffy nose, whet the appetite/reinforce stomach, strengthen and reinforce spleen, improve digestion, prevent or arrest vomiting, arrest epilepsy, astringe intestine, stop diarrhea, regulate or enhance energy flow (Qi), nourish or warm spleen/stomach/Qi, relieve pain, dispel endogenous cold, eliminate impediment;

Category 5:
Remove obstruction in meridians and collaterals/relax the muscles and joints, dispel endogenous damp, dispel endogenous wind, induce perspiration, relieve external syndrome, relieve itching, kill or expel parasites, relieve muscular spasm, expose exanthema/promote eruption, anesthetic;

Category 6:
Reduce aminotransferase, promote granulation, cool blood, stop bleeding, clear away heat, eliminate dampness, detoxification, relieve summer-heat, promote astringent function;

Category 7:
Activate water metabolism or excrete water, eliminate or relieve tuberculosis/soften hardness or dissolve masses, reduce swelling, antiatherosclerosis, antihypertension, anticancer, discharge pus/diminish inflammation/antiinfection, dispel evil spirit;

Category 8:
Invigorate blood circulation, absorb clots/eliminate stasis/resolve carbuncle/promote wound healing, Regulate menstruation/promote blood flow, promote lactation/stimulate milk secretion.

References

ABCAM. 2012. Available at: http://www.abcam.com/index.html?pageconfig=productmap&cl=2282. Accessed on October 10, 2016.

Achlioptas D, D'Souza RM, Spencer J. 2009. Explosive percolation in random networks. *Science*, 323: 1453–1455.

Adamic LA, Adar E. 2003. Friends and neighbors on the web. *Social Networks*, 25(3): 211–230.

Aggarwal CC. 2011. *An Introduction To Social Network Data Analytics*. Springer, Berlin, Germany.

Ahn YY, Bagrow JP, Lehmann S. 2010. Link communities reveal multiscale complexity in networks. *Nature*, 466(7307): 761–764.

Airoldi EM, Blei DM, Fienberg SE, *et al.* 2008. Mixed membership stochastic blockmodels. *Journal of Machine Learning Research*, 9: 1981–2014.

Albert R. 2004. Boolean modeling of genetic regulatory networks. *Lecture Notes Physics*, 650: 459–481.

Ali Azam T, Iwata A, Nishimura A, *et al.* 1999. Growth phase-dependent variation in protein composition of the *Escherichia coli* nucleoid. *The Journal of Bacteriology*, 181: 6361–6370.

Alizadeh R. 2011. A dynamic cellular automaton model for evacuation process with obstacles. *Safety Science*, 49: 315–323.

Allesina S, Bodini A, Pascual M. 2005. Functional links and robustness in food webs. *Philosophical Transactions of the Royal Society of London B*, 364(1524): 1701–1709.

Almendral JA, Sendiña-Nadal I, Yu D, *et al.* 2009. Regulating synchronous states of complex networks by pinning interaction with an external node. *Physical Review E*, 80: 066111.

Alon U. 2006. *An Introduction to Systems Biology: Design Principles of Biological Circuits*. CRC Press, USA.

Alon U, Surette MG, Barkai N, *et al.* 1999. Robustness in bacterial chemotaxis. *Nature*, 397(6715): 168–171.

Aloy P, Russell RB. 2003. InterPreTS: Protein interaction prediction through tertiary structure. *Bioinformatics*, 19: 161–162.

Amaral LAN. 2008. A truer measure of our ignorance. *Proceedings of the National Academy of Sciences of the United States of America*, 105: 6795–6796.

Amaral LAN, Scala A, Barthélémy M, Stanley HE. 2000. Classes of small-world networks. *Proceedings of the National Academy of Sciences of the United States of America*, 97: 11149–11152.

Amoroso S, Patt YN. 1972. Decision procedures for subjectivity and injectivity of parallel maps for tessellation structures. *Journal of Computer and System Sciences*, 6: 448–464.

Anderson F. 1992. *Criminological Theory Summaries*. Cullen & Agnew. Accessed on November 3, 2011.

Apsel B, Blair JA, Gonzalez B, *et al.* 2008. Targeted poly pharmacology: Discovery of dual inhibitors of tyrosine and phosphoinositide kinases. *Nature Chemical Biology*, 4(11): 691–699.

Arnold J, Taha TR, Deligiannidis L. 2012. GKIN: A tool for drawing genetic networks. *Network Biology*, 2(1): 26–37.

Ash J, Newt HD. 2007. Optimizing complex networks for resilience against cascading failure. *Physical Statistical Mechanics and its Applications*, 380(7): 673–683.

Ashby WR. 1947. Principles of the self-organizing dynamic system. *Journal of General Psychology*, 37(2): 125–128.

Bakshy E, Wilensky U. 2007. NetLogo Team Assembly Model. Center for Connected Learning and Computer-Based Modeling, Northwestern University, Evanston, IL, USA.

Ball P. 2014. One rule of life: Are we poised on the border of order? Available at: http://www.newscientist.com/article/mg22229660.700-one-rule-of-life-are-we-poised-on-the-border-of-order.html?cmpid=RSS|NSNS|2012-GLOBAL|online-news. Accessed on April 30, 2014.

Ballestores F Jr, Qiu ZY. 2012. An integrated parcel-based land use change model using cellular automata and decision tree. *Proceedings of the International Academy of Ecology and Environmental Sciences*, 2(2): 53–69.

Barabasi AL. *The New Science of Networks*. Perseus Publishing, Cambridge, Massachusetts, USA, 2002, 79–92.

Barabasi AL. 2009. Scale-free networks: A decade and beyond. *Science*, 325: 412–413.

Barabasi AL, Albert R. 1999. Emergence of scaling in random networks. *Science*, 286(5439): 509.

Barabasi AL, Jeong H, Ravasz E, Neda Z, *et al.* 2002. Evolution of the social network of scientific collaborations. *Physica A*, 311: 590–614.

Barabasi AL, Oltvai ZN. 2004. Network biology: Understanding the cell's functional organization. *Nature Reviews Genetics*, 5: 101–113.

Barranco M, Proenza J, Almeida L. 2011. Quantitative comparison of the error-containment capabilities of a bus and a star topology in CAN networks. *IEEE Transactions on Industrial Electronics*, 53(3): 802–803.

Barzel B, Barabási AL. 2013. Network link prediction by global silencing of indirect correlations. *Nature Biotechnology*, 31: 720–725.

Bascompte J. 2009. Disentangling the web of life. *Science*, 325: 416–419.

Bastiaens P, Birtwistle MR, Blüthgen N, *et al.* 2015. Silence on the relevant literature and errors in implementation. *Nature Biotechnology*, 33: 336–339.

Bastolla U, Fortuna MA, Pascual-Garcia A, *et al.* 2009. The architecture of mutualistic networks minimizes competition and increases biodiversity. *Nature*, 458: 1018–1020.

Bednar JA, Choe Y, Paula JD, *et al.* 2004. Modeling cortical maps with topographica. *Neurocomputing*, 58: 1129–1135.

Beirne P. 1987. Adolphe quetelet and the origins of positivist criminology. *American Journal of Sociology*, 92(5): 1140–1169.

Bellman RE. 1957. *Dynamic Programming*. Princeston University Press, Princeston, USA.

Berlow EL, Jennifer A, Dunne A, *et al.* 2009. Simple prediction of interaction strengths in complex food webs. *Proceedings of the National Academy of Sciences of the United States of America*, 106: 187–191.

Berlow EL, Navarrete SA, Briggs CJ, *et al.* 1999. Quantifying variation in the strengths of species interactions. *Ecology*, 80: 2206–2224.

Bian ZQ, Zhang XG. 2000. *Pattern Recognition* (2nd Edn.). Tsinghua University Press, Beijing, China.

Bock JR, Gough DA. 2001. Predicting protein-protein interactions from primary structure. *Bioinformatics*, 17: 455–460.

Bohannon J. 2009. Counter terrorism's new tool: "metanetwork" analysis. *Science*, 325(5939): 409–411.

Bohman T. 2009. Emergence of connectivity in networks. *Science*, 323: 1438–1439.

Bollobás B. 1985. *Random Graphs*. Academic Press, USA.

Bollobás B. 2001. *Random Graphs*. Cambridge University Press, Cambridge, UK.

Bonabeau E. 2002. Agent-based modeling: Methods and techniques for simulating human systems. *Proceedings of the National Academy of Sciences of the United States of America*, 99(Suppl. 3): 7280–7287.

Bonabeau E, Dorigo M, Theraulaz G. 1999. *Swarm Intelligence: From Natural to Artificial Systems*. Oxford University Press, UK.

Bonacich P, Lloyd P. 2001. Eigenvector-like measures of centrality for asymmetric relations. *Social Networks*, 23(3): 191–201.

Bond WJ. 1989. The tortoise and the hare: Ecology of angiosperm dominance and gymnosperm persistence. *Biological Journal of the Linnean Society*, 36: 227–249.

Borgatti SP. 2005. Centrality and network flow. *Social Networks*, 27: 55–71.

Borgatti S. 2011. *NetDraw 2.099*. Lexington, USA.

Borgatti SP, Mehra AJ, Brass DJ, *et al.* 2009. Network analysis in the social sciences. *Science*, 323: 892–895.

Bossuyt B, Honnay O, Hermy M. 2005. Evidence for community assembly constraints during succession in dune slack plant communities. *Plant Ecology*, 178: 201–209.

Breiman L. 2001. Statistical modeling: The two cultures (with discussion). *Statistical Science*, 16: 199–231.

Brogliato B, Lozano R, Maschke B, *et al.* 2007. *Dissipative Systems Analysis and Control: Theory and Applications* (2nd edn.). Springer Verlag, London, UK.

Brose U, Williams RI, Martinez ND. 2006. Allometric scaling enhances stability in complex food webs. *Ecology Letters*, 9: 1228–1236.

Budovsky A, Fraifeld VE. 2012. Medicinal plants growing in the Judea region: Network approach for searching potential therapeutic targets. Network Biology, 2(3): 84–94.

Bullmore E, Sporns O. 2009. Complex brain networks: Graph theoretical analysis of structural and functional systems. *Nature Review Neuroscience*, 10: 186–198.

Burnham KP, Overton WS. 1978. Estimation of the size of a closed population when capture probabilities vary among animals. *Biometrika*, 65: 623–633.

Burnham KP, Overton WS. 1979. Robust estimation of population when capture probabilities vary among animals. *Ecology*, 60: 927–936.

Bursik Jr, Robert J. 1988. Social disorganization and theories of crime and delinquency: Problems and prospects. *Criminology*, 26(4): 519–539.

Butts CT. 2009. Revisiting the foundations of network analysis. *Science*, 325: 414–416.

Byron CJ, Tennenhouse C. 2015. Commonality in structure among food web networks. *Network Biology*, 5(4): 146–162.

Cakir T, Kirdar B, Ulgen KO. 2004. Metabolic pathway analysis of yeast strengthens the bridge between transcriptomics and metabolic networks. *Biotechnology and Bioengineering*, 86(3): 251–260.

Camazine S. 2003. *Self-Organization in Biological Systems*. Princeton University Press, USA.

Campillos M, *et al.* 2008. Drug target identification using side effect similarity. *Science*, 321(5886): 263–266.

Cancho RF, Sole RV. 2001. Optimization in complex networks. Santafe Institute, USA.

Case TJ. 1990. Invasion resistance arises in strongly interacting species-rich model competition communities. *Proceedings of the National Academy of Sciences of the United States of America*, 87: 9610–9614.

Casti J. 1997. *Would-Be Worlds: How Simulation Is Changing the World of Science*. Wiley, New York, USA.

Cattin MF, Bersier LF, Banasek-Richter C, Baltensperger R, Gabriel JP. 2004. Phylogenetic constraints and adaptation explain food-web structure. *Nature*, 427: 835–839.

Cemanovic A, Sutkovic J, Gawwad MRA. 2014. Comparative structural analysis of HAC1 in *Arabidopsis thaliana*. *Network Biology*, 4(2): 67–73.

Cena G, Valenzano A. 2013. A star topology to improve CAN performance. Work in Progress Session of IEEE WFCS' 2000. 3rd IEEE International Workshop on Factory Communication Systems. 19–22, Porto, Portugal.

Chan SB, *et al.* 1982. *Graph Theory and Its Applications*. Science Press, Beijing, China.

Chan JX. 1987. *Lectures on Foundations of Algebraic Topology.* Higher Education Press, Beijing, China.

Chao A. 1984. Non-parametric estimation of the number of classes in a population. *Scandinavian Journal of Statistics*, 11: 265–270.

Chao A, Lee SM. 1992. Estimating the number of classes via sample coverage. *Journal of American Statistician Association*, 87: 210–217.

Chase JM. 2003. Community assembly: When should history matter? *Oecologia*, 136: 489–498.

Chen C. 2004. Searching for intellectual turning points: Progressive Knowledge Domain Visualization. *Proceedings of the National Academy of Sciences of the United States of America*, 101 (Suppl. 1): 5303–5310.

Chen C. 2006. CiteSpace II: Detecting and visualizing emerging trends and transient patterns in scientific literature. *Journal of the American Society for Information Science and Technology*, 57(3): 359–377.

Chen JX. 1987. *Course Notes on Algebraic Topology.* Higher Education Press, Beijing, China.

Chen LN. 2016. Teach you how to conduct gene co-expression network analysis. Available at: http://blog.sciencenet.cn/blog-3227893-1010792. html. Accessed on October 25, 2016.

Chen Q, Wang Z, Wei DQ. 2010. Progress in the applications of flux analysis of metabolic networks. *Chinese Science Bulletin*, 55(22): 2315–2322.

Chen X, Cohen JE. 2001. Global stability, local stability and permanence in model food webs. *Journal of Theoretical Biology*, 212: 223–235.

Chen Y, Hu XC, Liu BB, *et al.* 2008. Modeling and simulation of economic system based on UML and Swarm. *Computer Engineering and Design*, 29(15): 4040–4042, 4060.

Christensen V, Pauly D. 1992. ECOPATH II — a software forbalancing steady-state ecosystem models and calculating network characteristics. *Ecological Modelling*, 61(3–4): 169–185.

Christensen V, Walters CJ. 2004. Ecopath with Ecosim: Methods, capabilities and limitations. *Ecological Modelling*, 172: 109–139.

Clarke KR. 1993. Non-parametric multivariate analyses of changes in community structure. *Australia Journal of Ecology*, 18: 117–143.

Clauset A, Moore C, Newman MEJ. 2008. Hierarchical structure and the prediction of missing links in networks. *Nature*, 453: 98–101.

Clyde DE, Corado MS, Wu X, *et al.* 2003. A self-organizing system of repressor gradients establishes segmental complexity in Drosophila. *Nature*, 426(6968): 849–853.

Cohen JE. 1978. *Food Webs and Niche Space.* Princeton University Press, Princeton, NJ, USA.

Cohen JE, Briand F. 1984. Trophic links of community food webs. *Proceedings of the National Academy of Sciences of the United States of America*, 81: 4105–4109.

Cohen JE, Briand F, Newman CM. 1990. *Community Food Webs: Data and Theory.* Springer, Berlin, Germany.

Cohen JE, *et al.* 1993. Improving food webs. *Ecology*, 74: 252–258.

Cohen JE, Newman CM. 1985. A stochastic theory of community food webs I. models and aggregated data. *Proceeding of the Royal Society of London Series B: Biological Sciences*, 224(1237): 421–448.

Coleman BD, Mares MA, Willig MR, *et al.* 1982. Randomness, area, and species richness. *Ecology*, 63: 1121–1133.

Colwell RK, Coddington JA. 1994. Estimating terrestrial biodiversity through extrapolation. *Philosophical Transactions of the Royal Society of London B*, 345: 101–108.

Coombes S. 2009. *The Geometry and Pigmentation of Seashells*. Department of Mathematical Sciences, University of Nottingham, Nottingham, UK.

Csermely P, Agoston V, Pongor S. 2005. The efficiency of multi-target drugs: The network approach might help drug design. *Trends in Pharmacological Sciences*, 26: 178–182.

Damgaard C. 2011. Measuring competition in plant communities where it is difficult to distinguish individual plants. *Computational Ecology and Software*, 1(3): 125–137.

Dandekar T, Shel B, Huynen M, Bork P. 1998. Conservation of gene order: A fingerprint of proteins that physically interact. *Trends in Biochemical Sciences*, 23: 324–328.

Davies J. 2014. *Life Unfolding: How the Human Body Creates Itself*. Oxford University Press, UK.

de Araujo WS, Grandez-Rios JM, Bergamini LL, *et al.* 2017. Exotic species and the structure of a plant-galling network. *Network Biology*, 7(2): 21–32.

Deng M, Mehta S, Sun F, *et al.* 2002. Inferring domain–domain interactions from protein–protein interactions. *Genome Research*, 12: 1540–1548.

Diamond JM. 1975. Assembly of species communities. In: Ecology and Evolution of Communities, Cody ML, Diamond JM, (eds.), Belknap Press, Harvard University Press, Cambridge, USA, pp. 342–444.

Dickerson JE, Robinson JV. 1985. Microcosms as islands: A test of the MacArthur–Wilson equilibrium theory. *Ecology*, 66: 966–980.

Dijkstra EW. 1959. A note on two problems in connection with graphs. *Numerche Math*, 1: 269–271.

Din Q. 2014. Stability analysis of a biological network. *Network Biology*, 4(3): 123–129.

Ding DW. 2012. Identification of crucial nodes in biological networks. *Network Biology*, 2(3): 118–120.

Dodds SP, Watts DJ, Sabel FC. 2003. Information exchange and robustness of organizational networks. *Proceedings of the National Academy of Sciences of the United States of America*, 100(21): 12516–12521.

Doherty IA, Padian NS, Marlow C, *et al.* 2005. Determinants and consequences of sexual networks as they affect the spread of sexually transmitted infections. *Journal of Infection Diseases*, 191: S42–S54.

Dolev S, Elovici Y, Puzis R. 2010. Routing betweenness centrality. *Journal of the ACM*, 57(4): 25.

Dormann CF. 2011. How to be a specialist? Quantifying specialisation in pollination networks. *Network Biology*, 1(1): 1–20.

Downing DJ, Gardner RH, Hoffman FO. 1985. An examination of response surface methodologies for uncertainty analysis in assessment models. *Technometrics*, 27: 151–163.

Du W, Cai M, Du HF. 2010. Study on indices of network structure robustness and their application. *Journal of Xi'an Jiao Tong University*, 44(4): 93–97.

Dunne JA, Williams RJ, Martinez ND. 2002. Food-web structure and network theory: The role of connectance and size. *Ecology*, 99(20): 12917–12922.

EACH Unit. 1997. Evolution of altruistic and cooperative habits: Learning about complexity in evolution. Available at: http://ccl.northwestern.edu/cm/EACH/ for more information. Accessed on September 26, 2017.

Ebel H, Mielsch LI, Bornholdt S. 2002. Scale-free topology of e-mail networks. *Physical Review E*, 66: 035103.

Editorial. 2007. Ecological network theory. *Ecological Modelling*, 208: 1–2.

Edwards JS, Ibarra RU, Palsson BO. 2001. In silico predictions of *Escherichia coli* metabolic capabilities are consistent with experimental data. *Nature Biotechnology*, 19(2): 125–130.

Edwards JS, Palsson BO. 2000. Metabolic flux balance analysis and the in silico analysis of *Escherichia coli* K-12 gene deletions. *BMC Bioinformatics*, 1: 1–10.

Elena E, Grammauro M, Venturino E. 2013. Predator's alternative food sources do not support ecoepidemics with two-strains-diseased prey. *Network Biology*, 3(1): 29–44.

Erdos P, Renyi A. 1959. On random graphs. *Publicationes Mathematicae Debrecen*, 6: 290–297.

Estrada E. Rodriguez-Velazquez JA. 2005. Subgraph centrality in complex networks. *Physical Review E*, 71(5): 056103.

Fararo TJ, Sunshine N. 1964. *A Study of a Biased Friendship Network*. Syracuse University Press, Syria.

Fath BD, Patten BC. 1999a. Quantifying resource homogenization using network flow analysis. *Ecological Modelling*, 123: 193–205.

Fath BD, Patten BC. 1999b. Review of the foundations of network environ analysis. *Ecosystems*, 2: 167–179.

Fath BD, Scharler UM, Ulanowicz RE, *et al.* 2007. Ecological network analysis: Network construction. *Ecological Modelling*, 208(1): 49–55.

Fawcett T. 2006. An introduction to ROC analysis. *Pattern Recognition Letters*, 8(27): 861–874.

Felson M. 1994. *Crime and Everyday Life*. Pine Forge, Pennsylvania, USA.

Ferrarini A. 2011a. Some thoughts on the controllability of network systems. *Network Biology*, 1(3–4): 186–188.

Ferrarini A. 2011b. Some steps forward in semi-quantitative network modelling. *Network Biology*, 1(1): 72–78.

Ferrarini A. 2011c. Some thoughts on the control of network systems. *Network Biology*, 2011, 1(3–4):186–188.

Ferrarini A. 2013a. Controlling ecological and biological networks via evolutionary modelling. *Network Biology*, 3(3): 97–105.

Ferrarini A. 2013b. Exogenous control of biological and ecological systems through evolutionary modelling. *Proceedings of the International Academy of Ecology and Environmental Sciences*, 3(3): 257–265.

Ferrarini A. 2013c. Network modelling is strictly required for predicting climate change impacts on biodiversity. *Network Biology*, 3(2): 67–73.

Ferrarini A. 2014. Local and global control of ecological and biological networks. *Network Biology*, 4(1): 21–30.

Ferrarini A. 2015. Evolutionary network control also holds for nonlinear networks: Ruling the Lotka–Volterra model. *Network Biology*, 5(1): 34–42.

Finn JT. 1976. Measures of ecosystem structure and function derived from analysis of flows. *Journal of Theoretical Biology*, 56: 363–380.

Finn JT. 1978. Cycling index: A general definition for cycling incompartment models. In: Environmental Chemistry and Cycling Processes, Adriano DC, Brisbin IL (eds.), DOE Proceedings 45, Conf. 760429. National Technical Information Service, Springfield, VA, USA, pp. 138–165.

Finn JT. 1980. Flow analysis of models of the Hubbard Brook ecosystem. *Ecology*, 61: 562–571.

Floyd RW. 1962. Algorithm 97: Shortest path. *Communications of the Association for Computing Machinery*, 5: 345.

Foerster HV. 1960. On self-organising systems and their environments. *Self-Organising Systems*, 31–50.

Foerster HV. 1996. Self-organization, the progenitor of conversation and interaction theories. *Systems Research and Behavioral Science*, 13(3): 349–362.

Ford LR Jr, Fulkerson DR. l956. Maximal flow through a network. *Canadian Journal of Mathematics*, 8: 399–404.

Ford LR Jr, Fulkerson DR. l957. A simple algorithm for finding maximal network flow and application to the Hitchcock problem. *Canadian Journal of Mathematics*, 9: 210–218.

Förster J, Gombert AK, Nielsen J. 2002. A functional genomics approach using metabolomics and in silico pathway analysis. *Biotechnology and Bioengineering*, 79(7): 703–712.

Frantz S. 2005. Drug discovery: Playing dirty. *Nature*, 437(7061): 942–943.

Fraser HB, Hirsh AE, Steinmetz LM, *et al.* 2002. Evolutionary rate in the protein interaction network. *Science*, 296: 750–752.

Freeman L. 2006. *The Development of Social Network Analysis*. Empirical Press, Vancouver, Canada.

Fryxell KJ. 1996. The coevolution of gene family trees. *Trends in Genetics*, 12: 364–369.

Fukami T. 2010. Community assembly dynamics in space. In: *Community Ecology: Processes, Models, and Applications*, Verhoef HA, Morin PJ, (eds.), Oxford University Press, UK, pp. 45–54.

Gaasterland T, Ragan MA. 1998. Microbial geneseapes phyletic and functional patterns of ORF distribution among prokaryotes. *Microbial & Comparative Genomics*, 3: 199–217.

Gao L, Guo JL. 2011. Review in research on biological networks. *China Journal of Bioinformation*, 9(2): 113–119.

Gao L, Li MH, Wu JS. 2006. Betweenness-based attacks on nodes and edges of food webs dynamics of continuous. *Discrete and Impulsive Systems Series B*, 13(3): 421–428.

Gardner M. 1970. The fantastic combinations of John Conway's new solitaire game "Life". *Scientific American*, 223: 120–123.

Gawwad MRA, Sutkovic J, Matakovic L, *et al.* 2013a. Functional interactome of Aquaporin 1 sub-family reveals new physiological functions in *Arabidopsis Thaliana*. *Network Biology*, 3(3): 87–96.

Gawwad MRA, Sutkovic J, Zahirovic E, *et al.* 2013b. 3D structure prediction of replication factor C subunits (RFC) and their interactome in *Arabidopsis thaliana*. *Network Biology*, 3(2): 74–86.

Geipel MM. 2007. Self-organization applied to dynamic network layout. *International Journal of Modern Physics C*, 18(10): 1537–1549.

Gentle JE. 2002. *Elements of Computational Statistics*. Springer Science + Business Media Inc., Netherlands.

Ghanbarnejad F, Klemm K. 2012. Impact of individual nodes in Boolean network dynamics. *Europhysics Letters*, 99(5): 58006.

Glansdorff P, Prigogine I. 1971. *Thermodynamic Theory of Structure, Stability and Fluctuations*. Wiley-Interscience, New York, USA.

Glass L, Kauffman SA. 1973. The logical analysis of continuous, non-linear biochemical control networks. *Journal of Theoretical Biology*, 39(1): 103–129.

Göbel U, Sander C, Schneider R, Valencia A. 1994. Correlated mutations and residue contacts in proteins. *Proteins*, 18: 309–317.

Goemann B, Wingender E, Potapov AP. 2011. Topological peculiarities of mammalian networks with different functionalities: Transcription, signal transduction and metabolic networks. *Network Biology*, 1(3–4): 134–148.

Goh CS, Bogan AA, Joachimiak M, *et al.* 2000. Co-evolution of proteins with their interactions partners. *Journal of Molecular Biology*, 299: 283–293.

Goh KI, Eom YH, Jeong H, *et al.* 2006. Structure and evolution of online social relationships: Heterogeneity in unrestricted discussions. *Physical Review E*, 73: 066123.

Golder SA, Wilkinson D, Huberman BA. 2007. Rhythms of social interaction: Messaging within a massive online network. In: Proceedings of the 3rd Communication Technology Conference. Springer, pp. 41–66.

Gonenc IE, Koutitonsky VG, Rashleigh B. 2007. Assessment of the fate and effects of toxic agents on water resources. In: Proceedings of the NATO Advanced Study Institute on Advanced Modeling Techniques for Rapid Diagnosis and Assessment of CBRN Agents Effects on Water Resources, Istanbul, Turkey.

González RER, Coutinho S, dos Santos RMZ, de Figueirêdo PH. 2013. Dynamics of the HIV infection under antiretroviral therapy: A cellular automata approach. *Physica A*, 392(19): 4701–4716.

Graham R, Wunderlin A. 1987. *Lasers and Synergetics*. Springer-Verlag, New York, USA.

Griebeler EM. 2011. Are individual based models a suitable approach to estimate population vulnerability? — a case study. *Computational Ecology and Software*, 1(1): 14–24.

Grimma V, Berger U, Bastiansen F. 2006. A standard protocol for describing individual-based and agent-based models. *Ecological Modelling*, 198: 115–126.

Gross JL, Yellen J. 2005. *Graph Theory and Its Applications* (2nd edn.). Chapman & Hall/CRC, USA.

Guimera R, Sales-Pardo M. 2009. Missing and spurious interactions and the reconstruction of complex networks. *Proceedings of the National Academy of Sciences of the United States of America*, 106: 22073–22078.

Guimera R, Uzzi B, Spiro J, Amaral L. 2005. Team assembly mechanisms determine collaboration network structure and team performance. *Science*, 308(5722): 697–702.

Habib N, Ahmed K, Jabin I, *et al.* 2016. Application of R to investigate common gene regulatory network pathway among bipolar disorder and associate diseases. *Network Biology*, 6(4): 86–100.

Habib N, Ahmed K, Jabin I, *et al.* 2017. Drug design and analysis for bipolar disorder and associated diseases: A bioinformatics approach. *Network Biology*, 7(2): 41–56.

Hakamada K, Hanai T, Honda H. 2001. Identifying genetic network using experimental time series data by Boolean algorithm. *Genome Informatics*, 12: 272–273.

Haken H. 1978. Synergetics: *An Introduction*. Springer, Berlin, Germany.

Haken H. 1983. *Synergetics, an Introduction: Nonequilibrium Phase Transitions and Self-Organization in Physics, Chemistry, and Biology* (3rd rev. enl. edn.). Springer-Verlag, New York, USA.

Haken H. 2004. Future trends in synergetics. *Solid State Phenomena*, 97–98: 3–10.

Haliki E, Kazanci N. 2017. Effects of a silenced gene in Boolean network models. *Network Biology*, 7(1): 10–20.

Han LS, Li XY, Yan DK. 2008. A discussion on several mathematical methods of sensitivity analysis. *China Water Transport*, 8(4): 177–178.

Hand DJ. 2007. Principles of data mining. *Drug safety*, 30(7): 621–622.

Havens K. 1992. Scale and structure in natural food webs. *Science*, 257: 1107–1109.

Hess B, Mikhailov A. 1994. Self-organization in living cells. *Science*, 264(5156): 223–224.

Hester S, Eglin P. 1992. *A Sociology of Crime*. Routledge, London, UK.

Hickman GJ, Hodgman TC. 2009. Inference of gene regulatory networks using boolean-network inference methods. *Journal of Bioinformatics and Computational Biology*, 7(6): 1013–1029.

Higashi M, Patten BC. 1986. Further aspects of the analysis of indirect effects in ecosystems. *Ecological Modelling*, 31: 69–77.

Higashi M, Patten BC. 1989. Dominance of indirect causality in ecosystems. *American Naturalist*, 133: 288–302.

Hirschi T. 1969. *Causes of Delinquency*. Transaction Publishers, USA.

Holland JH. 1975. *Adaptation in Natural and Artificial Systems. An Introductory Analysis with Applications to Biology, Control and Artificial Intelligence.* University of Michigan Press, Ann Arbor, MI, USA.

Honey CJ, Sporns O, Cammoun L, *et al.* 2009. Predicting human resting-state functional connectivity from structural connectivity. *Proceedings of the National Academy of Sciences of the United States of America,* 106(6): 2035–2040.

Hood L. 1998. Systems Biology: New opportunities arising from genomics, proteomics and beyond. *Experimental Hematology,* 26: 681.

Hood L. 2002. A personal view of molecular technology and how it has changed biology. *Journal of Proteome Research,* 1(5): 399–409.

Hood L. 2003. Systems Biology: Integrating technology, biology, and computation. *Mechanisms of Ageing and Development,* 124(1): 9–16.

Hopkins AL. 2007. Network pharmacology. *Nature Biotechnology,* 25(10): 1110–1111.

Hopkins AL. 2008. Network pharmacology: The next paradigm in drug discovery. *Nature Chemical Biology,* 4(11): 682–690.

Hraber PT, Milne BT. 1997. Community assembly in a model ecosystem. *Ecological Modelling,* 103: 267–285.

Huang JQ, Zhang WJ. 2012. Analysis on degree distribution of tumor signaling networks. *Network Biology,* 2(3): 95–109.

Ide K, Zamami R, Namatame A. 2013. Diffusion centrality in interconnected networks. *Procedia Computer Science,* 24: 227–238.

Ideker T, Krogan NJ. 2012. Differential network biology. *Molecular Systems Biology,* 8: 565.

Igor U, Ron S. 2005. Pathway redundancy and protein essentiality revealed in the *Saccharomyces cerevisiae* interaction networks. *Molecular Systems Biology,* 3: 104.

Ilachinski A. *Cellular Automata: A Discrete Universe.* World Scientific, Singapore, 2001.

Iqbal S, Ejaz H, Nawaz MS, *et al.* 2014. Meta-analysis of cancer transcriptomes: A new approach to uncover molecular pathological events in different cancer tissues. *Network Biology,* 4(1): 1–20.

Jablonka E, Lamb MJ. *Evolution in Four Dimensions.* 47–79, Massachusets Institute of Technology Press, USA, 2005.

Jahiruddin AM, Dey L. 2010. A concept-driven biomedical knowledge extraction and visualization framework for conceptualization of test copora. *Journal of Biomedical Informatics,* 43: 1020–1035.

Jain S. 2016a. Compendium model using frequency/cumulative distribution function for receptors of survival proteins: Epidermal growth factor and insulin. *Network Biology,* 6(4): 101–110.

Jain S. 2016b. Regression analysis on different mitogenic pathways. *Network Biology,* 6(2): 40–46.

Jain S, Bhooshan SV, Naik PK. 2011. Mathematical modeling deciphering balance between cell survival and cell death using insulin. *Network Biology,* 1(1): 46–58.

Janga SC, Tzakos A. 2009. Structure and organization of drug-target networks: Insights from genomic approaches for drug discovery. *Molecular Biosystems*, 5(12): 1536–1548.

Janson S, Knuth DE, Luczak T, Pittel B. 1993. The birth of the giant component. *Random Structures and Algorithms*, 4(3): 233–358.

Jennings NR. 2000. On agent-based software engineering. *Artificial Intelligence*, 117: 277–296.

Jesmin T, Waheed S, Emran AA. 2016. Investigation of common disease regulatory network for metabolic disorders: A bioinformatics approach. *Network Biology*, 6(1): 28–36.

Jiang LQ, Zhang WJ, Li X. 2015. Some topological properties of arthropod food webs in paddy fields of South China. *Network Biology*, 5(3): 95–112.

Jiang LQ, Zhang WJ. 2015a. Determination of keystone species in CSM food web: A topological analysis of network structure. *Network Biology*, 5(1): 13–33.

Jiang LQ, Zhang WJ. 2015b. Effects of parasitism on robustness of food webs. *Selforganizology*, 2(2): 21–34.

Jiao Y, Torquato S. 2011. Emergent behaviors from a cellular automaton model for invasive tumor growth in heterogeneous microenvironments. *PLOS Computational Biology*, 7(12): e1002314.

Jordán F, Takacs-Santa A, Molnar I. 1999. Are liability theoretical quest for key stones. *Oikos*, 86: 453–462.

Jordán F. 2001. Trophic fields. *Community Ecology*, 2: 181–185.

Jordán F, Liu W, Davis AJ. 2006. Topological keystone species: Measures of positional importance in food webs. *Oikos*, 112: 535–546.

Jørgensen SE, Fath B. 2006. Examination of ecological networks. *Ecological Modelling*, 196: 283–288.

Junker BH, Koschutzki D, Schreiber F. 2006. Exploration of biological network centralities with CentiBiN. *BMC Bioinformatics*, 7: 219.

Kanehisa Laboratories. 2016. KEGG. Available at: http://www.kegg.jp/kegg/pathway.html. Accessed on December 10, 2016.

Kanehisa M, Araki M., Goto S, *et al.* 2008. KEGG for linking genomes to life and the environment. *Nucleic Acids Research*, 36: D480–D484.

Katz L. 1953. A new status index derived from sociometric analysis. *Psychometrika*, 18: 39–43.

Kauffman SA. 1969. Metabolic stability and epigenesis in randomly constructed genetic nets. *Journal of Theoretical Biology*, 22(3): 437–467.

Kennedy J, Eberhart, R. 1995. Particle Swarm Optimization. In: Proceedings of IEEE International Conference on Neural Networks IV. Perth, Australia, 1942–1948.

Kerner BS, Klenov SL, Schreckenberg M. 2011. Simple cellular automaton model for traffic breakdown, highway capacity, and synchronized flow. *Physical Review E*, 84(4 pt 2): 046110.

Kevin MB, Anthony W. 2011. Biosocial Criminology. In: *The Ashgate Research Companion to Biosocial Theories of Crime*, Beaver KM, (ed.), Routledge, USA.

Khansari M, Kaveh A, Heshmati Z, *et al.* 2016. Centrality measures for immunization of weighted networks. *Network Biology*, 6(1): 12–27.

Kim TH, Monsefi N, Song JH, *et al.* 2015. Network-based identification of feedback modules that control RhoA activity and cell migration. *Journal of Molecular Cell Biology*, 7(3): 242–252.

Kitsak M, Gallos LK, Havlin S, *et al.* 2010. Identification of influential spreaders in complex networks. *Nature Physics*, 6(11): 888–893.

Klein DJ, Randi M. 1993. Resistance distance. *Journal of Mathematical Chemistry*, 12(1): 81–95.

Klemm K, Serrano M, Eguíluz VM, San Miguel M. 2012. A measure of individual role in collective dynamics. *Scientific Reports*, 2: 292.

Kola I, Landis J. 2004. Can the pharmaceutical industry reduce attrition rates. *Nature Reviews Drug Discovery*, 3(8): 711–715.

Kolaczyk ED. Statistical Analysis of Network Data; Methods and Models. Springer Science + Business Media, NY, USA, 2009.

Kovalevsky V. Algorithms in digital geometry based on cellular topology. In: *IWCIA 2004*, Klette R, Zunic J, (eds.), LNCS, Springer, 2004, pp. 366–393.

Kuang WP, Zhang WJ. 2011. Some effects of parasitism on food web structure: A topological analysis. *Network Biology*, 1(3–4): 171–185.

Kwon YK, Cho KH. 2007. Analysis of feedback loops and robustness in network evolution based on Boolean models. *BMC Bioinformatics*, 8(9): 430–438.

Lando DY, Teif VB. 2000. Long-range interactions between ligands bound to a DNA molecule give rise to adsorption with the character of phase transition of the first kind. *Journal of Biomolecular Structure and Dynamics*, 17(5): 903–911.

Lange RP, Locher HH, Wyss PC, Then RL. 2007. The targets of currently used antibacterial agents: Lessons for drug discovery. *Current Pharmaceutical Design*, 13(30): 3140–3150.

Langfelder P, Horvath S. 2008. WGCNA: An R package for weighted correlation network analysis. *BMC Bioinformatics*, 9: 559.

Laska MS, Wooton JT. 1998. Theoretical concepts and empirical approaches to measuring interaction strength. *Ecology*, 79: 461–476.

Latham LG. 2006. Network flow analysis algorithms. *Ecological Modelling*, 192: 586–600.

Latham LG, Scully EP. 2002. Quantifying constraint to assess development in ecological networks. *Ecological Modelling*, 154: 25–44.

Lau KY, Ganguli S, Tang C. 2007. Function constrains network architecture and dynamics: A case study on the yeast cell cycle Boolean network. *Physical Review E*, 75: 1–9.

Law R, Blackford JC. 1992. Self-assembling food webs. A global view-point of coexistence of species in Lotka–Volterra communities. *Ecology*, 73: 567–578.

Lehn JM. 1988. Perspectives in supramolecular chemistry-from molecular recognition towards molecular information processing and self-organization. *Angewandte Chemie International Edition in English*, 27(11): 89–121.

Lehn JM. 1990. Supramolecular chemistry-scope and perspectives: Molecules, supermolecules, and molecular devices (Nobel Lecture). *Angewandte Chemie International Edition in English*, 29(11): 1304–1319.

Lemke N, Heredia F, Barcellos CK, *et al.* 2004. Essentiality and damage in metabolic networks. *Bioinformatics*, 20(1): 115–119.

Li D. 1982. *Operational Research*. Tsinghua University Press, Beijing, China.

Li DC, Zheng C, Han CY, Liu YX. 2012. Research and application of multiple spanning tree network topology discovery algorithm. *Advances in Computer, Communication, Control and Automation*, 121: 165–172.

Li HL, Chen H, Jin SY. 2007. Research of modeling method for agent-based complex systems distributed simulation. *Computer Engineering and Applications*, 43(8): 209–213.

Li JR, Zhang WJ. 2013. Identification of crucial metabolites/reactions in tumor signaling networks. *Network Biology*, 3(4): 121–132.

Li P, Zhang J, Xu XK, Small M. 2012. Dynamical influence of nodes revisited: A Markov chain analysis of epidemic process on networks. *Chinese Physics Letters*, 29(4): 048903.

Li Y, Ma SF. 2006. Modeling of agent-based simulation system. *Journal of Systems Engineering*, 21(3): 225–231.

Liang X, Wang JH, Yong HE. 2010. Human connectome: Structural and functional brain networks. *Science Bulletin*, 55(16): 1565–1583.

Liao X, Yuan L, Zhao T, *et al.* 2015. Spontaneous functional network dynamics and associated structural substrates in the human brain. *Frontiers in Human Neuroscience*, 9: 478.

Liben-Nowell D, Novak J, Kumar R, Raghavan P, *et al.* 2005. Geographic routing in social networks. *Proceedings of the National Academy of Sciences of the United States of America*, 102: 11623–11628.

Libralato S, Christensen V, Pauly D. 2006. A method for identifying keystone species in food web models. *Ecological Modeling*, 195: 153–171.

Liljeros F, Edling CR, Amaral LAN. 2003. Sexual networks: Implications for the transmission of sexually transmitted infections. *Microbes and Infection*, 5: 189–196.

Lin JK. 1998. *Foundations of Topology*. Science Press, Beijing, China.

Liu BC. 2000. *Functional Analysis*. Science Press, Beijing, China.

Liu JG, Lin JH, Guo Q, Zhou T. 2016. Locating influential nodes via dynamics-sensitive centrality. *Scientific Reports*, 6: 21380.

Liu YY, Slotine JJ, Barabasi AL. 2011. Controllability of complex networks. *Nature*, 473: 167–173.

Lochner L. 2004. The effect of Education on crime: Evidence from prison inmates, arrests, and self-reports. *The American Economic Review*, 94(1): 155–189.

Lockwood JL, Powell RD, Nott MP, Pimm SL. 1997. Assembling ecological communities in space and time. *Oikos*, 80: 549–553.

Lorrain F, White HC. 1971. Structural equivalence of individuals in social networks. *Annual Review of Sociology*, 1(1): 49–80.

Lu KC, Lu HM. 1995. *Graph Theory and Its Applications*, (2nd edn.), Tsinghua University Press, Beijing, China.

Lü LY, Medo M, Yeung CH, *et al.* 2012. Recommender systems. *Physics Reports*, 519: 1–49.

Lü L, Pan L, Zhou T, *et al.* 2015. Toward link predictability of complex networks. *Proceedings of the National Academy of Sciences of the United States of America*, 112(8): 2325–2330.

Lü LY, Zhou T. 2011. Link prediction in complex networks: A survey. *Physica A*, 390: 1150–1170.

Lüi XR. 2011. Quantitative risk analysis and prediction of potential distribution areas of common lantana (*Lantana Camara*) in China. *Computational Ecology and Software*, 1(1): 60–65.

Luna B, Galan-Vasquez E, Ugalde E, *et al.* 2013. Structural comparison of biological networks based on dominant vertices. *Molecular BioSystems*, 9: 1765.

Luo HF. 2007. An Optimization Analysis on Cellular Metabolic Network Model. MD Thesis (Supervisor: WenJun Zhang). Sun Yat-sen University, Guangzhou, China.

Macal CM, North MJ. 2005. Tutorial on agent-based modeling and simulation. In: *Proceedings of the 37th Winter Simulation Conference*, Kuhl ME, Steiger NM, Armstrong FB, Joines JA (eds.), Orlando, FL, USA, pp. 2–15.

MacArthur R. 1955. Fluctuation of animal populations and a measure of community stability. *Ecology*, 36(3): 533–536.

Mackey DJC. 1992. Bayesian interpolation. *Neural Computation*, 4: 415–447.

Manly BFJ. 1997. *Randomization, Bootstrap and Monte Carlo Methods in Biology* (2nd edn.), Chapman & Hall, London, UK.

Manyika J, Chui M, Brown B, *et al.* 2011. Big data: The next frontier for innovation, competition, and productivity. McKinsey Global Institute, 1–137.

Marcot EM, Pellegrini M, He LN, Rice DW, *et al.* 1999. Detecting protein function and protein–protein interactions from genome sequences. *Science*, 285: 751–753.

Marina K, Das SR, Subramanian AP. 2010. A topology control approach for utilizing multiple channels in multi-radio wireless mesh networks. *Computer Networks*, 54: 241–256.

Marr C, Geertz M, Hutt MT, *et al.* 2008. Dissecting the logical types of network control in gene expression profiles. *BMC Systems Biology*, 2: 18.

Martínez-Antonio A. 2011. *Escherichia coli* transcriptional regulatory network. *Network Biology*, 1(1): 21–33.

Martínez-Antonio A, Collado-Vides J. 2003. Identifying global regulators in transcriptional regulatory networks in bacteria. *Current Opinion in Microbiology*, 6: 482–489.

Martinez ND. 1991. Artifacts or attributes? Effects of resolution on the Little Rock Lake food web. *Ecological Monographs*, 61: 367–392.

Martinez ND. 1992. Constant connectance in community food webs. *American Naturalist*, 139: 1208–1218.

Martinez ND. 1994. Scale-dependent constraints on food web structure. *American Naturalist*, 144: 935–953.

Martinez ND, Hawkins BA, Dawah HA, and Feifarek B. 1999. Effects on sampling effort on characterization of food-web structure. *Ecology*, 80: 1044–1055.

Mathworks. 2015. MATLAB 2015. Mathworks, Natick, USA.

May RM. 1973. *Stability and Complexity in Model Systems*. Princeton University Press, USA.

May RM. 1972. Will a large complex system be stable. *Nature*, 238: 413.

McAdams HH, Arkin A. 1999. It's a noisy business! Genetic regulation at the nanomolar scale. *Trends in Genetics*, 15(2): 65–69.

McCann KS. 2000. The diversity–stability debate. *Nature*, 405: 228–233.

McCann KS, et al. 1998. Weak trophic interactions and the balance of nature. *Nature*, 395: 794–798.

McKay MD, Beckman RJ, Conover WJ. 1979. A comparison of three methods for selecting values of input variables in the analysis of output from a computer code. *Technometrics*, 21: 239–245.

Mellouli S, Mineau G et al. 2003. Laying the foundations for an agent modelling methodology for fault tolerant multi-agent systems. In: *Fourth International Workshop Engineering Societies in the Agents World*. Imperial College London, UK.

Memmott J, Waser NM, Price MV. 2004. Tolerance of pollination networks to species extinctions. *Proceedings of the Royal Society of London Series B (Biological Sciences)*, 271: 2605–2611.

Men SP, Feng JH. 2005. *Applied Functional Analysis*. Science Press, Beijing, China.

Milo R, et al. 2002. Network motifs: Simple building blocks of complex networks. *Science*, 298: 824–827.

Minty GJ. 1965. A simple algorithm for listing all the trees of a graph. *IEEE Transactions on Circuit Theory*, CT-12(1): 120.

Mirjalili S, Mirjalili SM, Lewis A. 2014. Grey Wolf Optimizer. *Advances in Engineering Software*, 69: 46–61.

Misteli T. 2001. The concept of self-organization in cellular architecture. *Journal of Cell Biology*, 155(2): 181–185.

Montoya JM, Pimm SL, Sole RV. 2006. Ecological networks and their fragility. *Nature*, 442: 259–264.

Montoya JM, Sole RV. 2002. Small world patters in food webs. *Journal of Theoretical Biology*, 214: 405–412.

Montoya JM, Sole RV. 2003. Topological properties of food webs: From real data to community assembly models. *Oikos*, 102: 614–622.

Moreno JL. 1934. *Who Shall Survive?* Beacon House, USA.

Morenoff J, Robert S, Stephen R. 2001. Neighborhood inequality, collective efficacy and the spatial dynamics of urban violence. *Criminology*, 39 (3): 517–560.

Morin PJ, Lawler SP. 1996. Effects of food chain length and omnivory on population dynamics in experimental food webs. In: *Food Webs: Integration of Patterns and Dynamics*, Polis GA, Winemiller KO (eds.), Chapman and Hall, USA, pp. 218–230.

Morris MD. 1991. Factorial sampling plans for preliminary computational experiments. *Technometrics*, 33: 161–174.

Morton D, Law R, Pimm SL, et al. 1996. On models for assembling ecological communities. *Oikos*, 75: 493–499.

Motegi F, Zonies S, Hao Y, *et al.* 2011. Microtubules induce self-organization of polarized PAR domains in *Caenorhabditis elegans zygotes*. *Nature Cell Biology*, 13(11): 1361–1367.

Narad P, Upadhyaya KC, Som A. 2017. Reconstruction, visualization and explorative analysis of human pluripotency network. *Network Biology*, 7(3): 57–75.

Navia AF, Cortés E, Mejía-Falla PA. 2010. Topological analysis of the ecological importance of elasmobranch fishes: A food web study on the Gulf of Tortugas, Colombia. *Ecological Modelling*, 221: 2918–2926.

Nedorezov LV. 2011. About a dynamic model of interaction of insect population with food plant. *Computational Ecology and Software*, 1(4): 208–217.

Nedorezov LV. 2012. Continuous-discrete model of population dynamics with time lag in a reaction of intra-population self-regulative mechanisms. *Network Biology*, 2(4): 139–147.

Nedorezov LV, Neklyudova. 2014. Continuous-discrete model of parasite-host system dynamics: Trigger regime at simplest assumptions. *Computational Ecology and Software*, 4(3): 163–169.

Neidhardt FC, Curtiss III R, Ingraham JL, Lin ECC, Low KB, *et al.* 1996. *Escherichia coli* and *Salmonella*: Cellular and molecular biology (2nd edn.). ASM Press, Washington, USA.

Newman MEJ. 2001a. Scientific collaboration networks: I. Network construction and fundamental results. *Physical Review E*, 64: 016131.

Newman MEJ. 2001b. Scientific collaboration networks: II. Shortest paths, weighted networks, and centrality. *Physical Review E*, 64: 016132.

Newman MEJ. 2001c. The structure of scientific collaboration networks. *Proceedings of the National Academy of Sciences of the United States of America*, 98: 404–409.

Newman N, Barabasi A, Warrs DJ. 2006. *The Structure and Dynamics of Networks*. Princeton University Press, Princeton, NJ, USA.

Nicolis G, Prigogine I. 1977. *Self-Organization in Non-Equilibrium Systems: From Dissipative Structures to Order Through Fluctuations*. Wiley Interscience, New York, USA.

Nissen TL, Schulze U, Nielsen J, Villadsen J. 1997. Flux distributions in anaerobic, glucose-limited continuous cultures of *Saccharomyces cerevisiae*. *Microbiology*, 143: 203–218.

Nowak MA. 2006. *Evolutionary Dynamics*. Harvard University Press, USA, pp. 209–249.

Norton M. 1972. *Modern Control Engineering*. Pergamon Press, UK.

Odum EP. 1983. *Basic Ecology*. Saunders College Publishing, Philadelphia, USA.

Oleksiuk O, Jakovljevic V, Vladimirov N, *et al.* 2011. Thermal robustness of signaling in bacterial chemotaxis. *Cell*, 145(2): 312–321.

Olmea O, Rost B, Valencia A. 1999. Effective use of sequence correlation and conservation in fold recognition. *Journal of Molecular Biology*, 293: 1221–1239.

Onnela JP, Saramaki J, Hyvonen J, *et al.* 2007. Structure and tie strengths in mobile communication networks. *Proceedings of the National Academy of Sciences of the United States of America*, 104: 7332–7336.

Opsahl T, Agneessens F, Skvoretz J. 2010. Node centrality in weighted networks: Generalizing degree and shortest paths. *Social Networks*, 32(3): 245–251.

Overbeek R, Fonstein M, D'Souza M, Pusch CD, *et al.* 1999. Use of contiguity on the chromosome to predict functional coupling. *In Silico Biology*, 1: 93–108.

Pages S, Belaich A, Belaich JP, Morag E, *et al.* 1997. Species-specificity of the cohesin–dockerin interaction between *Clostridium thermocellum* and *Clostridium cellulolyticum*: Prediction of specificity determinants of the dockerin domain. Proteins, 29: 517–527.

Paine RT. 1980. Food webs: Linkage, interaction strength and community infrastructure. *Journal of Animal Ecology*, 49: 667–686.

Paine RT. 1992. Food-web analysis through field measurement of per capita interaction strength. *Nature*, 355: 73–75.

Pal S, Talwar V, Mitra P. 2002. Web mining in soft computing framework: Relevance, state of the art and future directions. *IEEE Transactions on Neural Networks*, 13(5): 1163–1177.

Pan Y, Li DH, Liu JG, *et al.* 2010. Detecting community structure in complex networks via node similarity. *Physica A: Statistical Mechanics and its Applications*, 389(14): 2849–2857.

Park YS, Sovan L, Scardi M, Versonshot PFM, Jorgensen SE. 2006. Patterning exergy of benthic macroinvertebrate communities using self-organising maps. *Ecological Modelling*, 190: 105–113.

Pascual M, Dunne JA. 2006. *Ecological Networks. Linking Structure to Dynamics in food Webs.* Oxford University Press, Oxford, UK.

Pathway Central. 2012. SABiosciences. Available at: http://www.sabiosciences.com/pathwaycentral.php. Accessed on January 4, 2013.

Paton K. 1969. An algorithm for finding a fundamental set of cycles of a graph. *Communications of the Association for Computing Machinery*, 12(9): 514–518.

Patten BC, Bosserman RW, Finn JT, Cale WG. 1976. Propagation of cause in ecosystems. In: *Systems Analysis and Simulation in Ecology*, Vol. 4, Patten BC (ed.), Academic Press, New York, USA, pp. 457–579.

Pauly D, Christensen V, Walters C. 2000. Ecopath, Ecosim, and Ecospace as tools for evaluating ecosystem impacts on marine ecosystems. *ICES Journal of Marine Science*, 57: 697–706.

Pazos F, Valencia A. 2001. Similarity of phylogenetic trees as indicators of protein-protein interaction. *Protein Engineering*, 14: 609–614.

Peak D. 2004. Evidence for complex, collective dynamics and emergent, distributed computation in plants. *Proceedings of the National Academy of Sciences of the United States of America*, 101(4): 918–922.

Pei S, Makse HA. 2013. Spreading dynamics in complex networks. *Journal of Statistical Mechanics: Theory and Experiment*, 12: P12002.

Pellegrini M, Marcotte EM, Thompson MJ, *et al.* 1999. Assigning protein functions by comparative genome analysis: Protein phylogenetic profiles. *Proceedings of the National Academy of Sciences of the United States of America*, 96: 4285–4288.

Petchey OL, Beckerman AP, Riede JO, *et al.* 2008. Size, foraging, and food web structure. *Proceedings of the National Academy of Sciences of the United States of America*, 105(11): 4191–4196.

Pimm SL. 1980. Food web design and the effect of species deletion. *Oikos*, 35: 139–149.

Pimm SL. 1982. *Food Webs*. Chapman & Hall, London, UK.

Pimm SL. 1991. *The Balance of Nature? Ecological Issues in the Conservation of Species and Communities*. University of Chicago Press, USA.

Pimm SL, Lawton JH. 1980. Are food webs divided into compartments? *Journal of Animal Ecology*, 49: 879–898.

Pimm SL, Lawton JH, Cohen JE. 1991. Food web patterns and their consequences. *Nature*, 350: 669–674.

Pinnegar JK, Blanchard JL, Mackinson S, *et al.* 2005. Aggregation and removal of weak-links in food-web models: System stability and recoveryfrom disturbance. *Ecological Modelling*, 184: 229–248.

Piraveenan M, Prokopenko M, Hossain L. 2013. Percolationcentrality: Quantifying graph-theoretic impact of nodes during percolation innetworks. *PloS one*, 8(1): e53095.

Piyaratne M, Zhao HY, Hu ZQ, *et al.* 2014. Catastrophic behavior of aphid population dynamics: An analysis of swallowtail model. *Computational Ecology and Software*, 4(3): 135–146.

Polovina JJ. 1984. Model of a coral-reef ecosystem. 1. The Ecopath model and its application to French Frigate Shoals. *Coral Reefs*, 3: 1–11.

Power ME, Tilman D, Estes JA, *et al.* 1996. Challenges in the quest for keys. *Bioscience*, 46: 609–620.

Psilegrini M, Marcotte EM, Thompson MJ, Eisenberg D, Yeates TO. 1999. Assigning protein functions by comparative genome analysis: Protein phylogenetic profiles. *Proceedings of the National Academy of Sciences of the United States of America*, 96: 4285–4288.

Puccia CJ, Levins R. 1985. *Qualitative Modelling of Complex Systems: An Introduction to Loop Analysis and Time Averaging*. Harvard University Press, Cambridge, UK.

Qi YH. 2003. Habitat Heterogeneity and Pest Diffusion in Different Habitats: Models and Algorithms. MSc Thesis. Sun Yat-sen University, China.

Qi YH. 2004. A network software of multivariable regression model used in statistical analysis of information. *Information Science*, 22(1): 104–106, 114.

Qi YH. 2005. The web software for ordered cluster analysis of sequential information. *Information Science*, 23(Suppl.): 99–101.

Qi YH. 2007. Simulation model of ordinary linear system and java based network implementation. *Computer Applications and Software*, 24(2): 130–131, 142.

Qi YH, Liu GH, Zhang WJ. 2016. A Matlab program for stepwise regression. *Network Pharmacology*, 1(1): 36–41.

Qi YH, Zhang WJ. 2002. A percolation model for pest perturbation in diverse habitat and network computing software. *Modern Computer*, 133: 16–19.

Qi YH, Zhang ZG, Zhang WJ. 2004. Simulation models and network software for describing pest diffusion in homogeneous habitat. *Computer Applications Research*, 24(4): 131–134.

Qian XS, Song J. 1983. *Engineering Cybernetics*. Science Press, Beijing, China.

Raffaelli DG, Hall SJ. 1996. Assessing the relative importance of trophic links in food webs. In: *Food Webs: Integration of Patterns and Dynamics*, Polis GA and Winemiller KO (eds.), Chapman and Hall, USA, pp. 185–191.

Rager J, Fry R. 2013. Systems biology and environmental exposures. In: *Network Biology: Theories, Methods and Applications*, WenJun Zhang, (ed.). Nova Science Publishers, New York, USA, pp. 81–132.

Rahman KMT, Md. Islam F, Banik RS, *et al.* 2013. Changes in protein interaction networks between normal and cancer conditions: Total chaos or ordered disorder? *Network Biology*, 3(1): 15–28.

Rai V, Upadhyay RK, Raw SN, *et al.* 2011. Some aspects of animal behavior and community dynamics. *Computational Ecology and Software*, 1(3): 153–182.

Ramakrishna R, Edwards JS, McCulloch A, Palsson BO. 2001. Flux-balance analysis of mitochondrial energy metabolism: consequences of systemic stoichiometric constraints. *American Journal of Physiology. Regulatory, Integrative and Comparative Physiology.*, 280: R695–R704.

Rapoport A, Horvath WJ. 1961. A study of a large sociogram. *Behavioral Science*, 6: 279–291.

Ravasz E, Barabási L. 2003. Hierarchical organization in complex networks. *Physical Review E*, 67: 026112.

Ravasz E, Somera AL, Mongru DA, Oltvai ZN, Barabasi AL. 2002. Hierarchical organization of modularity in metabolic networks. *Science*, 297: 1551–1555.

Resnick M. 1994. *Turtles, Termites and Traffic Jams*. MIT Press, USA.

Rezende EL, Lavabre JE, Guimaraes PR, *et al.* 2007. Nonrandom coextinctions in phylogenetically structured mutualistic networks. *Nature*, 448: 925–928.

Rhodes R. 2000. *Why They Kill: The Discoveries of a Maverick Criminologist*. Vintage Books, New York, USA.

Rossberg AG, Matsuda H, Amemiya T, *et al.* 2005. An explanatory model for food-web structure and evolution. *Ecological Complexity*, 2: 312–321.

Rudin W. 1991. *Functional Analysis* (2nd edn.). McGraw-Hill, Columbus, USA.

Rutledge RW, Basorre BL, Mulholland RJ. 1976. Ecological stability: An information theory viewpoint. *Journal of Theoretical Biology*, 57: 355–371.

Saltelli A, Chan K, Scott M. 2000. *Sensitivity Analysis, Probability and Statistics Series*. John Wiley, New York, USA.

Saltelli A, Tarantola S, Chan KPS. 1999. A quantitative modelindependent method for global sensitivity analysis of model output. *Technometrics*, 41: 39–56.

Sante I, Gacia AM, Miranda D, *et al.* 2010. Cellular automata models for the simulation of real-world urban processes: A review and analysis. *Landscape and Urban Planning*, 96: 108–122.

Sarma AD, Gollapudi S, Panigrahy R. 2011. Estimating pagerank on graph streams. *Journal of the ACM*. 58(3): 13.

SAS. Big Data Analytics. Available at: http://www.sas.com/en_us/insights/analytics/big-data-analytics.html. Accessed on February 2, 2016.

Saunders PT. 1980. *An Introduction to Catastrophe Theory.* Cambridge University Press, UK.

Savageau MA. 1977. Design of molecular control mechanisms and the demand for gene expression. *Proceedings of the National Academy of Sciences of the United States of America,* 74: 5647–5651.

Savinell JM, Palsson BO. 1992. Network analysis of intermediary metabolism using linear optimization. I. Development of mathematical formalism. *Journal of Theoretical Biology,* 154(4): 421–154.

Scardoni G, Laudanna C. 2012. Centralities based analysis of complex networks. In: *New Frontiers in Graph Theory,* Zhang YG, (ed.), InTech, Crotia, pp. 323–348.

Schellenberger J, Que R, Fleming RMT, *et al.* 2011. Quantitative prediction of cellular metabolism with constraint based models: The COBRA Toolbox v2.0. *Nature Protocols,* 6(9): 1290–1307.

Schilling G, Becher MW, Sharp AH, *et al.* 1999. Intranuclear inclusions and neurotic aggregates in transgenic mice expressing a mutant N-terminal fragment of huntingtin. *Human Molecular Genetics,* 8(3): 397–407.

Schoenharl TW. 2005. An Agent Based Modeling Approach for the Exploration of Self-organizing Neural Networks. MS Thesis. University of Notre Dame, USA.

Schoenly K, Beaver RA, Heumier TA. 1991. On the trophic relations of insects: A food web approach. *The American Naturalist* 137: 597–638.

Schoenly KG, Zhang WJ. 1999. IRRI Biodiversity Software Series. V. RARE, SPPDISS, and SPPANK: Programs for detecting between-sample difference in community structure. IRRI Technical Bulletin No. 5. International Rice Research Institute, Manila, Philippines.

Schrattenholz A, Groebe K, Soskic V. 2010. Systems biology approaches and tools for analysis of interactomes and multi-target drugs. *Methods in Molecular Biology,* 662: 29–58.

Scott J. 2000. *Social Network Analysis: A Handbook.* Sage Publications, USA.

Segre D, Vitkup D, Church GM. 2002. Analysis of optimality in natural and perturbed metabolic networks. *Proceedings of the National Academy of Sciences of the United States of America,* 99: 15112–15117.

Shakil M, Wahab HA, Naeem M, *et al.* 2015a. The modeling of predator–prey interactions. *Network Biology,* 2015, 5(2): 71–81.

Shakil M, Wahab HA, Naeem M, *et al.* 2015b. The predator–prey models for the mechanism of autocatalysis, pairwise interactions and movements to free places. *Network Biology,* 2015, 5(4): 169–179.

Shams B, Khansari M. 2014. Using network properties to evaluate targeted immunization algorithms. *Network Biology,* 4(3): 74–94.

Shen-Orr SS, Milo R, Mangan S, *et al.* 2002. Network motifs in the transcriptional regulation network of *Escherichia coli. Nature Genetics,* 31: 64–68.

Shi SH, Cai YP, Cai XJ, *et al.* 2014. A network pharmacology approach to understanding the mechanisms of action of traditional medicine: Bushen-huoxue formula for treatment of chronic kidney disease. *PLoS ONE* 9(3): e89123.

Simko GI, Csermely P. 2013. Nodes having a major influence to break cooperation define a novel centrality measure: Game centrality. *PloS one*, 8(6): e67159.

Smith EP, van Belle G. 1984. Nonparametric Estimation of Species Richness. *Biometrics*, 40(1): 119–129.

Sobol IM. 1993. Sensitivity Estimates for nonlinear mathematical models. *Mathematical Model and Computational Experiment*, 1: 407–414.

Sole RV, Montoya JM. 2001. Complexity and fragility in ecological networks. *Proceedings of the Royal Society of London Series B (Biological Sciences)*, 268: 2039–2045.

Solow AR. 1993. A simple test for change in community structure. *Journal of Animal Ecology*, 62: 191–193.

Spanier EH. 1966. *Algebraic Topology*. Springer-Verlag, New York, USA.

Sparks RF. 1980. A critique of marxist criminology. *Crime and Justice*, 2: 170–171.

Spearman C. 1904. The proof and measurement of association between two things. *American Journal of Psychology*, 15: 72–101.

Stonedahl F, Wilensky U. 2008a. NetLogo Diffusion on a Directed Network model. Center for Connected Learning and Computer-Based Modeling, Northwestern University, Evanston, IL, USA.

Stonedahl F, Wilensky U. 2008b. NetLogo Particle Swarm Optimization Model. Center for Connected Learning and Computer-Based Modeling, Northwestern University, Evanston, IL, USA.

Stonedahl F, Wilensky U. 2008c. NetLogo Simple Genetic Algorithm Model. Center for Connected Learning and Computer-Based Modeling, Northwestern University, Evanston, IL, USA.

Stuart JM, Segal E, Koller D, Kim SK. 2003. A gene-coexpression network for global discovery of conserved genetic modules. *Science*, 302: 249–255.

Sugihara G, Schoenly K, Trombla A. 1989. Scale invariance in food web properties. *Science*, 245: 48–52.

Sun DC, Lin FY. 2004. *An Introduction to Systems Engineering*. Tsinghua University Press, Beijing, China.

Sun Y, Liu Q, Cao ZW. 2014. Network Based Deciphering of the Mechanism of TCM. In: *Data Analytics for Traditional Chinese Medicine Research*, Poon J, Poon SK, (eds.), Springer, Switzerland, pp. 81–96.

Sutkovic J, Gawwad MRA. 2014. In silico prediction of three-dimensional structure and interactome analysis of Tubulin α subfamily of *Arabidopsis thaliana*. *Network Biology*, 4(2): 47–57.

Tacutu R, Budovsky A, Yanai H, *et al.* 2011. Immunoregulatory network and cancer-associated genes: Molecular links and relevance to aging. *Network Biology*, 1(2): 112–120.

Tamames J, Casari O, Ouzounis C. 1997. A Conserved clusters of functionally related genes in two bacterial genomes. *Journal of Molecular Evolution*, 44: 66–73.

Tanaka H, Ogishima S. 2015. Network biology approach to epithelial–mesenchymal transition in cancer metastasis: Three stage theory. *Journal of Molecular Cell Biology*, 7(3): 253–266.

Tarantola S, Giglioli N, Jesinghaus J, *et al.* 2002. Can global sensitivity analysis steer the implementation of models for environmental assessments and decision-making? *Stochastic Environmental Research and Risk Assessment,* 16: 63–76.

Tarjan RE. 1972. Depth-first search and linear graph algorithms. *SIAM Journal on Computing,* 1(2): 146–160.

Thomas R. 1973. Boolean formalization of genetic control circuits. *Journal of Theoretical Biology,* 42: 563–585.

Thompson J, Michael T. 1982. *Instabilities and Catastrophes in Science and Engineering.* Wiley, New York, USA.

Tomassini M, Sipper M, Perrenoud M. 2000. On the generation of high-quality random numbers by two-dimensional cellular automata. *IEEE Transactions on Computers,* 40(10): 1146–1151.

Tomoyoshi K, Tsutomu M. 2002. High speed computation of three dimensional cellular automaton with FPGA. FPL, 2438: 1126–1130.

Topping CJ, Hansen TS, Jensen TS, *et al.* 2003. ALMaSS, an agent-based model for animals in temperate European landscapes. *Ecological Modeling,* 167: 65–82.

Tu WJ. 2006. Protein–protein interactions of lung cancer related proteins. MSc Thesis. Sun Yat-sen University, Guangzhou, China.

Ulanowicz RE. 1980. An hypothesis on the development of natural communities. *Journal of Theoretical Biology,* 85: 223–245.

Ulanowicz RE. 1983. Identifying the structure of cycling in ecosystems. *Mathematical Biosciences,* 65: 219–237.

Ulanowicz RE. 1986. *Growth and Development, Ecosystems Phenomenology.* Springer, New York, USA.

Ulanowicz RE. 1995. Ecosystem trophic foundations: Lindeman exonerata. In: *Complex Ecology: The Part-Whole Relation in Ecosystems,* Patten BC, Jørgensen SE (eds.), Prentice-Hall, Englewood Cliffs, New Jersey, USA, pp. 549–560.

Ulanowicz RE. 1997. Ecology, the ascendent perspective. In: *Complexity in Ecological Systems Series,* Allen TFH, Roberts DW (eds.), Columbia University Press, New York, USA.

Ulanowicz RE. 2004. A synopsis of quantitative methods for Ecological Network Analysis. *Computational Biology and Chemistry,* 28(5–6): 321–339.

Ulanowicz RE, Norden JS. 1990. Symmetrical overhead in flow networks. *International Journal of Systems Science,* 21: 429–437.

Ulanowicz RE, Puccia CJ. 1990. Mixed trophic impacts in ecosystems. *Coenoses,* 5(1): 7–16.

Valente TW, Foreman RK. 1998. Integration and radiality: Measuring the extent of an individual's connectedness and reachability in a network. *Social Networks,* 20(1): 89–105.

Vallance P, Smart TG. 2006. The future of pharmacology. *British Journal of Pharmacology,* 147(S1): S304–S307.

Varma A, Palsson BO. 1994. Metabolic flux balancing: Basic concepts, scientific and practical use. *Biotechnology,* 12: 994–998.

Varma A, Boesch BW, Palsson BO. 1993. Stoichiometric interpretation of Escherichia coli glucose catabolism under various oxygenation rates. *Applied and Environmental Microbiology*, 59: 2465–2473.

Vieira VMNCS. 2012. Permutation tests to estimate significances on Principal Components Analysis. *Computational Ecology and Software*, 2(2): 103–123.

Walhout AJM, Sordella R, Lu X, *et al.* 2000. Protein interaction mapping in *C elegans* proteins involved in vulval development. *Science*, 287: 116.

Walters CJ, Christensen V, Pauly D. 1997. Structuring dynamic models of exploited ecosystems from trophic mass-balance assessments. *Reviews in Fish Biology and Fisheries*, 7: 139–172.

Walters CJ, Kitchell JF, Christensen V, Pauly D. 2000. Representing density dependent consequences of life history strategies in aquatic ecosystems: Ecosim II. *Ecosystems*, 3: 70–83.

Wang B, Tang HW, Guo CH. 2006a. Optimization of network structure to random failures. *Physical Statistical Mechanics and its Applications*, 368(2): 607–614.

Wang SY. 2017. How to extract topics from mountainous text data using Python. Available at: http://www.jianshu.com/p/fdde9fc03f94. Accessed on August 20, 2017.

Wang XF, Li X, Chen GR. 2006b. *The Theory of Complex Network and Its Application*. Tsinghua University Press, Beijing, China.

Wang ZJ, Wang NC, Feng F, Tian WF. 2007. Research on evolved behavior of cellular automata. *Application Research of Computers*, 24(8): 38–41.

Warren PH. 1994. Making connections in food webs. *Trends in Ecology and Evolution*, 4: 136–140.

Warren PH, Law R, Weatherby AJ. 2003. Mapping the assembly of protest communities in microcosms. *Ecology*, 84(4): 1001–1011.

Wasserman S, Faust K. 1994. *Social Network Analysis: Methods and Applications*. Cambridge University Press, Cambridge, UK.

Watts D, Strogatz S. 1998. Collective dynamics of "small world" networks. *Nature*, 393: 440–442.

Watts MJ, Worner SP. 2011. Improving cluster-based methods for investigating potential for insect pest species establishment: Region-specific risk factors. *Computational Ecology and Software*, 1(3): 138–145.

Weerasinghe S. 2013. Analytical Frameworks of Social Network Analyses. In: *Network Biology: Theories, Methods and Applications*, WenJun Zhang (ed.), Nova Science Publishers, New York, USA, pp. 1–22.

Wei W. 2010. Biodiversity Analysis on Arthropod and Weed Communities in Paddy Rice Fields of Pearl River Delta. Master Degree Dissertation. Sun Yat-sen University, China.

Wikipedia. 2014a. Dissipative system. Available at: http://en.wikipedia.org/wiki/Dissipative_system. Accessed on March 13, 2014.

Wikipedia. 2014b. Self-organization. Available at: http://en.wikipedia.org/wiki/Self-organization. Accessed on March 12, 2014.

Wikipedia. 2014c. Catastrophe theory. Available at: http://en.wikipedia.org/wiki/Catastrophe_theory. Accessed on March 13, 2014.

Wikipedia. 2014d. Molecular self-assembly. Available at: http://en.wikipedia.org/wiki/Molecular_self-assembly. Accessed on March 14, 2014.

Wikipedia. 2016a. Network science. Available at: https://en.wikipedia.org/wiki/Network_science. Accessed on January 28, 2016.

Wikipedia. 2016b. Pharmacology. Available at: https://en.wikipedia.org/wiki/Pharmacology. Accessed on January 28, 2016.

Wikipedia. 2016c. Big data. Available at: https://en.wikipedia.org/wiki/Big_data. Accessed on February 2, 2016.

Wikipedia. 2016d. Criminology. Available at: https://en.wikipedia.org/wiki/Criminology. Accessed on November 30, 2016.

Wikipedia. 2017. Dunbar's number. Available at: https://en.wikipedia.org/wiki/Dunbar%27s_number. Accessed on September 17, 2017.

Wilensky U. 1997a. NetLogo Ants Model. Center for Connected Learning and Computer-based Modeling, Northwestern University, Evanston, IL, USA.

Wilensky U. 1997b. NetLogo Termites Model. Center for Connected Learning and Computer-based Modeling, Northwestern University, Evanston, IL, USA.

Wilensky U. 1998. NetLogo Tumor Model. Center for Connected Learning and Computer-based Modeling, Northwestern University, Evanston, IL, USA.

Wilensky U. 2002. NetLogo Crystallization Basic Model. Center for Connected Learning and Computer-based Modeling, Northwestern University, Evanston, IL, USA.

Wilensky U. 2005. NetLogo Giant Component Model. Center for Connected Learning and Computer-based Modeling, Northwestern University, Evanston, IL, USA.

Wilensky U. 2007. NetLogo Hex Cell Aggregation Model. Center for Connected Learning and Computer-based Modeling, Northwestern University, Evanston, IL, Evanston, IL, USA.

Wilensky U, Reisman K. 1999. Connected science: Learning biology through constructing and testing computational theories — an embodied modeling approach. *International Journal of Complex Systems*, 234: 1–12.

Williams RJ. 2010. Simple MaxEnt models explain food web degree distributions. *Theoretical Ecology*, 3: 45–52.

Williams RJ, Martinez ND. 2000. Simple rules yield complex food webs. *Nature*, 404: 180–183.

Wilson JB, Roxburgh SH. 1994. A demonstration of guild-based assembly rules for a plant community, and determination of intrinsic guilds. *Oikos*, 69: 267–276.

Wolfram S. 2002. *A New Kind of Science*. Wolfram Media, Champaign, IL, USA.

Woodcock AER, Davis M. 1978. *Catastrophe Theory*. E. P. Dutton, New York, USA.

Wu S, Hofman JM, Mason WA, Watts DJ. 2011. Who says what to whom on twitter. In: *Proceedings of the 20th International Conference on WWW*, ACM Press, USA, pp. 705–714.

Xia Y, Tse CK, Tam WM, *et al.* 2005. Small, scale-free user-network approach to telephone network traffic analysis. *Physical Review E*, 72: 026116.

Xiao X, Wang P, Chou KC. 2011. Cellular automata and its applications in protein bioinformatics. *Current Protein and Peptide Science*, 12(6): 508–519.

Xu CG, Hu YM, Chang Y, *et al.* 2004. Sensitivity analysis in ecological modeling. *Chinese Journal of Applied Ecology*, 15(6): 1056–1062.

Yan PF, Zhang CS. 2000. *Artificial Neural Networks and Computation of Simulated Evolution*. Tsinghua University Press, Beijing, China.

Yang XS. 2014. *Nature-Inspired Optimization Algorithms*. Elsevier, Netherlands.

Yang XS, Deb S, Loomes M, *et al.* 2013. A framework for self-tuning optimization algorithm. *Neural Computing and Applications*, 23(7–8): 2051–2057.

YashRoy RC. 1987. 13-C NMR studies of lipid fatty acyl chains of chloroplast membranes. *Indian Journal of Biochemistry and Biophysics*, 24(6): 177–178.

Yildirim MA, Goh KI, Cusick ME, Barabasi AL, Vidal M. 2007. Drug-target network. *Nature Biotechnology*, 25(10): 1119–1126.

Ying HJ, Ouyang PK. 2000. Inquiry into mechanism of FDP accumulating using model of metabolic pathway flux. *Journal of Chemical Industry and Engineering*, 51(3): 313–319.

Yodzis P. 1980. The connectance of real ecosystems. *Nature*, 284: 544–545.

Yong L, Xin L, Michael P, Yuanyuan C. 2010. Social network analysis of a criminal hacker community. *Journal of Computer Information Systems*, 51(2): 31–41.

Yu HY, Braun P, Yildirim MA, *et al.* 2008. High-quality binary protein interaction map of the yeast interactome network. *Science*, 322: 104–110.

Zeeman EC. 1976. Catastrophe theory. *Scientific American*, 234: 65–83.

Zeeman EC. 1977. *Catastrophe Theory-Selected Papers 1972-1977*. Addison-Wesley, MA, USA.

Zeitoun AH, Ibrahim SS, Bagowski CP. 2012. Identifying the common interaction networks of amoeboid motility and cancer cell metastasis. *Network Biology*, 2(2): 45–56.

Zhang B, Horvath S. 2005. A general framework for weighted gene co-expression network analysis. *Statistical Applications in Genetics and Molecular Biology*, 4: Article 17.

Zhang PP, Chen K, He Y, Zhou T, *et al.* 2006. Model and empirical study on some collaboration networks. *Physica A*, 360: 599–616.

Zhang SY, Zhang X. 2009. Biological networks and some of their developments. *Journal of System Simulation*, 17: 5300–5305.

Zhang WJ. 2007. Computer inference of network of ecological interactions from sampling data. *Environmental Monitoring and Assessment*, 124: 253–261.

Zhang WJ. 2010. *Computational Ecology: Artificial Neural Networks and Their Applications*. World Scientific, Singapore.

Zhang WJ. 2011a. A Java algorithm for non-parametric statistic comparison of network structure. *Network Biology*, 1(2): 130–133.

Zhang WJ. 2011b. A Java program to test homogeneity of samples and examine sampling completeness. *Network Biology*, 1(2): 127–129.

Zhang WJ. 2011c. Constructing ecological interaction networks by correlation analysis: Hints from community sampling. *Network Biology*, 1(2): 81–98.

Zhang WJ. 2011d. Network Biology: An exciting frontier science. *Network Biology*, 1(1): 79–80.

Zhang WJ. 2011e. A Java program for non-parametric statistic comparison of community structure. *Computational Ecology and Software*, 1(3): 183–185.

Zhang WJ. 2012a. A Java software for drawing graphs. *Network Biology*, 2(1): 38–44.

Zhang WJ. 2012b. How to construct the statistic network? An association network of herbaceous plants constructed from field sampling. *Network Biology*, 2(2): 57–68.

Zhang WJ. 2012c. Modeling community succession and assembly: A novel method for network evolution. *Network Biology*, 2(2): 69–78.

Zhang WJ. 2012d. Several mathematical methods for identifying crucial nodes in networks. *Network Biology*, 2(4): 121–126.

Zhang WJ. 2012e. *Computational Ecology: Graphs, Networks and Agent-based Modeling*. World Scientific, Singapore.

Zhang WJ. 2013. Selforganizology: A science that deals with self-organization. *Network Biology*, 3(1): 1–14.

Zhang WJ. 2013a. *Network Biology: Theories, Methods and Applications*. Nova Science Publishers, New York.

Zhang WJ. 2013b. *Self-organization: Theories and Methods*. Nova Science Publishers, New York.

Zhang WJ. 2014a. A framework for agent-based modeling of community assembly and succession. *Selforganizology*, 1(1): 16–22.

Zhang WJ. 2014b. Interspecific associations and community structure: A local survey and analysis in a grass community. *Selforganizology*, 1(2): 89–129.

Zhang WJ. 2014c. Research advances in theories and methods of community assembly and succession. *Environmental Skeptics and Critics*, 3(3): 52–60.

Zhang WJ. 2014d. Selforganizology: A more detailed description. *Selforganizology*, 1(1): 31–46.

Zhang WJ. 2015a. Complete Impact Factor (CIF): A full index to evaluate journal impact. *Selforganizology*, 2(1): 18–20.

Zhang WJ. 2015b. A generalized network evolution model and self-organization theory on community assembly. *Selforganizology*, 2(3): 55–64.

Zhang WJ. 2015c. A hierarchical method for finding interactions: Jointly using linear correlation and rank correlation analysis. *Network Biology*, 5(4): 137–145.

Zhang WJ. 2015d. Calculation and statistic test of partial correlation of general correlation measures. *Selforganizology*, 2(4): 65–77.

Zhang WJ. 2015e. Prediction of missing connections in the network: A node-similarity based algorithm. *Selforganizology*, 2(4): 91–101.

Zhang WJ. 2016a. A mathematical model for dynamics of occurrence probability of missing links in predicted missing link list. *Network Pharmacology*, 1(4): 86–94.

Zhang WJ. 2016b. A Matlab program for finding shortest paths in the network: Application in the tumor pathway. *Network Pharmacology*, 1(1): 42–53.

Zhang WJ. 2016c. A method for identifying hierarchical sub-networks/modules and weighting network links based on their similarity in sub-network/module affiliation. *Network Pharmacology*, 1(2): 54–65.

Zhang WJ. 2016d. A node-similarity based algorithm for tree generation and evolution. *Network Biology*, 6(3): 55–64.

Zhang WJ. 2016e. A node degree dependent random perturbation method for prediction of missing links in the network. *Network Biology*, 6(1): 1–11.

Zhang WJ. 2016f. A random network based, node attraction facilitated network evolution method. *Selforganizology*, 3(1): 1–9.

Zhang WJ. 2016g. Detecting connectedness of network: A Matlab program and application in tumor pathways and a phylogenic network. *Selforganizology*, 3(4): 117–120.

Zhang WJ. 2016h. Finding trees in the network: Some Matlab programs and application in tumor pathways. *Network Pharmacology*, 1(2): 66–73.

Zhang WJ. 2016i. Generate networks with power-law and exponential-law distributed degrees: with applications in link prediction of tumor pathways. *Network Pharmacology*, 1(1): 15–35.

Zhang WJ. 2016j. How to find cut nodes and bridges in the network? A Matlab program and application in tumor pathways. *Network Pharmacology*, 1(3): 82–85.

Zhang WJ. 2016k. Network chemistry, network toxicology, network informatics, and network behavioristics: A scientific outline. *Network Biology*, 6(1): 37–39.

Zhang WJ. 2016l. Network informatics: A new science. *Selforganizology*, 3(2): 43–50.

Zhang WJ. 2016m. Network pharmacology: A further description. *Network Pharmacology*, 1(1): 1–14.

Zhang WJ. 2016n. Network robustness: Implication, formulization and exploitation. *Network Biology*, 6(4): 75–85.

Zhang WJ. 2016o. Network toxicology: A new science. *Computational Ecology and Software*, 6(2): 31–40.

Zhang WJ. 2016p. Screening node attributes that significantly influence node centrality in the network. *Selforganizology*, 3(3): 75–86.

Zhang WJ. 2016q. Some methods for sensitivity analysis of systems/networks. *Network Pharmacology*, 1(3): 74–81.

Zhang WJ. 2016r. *Selforganizology: The Science of Self-Organization*. World Scientific, Singapore.

Zhang WJ. 2017a. Estimation of node richness by sampling: Application of nonparametric methods. *Selforganizology*, 4(1): 10–13.

Zhang WJ. 2017b. Finding fundamental circuits in the network: A Matlab program and application in tumor pathway. *Selforganizology*, 4(1): 14–17.

Zhang WJ. 2017c. Finding minimum cost flow in the network: A Matlab program and application. *Selforganizology*, 4(2): 30–34.

Zhang WJ. 2017d. Finding the shortest tree in the network: A Matlab program and application in tumor pathway. *Network Pharmacology*, 2(1): 13–16.

Zhang WJ. 2017e. MATASS: The software for multi-attribute assessment problems. *Computational Ecology and Software*, 7(2): 38–48.

Zhang WJ. 2017f. Measurement and identification of positive plant interactions: Overview and new perspective. *Network Biology*, 7(2): 33–40.

Zhang WJ. 2017g. Network criminology: The criminology based on network science. *Network Biology*, 7(1): 1–9.

Zhang WJ. 2017h. Network pharmacology of medicinal attributes and functions of Chinese herbal medicines: (I) Basic statistics of medicinal attributes and functions for more than 1100 Chinese herbal medicines. *Network Pharmacology*, 2(2): 17–37.

Zhang WJ. 2017i. Network pharmacology of medicinal attributes and functions of Chinese herbal medicines: (III) Canonical correlation functions between attribute classes and linear eigenmodels of Chinese herbal medicines. *Network Pharmacology*, 2(3): 67–81.

Zhang WJ. 2017j. Network pharmacology of medicinal attributes and functions of Chinese herbal medicines: (IV) Classification and network analysis of medicinal functions of Chinese herbal medicines. *Network Pharmacology*, 2(3): 82–104.

Zhang WJ. 2017k. Network pharmacology of medicinal attributes and functions of Chinese herbal medicines: (II) Relational networks and pharmacological mechanisms of medicinal attributes and functions of Chinese herbal medicines. *Network Pharmacology*, 2(2): 38–66.

Zhang WJ. 2017l. Finding maximum flow in the network: A Matlab program and application. *Computational Ecology and Software*, 8(2): 57–58.

Zhang WJ. 2017m. Phase recognition in network evolution. *Selforganizology*, 4(3): 35–40.

Zhang WJ. 2017n. A new model to describe the relationship between species richness and sample size. *Computational Ecology and Software*, 7(1): 1–7.

Zhang WJ, Bai CJ, Liu GD. 2007. Neural network modeling of ecosystems: A case study on cabbage growth system. *Ecological Modelling*, 201: 317–325.

Zhang WJ, Barrion AT. 2006. Function approximation and documentation of sampling data using artificial neural networks. *Environmental Monitoring and Assessment*, 122: 185–201.

Zhang WJ, Chen B. 2011. Environment patterns and influential factors of biological invasions: A worldwide survey. *Proceedings of the International Academy of Ecology and Environmental Sciences*, 1(1): 1–14.

Zhang WJ, Feng YT. 2017. Metabolic pathway of non-alcoholic fatty liver disease: Network properties and robustness. *Network Pharmacology*, 2(1): 1–12.

Zhang WJ, Gu DX. 1998. Mathematical models to describe mixture infection of multi-species pathogens to insects. *Journal of Biomathematics*, 13(2): 268–272.

Zhang WJ, Jiang LQ, Chen WJ. 2014a. Effect of parasitism on food webs: Topological analysis and goodness test of cascade model. *Network Biology*, 4(4): 170–178.

Zhang WJ, Li X. 2015a. General correlation and partial correlation analysis in finding interactions: With Spearman rank correlation and proportion correlation as correlation measures. *Network Biology*, 5(4): 163–168.

Zhang WJ, Li X. 2015b. Linear correlation analysis in finding interactions: Half of predicted interactions are undeterministic and one-third of candidate direct interactions are missed. *Selforganizology*, 2(3): 39–45.

Zhang WJ, Li X. 2016. A cluster method for finding node sets/sub-networks based on between- node similarity in sets of adjacency nodes: With application in finding sub-networks in tumor pathways. *Proceedings of the International Academy of Ecology and Environmental Sciences*, 6(1): 13–23.

Zhang WJ, Liu GH. 2012. Creating real network with expected degree distribution: A statistical simulation. *Network Biology*, 2(3): 110–117.

Zhang WJ, Liu GH, Dai HQ. 2008a. Simulation of food intake dynamics of holometabolous insect using functional link artificial neural network. *Stochastic Environmental Research and Risk Assessment*, 22(1): 123–133.

Zhang WJ, Pang Y, Qi YH, Chen QJ, 1997. Simulation model for epizootic disease of *Spodoptera litura* F. baculovirus. *Acta Scientiarum Naturalium Universitatis Sunyatseni*, 36(1): 54–59.

Zhang WJ, Qi YH. 2014. Pattern classification of HLA-DRB1 alleles, human races and populations: Application of self-organizing competitive neural network. *Selforganizology*, 1(3–4): 138–142.

Zhang WJ, Qi YH, Zhang ZG. 2014b. Two-dimensional ordered cluster analysis of component groups in self-organization. *Selforganizology*, 1(2): 62–77.

Zhang WJ, Qi YH, Zhang ZG. 2015. A cellular automaton for population diffusion in the homogeneous rectangular area. *Selforganizology*, 2(1): 13–17.

Zhang WJ, Schoenly KG. 1999a. IRRI Biodiversity Software Series. II. COLLECT1 and COLLECT2: Programs for Calculating Statistics of Collectors' Curves. IRRI Technical Bulletin No.2. International Rice Research Institute, Manila, Philippines.

Zhang WJ, Schoenly KG. 1999b. IRRI Biodiversity Software Series. IV. EXTSPP1 and EXTSPP2: programs for comparing and performance-testing eight extrapolation-based estimators of total taxonomic richness. IRRI Technical Bulletin No.4. International Rice Research Institute, Manila, Philippines.

Zhang WJ, Wang R, Zhang DL, *et al.* 2014c. Interspecific associations of weed species around rice fields in Pearl River Delta, China: A regional survey. *Selforganizology*, 1(3–4): 143–205.

Zhang WJ, Wei W. 2009. Spatial succession modeling of biological communities: A multi-model approach. *Environmental Monitoring and Assessment*, 158: 213–230.

Zhang WJ, Wopke van der Werf, Pang Y. 2011. A simulation model for vegetable-insect pest–insect nucleopolyhedrovirus epidemic system. *Journal of Environmental Entomology*, 33(3): 283–301.

Zhang WJ, Zhan CY. 2011. An algorithm for calculation of degree distribution and detection of network type: With application in food webs. *Network Biology*, 1(3–4): 159–170.

Zhang WJ, Zhang XY. 2008. Neural network modeling of survival dynamics of holometabolous insects: A case study. *Ecological Modelling*, 211: 433–443.

Zhang WJ, Zhong XQ, Liu GH. 2008b. Recognizing spatial distribution patterns of grassland insects: Neural network approaches. *Stochastic Environmental Research and Risk Assessment*, 22(2): 207–216.

Zhang Y, Chen M, Liao XF. 2013. Big Data Applications: A Survey. *Journal of Computer Research and Development*. 50(Suppl.): 216–233.

Zhang YT, Fang KT. 1982. *Introduction to Multivariate Statistics*. Science Press, Beijing, China.

Zhang ZL, Wang ZY. 2009. *Systems Biology*. Science Press, Beijing, China.

Zhao J, Miao LL, Yang Y, *et al.* 2015. Prediction of links and weights in networks by reliable routes. *Scientific Reports*, 5: 12261.

Zhao SN, Yu YX. 1987. *Catastrophe Theory and Its Application in Biomedical Science*. Science Press, Beijing, China.

Zhao XC. 2014. Swarm intelligence optimization algorithms and life. Available at: http://blog.sciencenet.cn/blog-86581-772563.html. Accessed on March 4, 2014.

Zhao Y, Zhang WJ. 2013. Organizational theory: With its applications in biology and ecology. *Network Biology*, 3(1): 45–53.

Zheng CS, Xu XJ, Ye HZ, *et al.* 2013. Network pharmacology-based prediction of the multi-target capabilities of the compounds in Taohong Siwu decoction, and their application in osteoarthritis. *Experimental and Therapeutic Medicine*, 6(1): 125–132.

Zhou C, Zemanov L, Zamora G, *et al.* 2006. Hierarchical organization unveiled by functional connectivity in complex brain networks. *Physical Review Letters*, 97(3): 238103.

Zhou T. 2012. A knowledge on social networks. Available at: http://blog.sciencenet.cn/blog-3075-531837.html. Accessed on January 26, 2012.

Zhou T. 2015. Why link prediction? Available at: http://blog.sciencenet.cn/blog-3075-912975.html. Accessed on August 14, 2015.

Zhou T. 2016a. Dynamics indices for assessing node importance. Available at: http://blog.sciencenet.cn/blog-3075-962600.html. Accessed on March 14, 2016.

Zhou T. 2016b. Available at: http://www.linkprediction.org/index.php/link/resource/code. Access on April 12, 2016.

Zhou T, Lu LY, Zhang YC. 2009. Predicting missing links via local information. *European Physical Journal B*, 71(4): 623–630.

Zhou T, Medo M, Cimini G, Zhang ZK, *et al.* 2011. Emergence of scale-free leadership structure in social recommender systems. *PLoS ONE*, 6: e20648.

Zhou WG. 2007. A Field Survey on Paddy Rice Arthropod Biodiversity in Northern Guangzhou. Master Degree Dissertation. Sun Yat-sen University, China.

Zhou WX, Cheng XR, Zhang YX. 2012. Network pharmacology-a new philosophy for understanding of drug action and discovery of new drugs. *Chinese Journal of Pharmacology and Toxicology*, 2(26): 4–9.

Zhu YH, Liu J. 2012. A research on the robustness of complex networks. *Science and Technology Information*, 32: 6.

Zorach AC, Ulanowicz RE. 2003. Quantifying the complexity of low networks: How many roles are there? *Complexity*, 8: 68–76.

Index

www.ingramcontent.com/pod-product-compliance
Lightning Source LLC
Chambersburg PA
CBHW052115230326
41598CB00079B/3701